Lecture Notes in Networks and Systems

Volume 737

The series "Lecture Notes in Networks and Systems" publishes the latest developments in Networks and Systems—quickly, informally and with high quality. Original research reported in proceedings and post-proceedings represents the core of LNNS.

Volumes published in LNNS embrace all aspects and subfields of, as well as new challenges in, Networks and Systems.

The series contains proceedings and edited volumes in systems and networks, spanning the areas of Cyber-Physical Systems, Autonomous Systems, Sensor Networks, Control Systems, Energy Systems, Automotive Systems, Biological Systems, Vehicular Networking and Connected Vehicles, Aerospace Systems, Automation, Manufacturing, Smart Grids, Nonlinear Systems, Power Systems, Robotics, Social Systems, Economic Systems and other. Of particular value to both the contributors and the readership are the short publication timeframe and the world-wide distribution and exposure which enable both a wide and rapid dissemination of research output.

The series covers the theory, applications, and perspectives on the state of the art and future developments relevant to systems and networks, decision making, control, complex processes and related areas, as embedded in the fields of interdisciplinary and applied sciences, engineering, computer science, physics, economics, social, and life sciences, as well as the paradigms and methodologies behind them.

Indexed by SCOPUS, INSPEC, WTI Frankfurt eG, zbMATH, SCImago.

All books published in the series are submitted for consideration in Web of Science.

For proposals from Asia please contact Aninda Bose (aninda.bose@springer.com).

Wojciech Zamojski · Jacek Mazurkiewicz ·
Jarosław Sugier · Tomasz Walkowiak ·
Janusz Kacprzyk
Editors

Dependable Computer Systems and Networks

Proceedings of the Eighteenth International Conference on Dependability of Computer Systems DepCoS-RELCOMEX, July 3–7, 2023, Brunów, Poland

Springer

Editors
Wojciech Zamojski
Department of Computer Engineering
Wrocław University of Science and Technology
Wrocław, Poland

Jacek Mazurkiewicz
Department of Computer Engineering
Wrocław University of Science and Technology
Wrocław, Poland

Jarosław Sugier
Department of Computer Engineering
Wrocław University of Science and Technology
Wrocław, Poland

Tomasz Walkowiak
Department of Computer Engineering
Wrocław University of Science and Technology
Wrocław, Poland

Janusz Kacprzyk
Polish Academy of Sciences
Systems Research Institute
Warsaw, Poland

ISSN 2367-3370 ISSN 2367-3389 (electronic)
Lecture Notes in Networks and Systems
ISBN 978-3-031-37719-8 ISBN 978-3-031-37720-4 (eBook)
https://doi.org/10.1007/978-3-031-37720-4

This Springer imprint is published by the registered company Springer Nature Switzerland AG
The registered company address is: Gewerbestrasse 11, 6330 Cham, Switzerland

Programme Committee

Wojciech Zamojski (Chairman), Wrocław University of Science and Technology, Poland
Ali Al-Dahoud, Al-Zaytoonah University, Amman, Jordan
Johnson I. Agbinya, Melbourne Institute of Technology, Australia
Michael Affenzeller, Upper Austria University of Applied Sciences, Austria
Patricio García Baez, Universidad de La Laguna, San Cristóbal de La Laguna, Spain
Andrzej Białas, Research Network ŁUKASIEWICZ-Institute of Innovative Technologies EMAG, Katowice, Poland
Ilona Bluemke, Warsaw University of Technology, Poland
Magdalena Bogalecka, Gdynia Maritime University, Poland
Wojciech Bożejko, Wrocław University of Science and Technology, Poland
Eugene Brezhniev, National Aerospace University "KhAI", Kharkov, Ukraine
Dariusz Caban, Wrocław University of Science and Technology, Poland
De-Jiu Chen, KTH Royal Institute of Technology, Stockholm, Sweden
Jacek Cichoń, Wrocław University of Science and Technology, Poland
Frank Coolen, Durham University, UK
Wiktor B. Daszczuk, Warsaw University of Science and Technology, Poland
Mieczysław Drabowski, Cracow University of Technology, Poland
Francesco Flammini, University of Linnaeus, Sweden
Peter Galambos, Óbuda University, Hungary
Manuel Gill Perez, University of Murcia, Spain
Franciszek Grabski, Gdynia Maritime University, Gdynia, Poland
Aleksander Grakowskis, Transport and Telecommunication Institute, Riga, Latvia
Laszlo Gulacsi, Óbuda University, Hungary
Atsushi Ito, Chuo University, Tokyo, Japan
Dariusz Jagielski, 4th Military Hospital, Wrocław, Poland
Łukasz Jeleń, Wroclaw University of Science and Technology, Poland
Ireneusz Jóźwiak, Wrocław University of Science and Technology, Poland
Igor Kabashkin, Transport and Telecommunication Institute, Riga, Latvia
Janusz Kacprzyk, Polish Academy of Sciences, Warsaw, Poland

Organizing Committee

Chair

Wojciech Zamojski, Wrocław University of Science and Technology, Poland

Members

Jacek Mazurkiewicz
Jarosław Sugier
Tomasz Walkowiak
Tomasz Zamojski
Mirosława Nurek
Wrocław University of Science and Technology, Poland

Preface

Dependability of computer processing means obtaining reliable (true and timely) results in the conditions of processing both quantitative and qualitative data, with application of precise and "fuzzy/imitating" models and algorithms

In this volume we are presenting the proceedings of the 18th International Conference on Dependability of Computer Systems DepCoS-RELCOMEX, which is scheduled to be held stationary in Brunów Palace, Poland, from 3rd to 7th July 2023.

It is the first time, after the 3-year disruption caused by the COVID pandemic, when the Conference will be organized in a normal way in a beautiful Brunów Palace, surrounded by many palaces and castles of the charming Valley. Three years ago, during preparation of the proceedings in February 2020, we couldn't predict that then-local pandemic would spread so fast also in Europe and our event, organized yearly since 2006, would have to be held exclusively online for the next 3 years. Although, despite technical difficulties, the remote sessions of 2020–2022 editions successfully accomplished conference goals, the limited possibilities of online contacts couldn't match benefits of a real-life meeting and this year we are pleased and relieved to return to the face-to-face discussions in our traditional, beautiful venue.

DepCoS-RELCOMEX scope has always been focused on diverse issues which are constantly arising in performability and dependability analysis of contemporary computer systems and networks. This year we are opening our Conference for topics related to computer and network solutions supporting applications in medicine (computer-aided medicine). It should be emphasized that Artificial Intelligence methods and tools are increasingly used in modern information technology and computer engineering, and therefore we are expanding our view on the dependability of systems that are increasingly using "algorithms" based on deep learning tools. In our opinion, this approach (**dependability** as the credibility of systems, **AI** tools, and **medical** applications) meets the needs of the modern computer science and technology, both in theory and in engineering practice. Ever-growing number of research methods being continuously developed for dependability analyses apply the newest results of artificial intelligence (AI) and computational intelligence (CI). Selection of papers in these proceedings illustrates the broad variety of multi-disciplinary topics which should be considered in these studies, but also proves that virtually all areas of contemporary computer systems and networks must take into account an aspect of dependability.

The Conference is now organized by Department of Computer Engineering at the Faculty of Information and Communication Technology, Wrocław University of Science and Technology, but its roots go back to the heritage of two other cycles of events: RELCOMEX (1977–89) and Microcomputer School (1985–95) which were organized by Institute of Engineering Cybernetics (predecessor of the Department) under the leadership of prof. Wojciech Zamojski, now also the DepCoS Chairman. These proceedings are the second ones published in the series "Lecture Notes in Networks and Systems", after last year's vol. 484. The previous volumes were printed, chronologically, first by the IEEE Computer Society (2006–09), then by Wrocław University of Science and Technology Publishing House (2010–12), and recently by Springer Nature in the "Advances in Intelligent Systems and Computing" volumes no. 97 (2011), 170 (2012), 224 (2013), 286 (2014), 365 (2015), 479 (2016), 582 (2017), 761 (2018), 987 (2019), 1173 (2020), and 1389 (2021). Springer Nature is one of the largest and most prestigious scientific publishers, with the LNNS titles being submitted for indexing in CORE Computing Research & Education database, Web of Science, SCOPUS, INSPEC, DBLP, and other indexing services.

We would like to express our thanks to everyone who participated in organization of the Conference and in preparation of this volume: members of the Program and the Organizing Committees, and to all who helped in making the Conference take place. Our special thanks go to the reviewers whose opinions and comments were invaluable in selecting and enhancing the contents of this volume. This year it was an effort of Andrzej Białas, Ilona Bluemke, Deju Chen, Manuel Gil Perez, Ireneusz Jóźwiak, Urszula Kużelewska, Alexey Lastovetsky, Jan Magott, Jacek Mazurkiewicz, Yiannis Papadopoulos, Czesław Smutnicki, Robert Sobolewski, Janusz Sosnowski, Jaroslaw Sugier, Kamil Szyc, Tomasz Walkowiak, Marek Woda, Min Xie, Irina Yatskiv (Jackiva), and Wojciech Zamojski. Their work, not mentioned anywhere else in this book, deserves to be highlighted and appreciated in this introduction.

Concluding the preface, we would like to thank all the authors who decided to publish and discuss the results of their work on the DepCoS-RELOCMEX platform. We express our hope that the papers of these proceedings will promote design, analysis, and engineering of dependable computer systems and networks, and will be a valuable and inspiring source material for scientists, researchers, engineers, and students who are working in this area.

Wrocław, Poland — Wojciech Zamojski
Wrocław, Poland — Jacek Mazurkiewicz
Wrocław, Poland — Jarosław Sugier
Wrocław, Poland — Tomasz Walkowiak
Warsaw, Poland — Janusz Kacprzyk

Contents

Line Segmentation of Handwritten Documents Using Direct Tensor Voting

Tomasz Babczyński and Roman Ptak

1 Introduction

The increasing exchange of electronic documents has still not replaced traditional paper documents. Moreover, a large number of archival materials and other publications are still available in paper form. Besides paper as a writing surface also papyrus and parchment were used. These documents are of interest to scientists from various branches, including historians. Handwritten manuscripts and paper documents, including primary sources of historical significance, are subject to frequent examination. Historical documents are kept in archives and libraries throughout the world. In many cases they are hardly available. Digitization process can make access to them easier. However, even after digitization, these documents remain unsearchable, underscoring the importance of rendering their contents in a machine-readable format. This implies the need to perform document image processing. There are various goals in the document processing. Handwriting is the main component of historical documents, so the recognition process must be performed on them. The reliability of this procedure relies on the quality of image segmentation. Clearly, the main objective is to recognize the text written in manuscript. It is crucial to perform text line localization and segmentation accurately and dependably, among other tasks, in the process of script recognition. Localizing the text is frequently the initial step in text line segmentation. Subsequently, line segmentation frequently occurs prior to partitioning the text into words. The text may be further divided into individual letters, after which character recognition can be performed.

T. Babczyński (✉) · R. Ptak
Department of Computer Engineering Wrocław University of Science and Technology, Wyb. Wyspiańskiego 27, 50-370 Wrocław, Poland
e-mail: tomasz.babczynski@pwr.edu.pl

R. Ptak
e-mail: roman.ptak@pwr.edu.pl

W. Zamojski et al. (eds.), *Dependable Computer Systems and Networks*, Lecture Notes in Networks and Systems 737, https://doi.org/10.1007/978-3-031-37720-4_1

In the present article, we focus on the problem of text line segmentation by means of Tensor Voting.

The rest of the paper is organized as follows. Section 2 describes some related works on text line segmentation. Our method and experimental results are presented in Sects. 3 and 4 respectively. Finally, Sect. 5 gives our conclusions.

2 Related Work

The text lines segmentation problem was solved using many various approaches. It is worth referring to articles: [14, 24]. Furthermore, a more recent review can be found in [18].

In segmentation, two categories of algorithms can be designed: top-down and bottom-up. Some top-down methods can be classified as accumulating data or voting. The review presented here is concentrated on this category. Additionally, some other methods are presented, especially those used in ICDAR 2009 *Handwriting Segmentation Contest* described with more details later. The result of our algorithm is compared with the ones of the attenders of this competition.

Projection profile methods accumulate pixels data along a given path. Projection consists of calculating the values of foreground pixels. For a binary image, it is summing all “black” pixels. Usually horizontal or vertical projection is used for segmentation – like in the selected publications [4, 10, 22]. Methods of this kind commonly maintain the global projection profile, but there are also methods applying the piece-wise projection profile of the document [2, 3]. The methods proposed there are robust against lines running into each other or slanted lines in handwriting documents.

The winner of the mentioned competition, the algorithm denoted there as CUBS is based on Adaptive Local Connectivity Map (ALCM), which is defined as a convolution, but is actually a projection in a moving window. Some directions of the projection are applied by means of a steerable directional filter [25]. The method from [11] utilizes a modification of the ALCM.

Another concept employed in the field of document analysis and belonging to the group of accumulating methods is the classical 1-pixel Hough transform. In this method, the coordinates of the pixels or the centroids of all connected components are accumulated – pixels of the whole domain act as voters [9]. It is a widely used technique for finding straight elements in images. It can be used to determine the slope of elements, to detect skew and slant and also to segment text lines (e.g. [1, 16]). Pixel- and block-based Hough transform can be employed to the task of text lines segmentation [15]. The Hough transform based methods can cope with documents with variations in the skew between lines [13, 23].

A similar idea lays behind the concept of Tensor Voting (TV) method. It is also a representative of the group of accumulating methods in image processing. The TV is based on tensor representation of image features and non-linear voting. Being similar to the Hough transform, it differs from it in a few details. The TV method works more locally – each voter casts its votes in a limited neighborhood. Non-linearity gives a possibility of finding not only straight lines but also 2nd order curves. In the problem of line segmentation the initial tensor field is usually built from central points of connected regions of foreground pixels. The TV is applied to obtain the set of points more likely belonging to actual lines. This set is then used to construct line chains and finally to segment the document.

Tensor Voting is used to estimate non-uniform skew of text lines of printed documents. The authors of the method [9] started from the centroids of the connected components to construct the initial tensor field. Then double voting was performed. The method performs very well on the documents with clearly separated letters even when they are distorted during the digitization process. Unfortunately, the handwritten texts are treated incorrectly by the procedure.

The algorithm was adapted to deal with handwritten text by the authors of [20]. It is based on 2D Tensor Voting. The separation of letters is ensured by artificial clearing pixels along some vertical lines.

A deep learning approach is also used to segmentation. For example, artificial neural networks can be used to recognize lines in the text. Segmentation of handwritten document image into lines applying fully convolutional neural network is presented in the papers [5, 27].

3 Proposed Method

3.1 Tensor Voting in 2D Space

In 1995, the Tensor Voting method was introduced and formally defined in the paper [12]. Since then, it has found various applications in pattern recognition. The method draws inspiration from Gestalt psychology, which asserts that humans tend to perceive shapes such as lines, circles, or ovals in images even when only some points are visible. The goal of the Tensor Voting method is to teach computers how to connect points in an image into shapes in a way that mimics human perception. As mentioned above, the Tensor Voting method is an accumulating method, similar to the Hough transformation. The Tensor Voting method employs symmetric, positive semi-definite tensors of second order as the main component of data processing. In 2D space, a tensor can be defined as shown in (1) and represented as a symmetrical 2×2 matrix.

$$T = [\mathbf{e_1}\ \mathbf{e_2}] \begin{bmatrix} \lambda_1 & 0 \\ 0 & \lambda_2 \end{bmatrix} \begin{bmatrix} \mathbf{e_1}^T \\ \mathbf{e_2}^T \end{bmatrix} \tag{1}$$

The tensor's orthonormal basis is formed by the eigenvectors $\mathbf{e_1}$ and $\mathbf{e_2}$. The eigenvalues λ_1 and λ_2 can be interpreted as the tensor's sizes in both directions, with $\lambda_1 \geq \lambda_2$ and both being non-negative. While the tensor can be decomposed into two

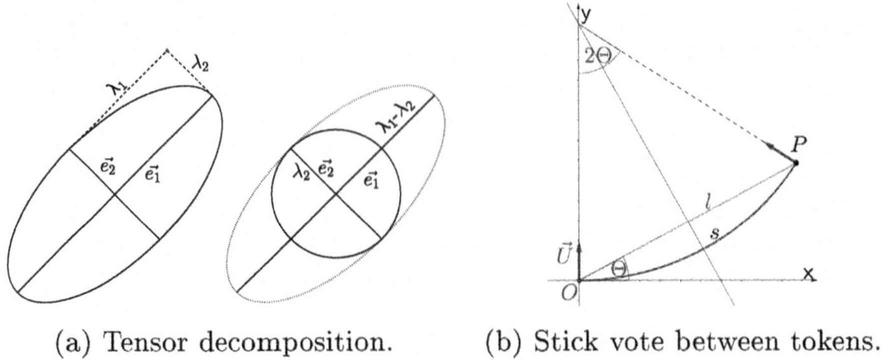

Fig. 1 Tensor features

orthogonal parts as shown in (2), the Tensor Voting method decomposes it into the *stick* and *ball* parts, as in (3). The left part of the (3) is referred to as the *stick tensor*, the right part as the *ball tensor*, and the actual tensor is a linear combination of the two.

$$T = \lambda_1 \mathbf{e_1}\mathbf{e_1}^T + \lambda_2 \mathbf{e_2}\mathbf{e_2}^T \tag{2}$$

$$T = (\lambda_1 - \lambda_2)\mathbf{e_1}\mathbf{e_1}^T + \lambda_2(\mathbf{e_1}\mathbf{e_1}^T + \mathbf{e_2}\mathbf{e_2}^T) \tag{3}$$

The eigenvalue λ_2 can be interpreted as a measure of the tensor's *ballness* or *anisotropic saliency*, which encodes information about junctions, areas, or noise in the image. The difference between eigenvalues λ_1 and λ_2 can be interpreted as a measure of the tensor's *curve saliency* or *stickness*, which represents the certainty that there is a line in the image passing through the given point and whose direction is normal to the $\mathbf{e_1}$ vector. Often, a 2D tensor is represented graphically as an ellipse in which the principal semiaxes have lengths proportional to the eigenvalues. This representation, along with the stick/ball decomposition, is shown in Fig. 1a.

The described method begins with encoding the input image as a tensor field, where the tensors are referred to as *tokens*. The encoding process varies depending on the problem to be solved, but in our case, only vertical unit stick tensors are utilized as described in the next subsection.

After generating the initial tensor field, the voting procedure is carried out. Tokens cast their votes on their neighborhood, either on the other tokens in the *sparse voting* or on all positions in the *dense voting*.

Direction of the vote Now, let us look at the Fig. 1b. There are two tokens presented – **O**, the *voter* and **P**, the *votee*. The most likely a smooth curve going through the two points lays on the osculating circle. In the example, it is the arc s. If the tensor at point **O** is a pure stick tensor with a nonzero eigenvector **U**, which is commonly referred to as the direction of the tensor, then the stick tensor at the votee position will also have a direction perpendicular to the arc s.

Strength of the vote The intensity of the vote is influenced by both the gap between the positions and the angle between the tensors. There are multiple variations of the Tensor Voting method, all of which rely on an exponential decay concerning distance, but differ in how they penalize curvature and measure the distance.

For instance, the Original Tensor Voting (OTV) measures the distance as the arc length, with curvature penalties calculated based on the curvature value κ, as shown in Eq. (4).

$$\mathrm{DF}(l|\sigma) = e^{-\frac{s^2+c\kappa^2}{\sigma^2}}, \tag{4}$$

where $s = 2\Theta r$ is the arc length between the positions of the analyzed tensors, $\kappa = 1/r$ is the curvature, $r = l/(2\sin\Theta)$ is the radius of the osculating circle, c is the constant calculated by the authors of the OTV. The σ value is the *scale of voting* and is the only free parameter of the method.

In the Steerable filters variant of the Tensor Voting (STV), which was introduced in [6], the distance metric is the Euclidean one instead of arc length, and the term that penalizes deviation from straight lines is modified to use a power of a trigonometric function instead of the curvature component. In STV, there is no actual voting performed. Instead, some convolutions of scalar fields in complex number space are performed and then three real valued scalar fields are calculated (*saliency*, *ballness*, *orientation*). No tensor field is calculated, but it can be formed from the three generated fields when needed. The decay function of the STV kernel looks like in (5).

$$\mathrm{DF}(l|\sigma) = e^{-\frac{l^2}{2\sigma^2}} \cos^{2n}(\Theta) \tag{5}$$

Here, l is the Euclidean distance between positions of the tokens, σ is the scale of the voting and n is the parameter used for curvature penalization and have the value 2 as proposed by the authors of this variant. Figure 1b presents all of these parameters. It's worth noting that STV is limited to stick voting and 2D space. There is no ball voting defined for this method. While these constraints may seem restrictive, they don't pose any issues for our algorithm as we only require stick voting in 2D space. One major advantage of the STV formulation is its speed; the voting process is significantly faster than that of the OTV. Moreover, the speed is independent of the number of tokens in the initial field, making the dense voting very efficient. It causes that we can perform the voting almost directly on the image without generating a sparse field, which is not the case for the OTV and the other known formulations of the Tensor Voting.

The reader interested in the details of the OTV formalism can find more information in two books – [17, 19]. The STV is deeply discussed in [6].

3.2 Algorithm

Our algorithm starts from the binary image. It does not address the binarization process. The procedure consists from the following steps.

Step 1 The average height of the character ($\overline{H}$) is calculated as the mean value of the heights of connected components. This parameter is used in the later steps, especially step 6. If the lines of text often touch each other, the value of $\overline{H}$ is too high but the algorithm is not very sensitive to this value.

Step 2 Each foreground pixel in the input image constitutes a tensor in the initial tensor field. All tensors are unit stick ones and are horizontal. This is the realization of the assumption that lines of text are nearly horizontal.

Step 3 The STV equivalent of the voting with the scale σ (varied in experiments) is performed generating the saliency, ballness, and orientation fields. There is no need of tensor field reconstruction because the saliency and orientation are sufficient for the next steps. The saliency field for the fragment of two lines of text can be found in the Fig. 2a. Darker points visualize regions with greater line saliency.

Step 4 The saliency field is smoothed using the Gaussian function with standard deviations $\sigma_x = 30$, $\sigma_y = 3$. This anisotropy is the consequence of the horizontal lines assumption. The values of the σ's were selected in rough experiments conducted on a sample of documents. The effect of the filtering can be seen in the Fig. 2b. Next, the vertical component of the gradient of the smoothed saliency field (s) is calculated using the central differences method ($G_{:,y} = 0.5(s_{:,y+1} - s_{:,y-1})$).

Step 5 Points that most likely belong to the text lines are selected as the ones for which all conditions in (6) are met.

$$\begin{cases} G_{x,y-1} < 0 \\ G_{x,y+1} > 0 \\ s_{x,y} > \rho\,\overline{s} \\ \left|o_{x,y}\right| < \phi \end{cases} \tag{6}$$

The s is the smoothed saliency field, $\overline{s}$ is the mean value of the s, ρ is the threshold of the saliency value, G is the vertical component of the gradient field, o is the orientation field, $|.|$ is for absolute value and ϕ is the angle limit (also varied during experiments). In the Fig. 3, there are shown areas where certain conditions are met. The result of this step can be seen in the Fig. 3b.

Step 6 Having a set of points, we can construct line chains for each text line. First, we look for starting points in the downward direction and then from the left to the right. Each point found (P) becomes a starting point of a polyline constructed as follows. Next points are selected in the moving window of the size $2 \cdot \overline{H} \times 70$ starting at the x position of P and centered on the P vertically. The line starting at P is calculated using linear regression. The

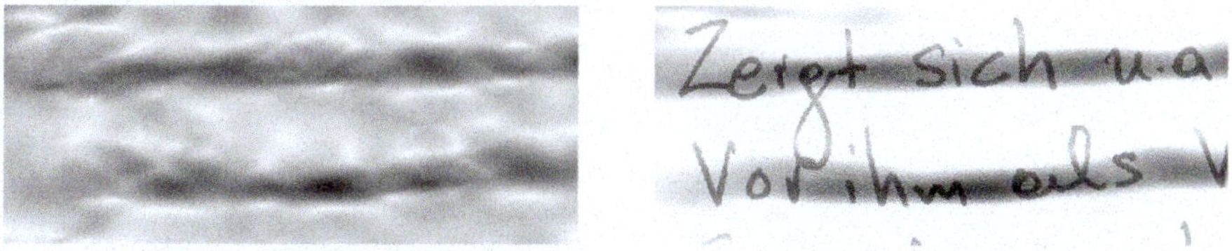

(a) (b)

Fig. 2 Fragment of the: **a** saliency field after voting, **b** smoothed saliency field after voting, overlaid with original text

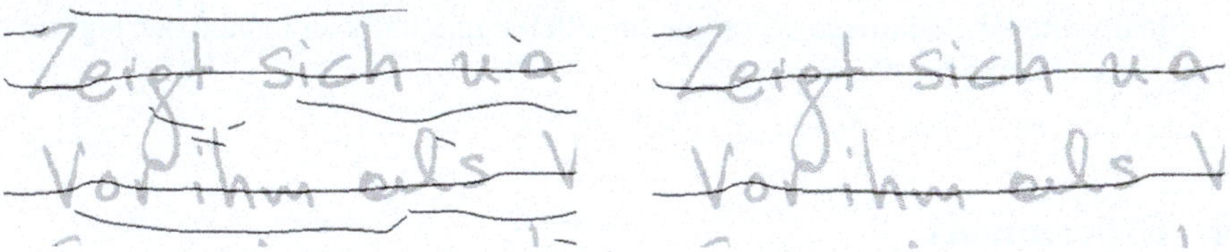

(a) (b)

Fig. 3 Fragment of the document with areas where the conditions in (6) are fulfilled, **a** 1 and 2, **b** all conditions

furthest point in the window such that its distance from the line is less than 5 is selected as the next point in the line chain, all points to the left of this one are cleared and the window is shifted to the new point. The procedure stops at thc right edge of the image or if there are no points in the window. Usually, too many line chains are detected, so the rectifying process should be applied. If the line does not touch any connected component of the original document or touches only components touched by another line chain – the line is removed. After the removal process, the line chains lying close to each other (with the maximal distance in the vertical direction lower than $\overline{H}$) are glued together getting the same order number. The result of this step is shown in the Fig. 4a.

Step 7 After the set of line chains is determined, the text is labeled. The connected regions in the image are identified and analyzed one by one. If the region is touched by one line chain, then all points in this region are labeled with the number of this chain. In the opposite case (i.e. no chain or more than one chain touches the region), each point in the region is labeled with the number of the nearest (using the Euclidean distance) chain. The results of the whole algorithm can be found in Fig. 4b. Among of the correctly classified points we can see also a fragment of the "g" letter incorrectly assigned to the second line. It is a weakness of the presented algorithm which can not separate overlapping letters of two consecutive lines where only one of them has a descender or an ascender.

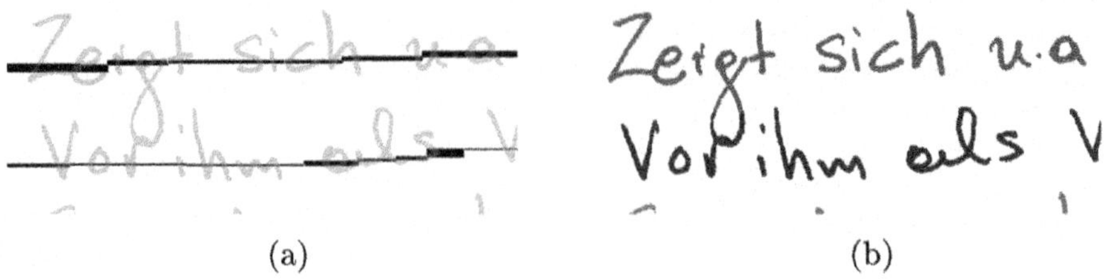

Fig. 4 **a** Line chains for two lines of text, **b** results of the segmentation

Finally, the labeled image is compared with the manually annotated one – ground truth – as described in the Sect. 4.

4 Experiments

4.1 Data-Set and Evaluation Methodology

Our algorithm was tested on a set of handwritten documents taken from the materials of the challenge *Handwriting Segmentation Contest* accompanying the ICDAR 2009 conference [7]. The data-set was divided into two parts – the training set and the benchmark set. The last one contained 200 1-page handwritten documents in four languages (English, French, German and Greek) written by many writers. The total number of lines in the documents was 4034. The training set included 100 documents, similar but less consistent in script, size and layout. This set was taken from the ICDAR 2007 competition. It was accepted as the training data during the ICDAR 2009 contest. All images were black-and-white. The competition organizers manually annotated each document to create the *ground truth* data-set which was used to evaluate the results of the participants. A label indicating which line the pixel belonged to was assigned to each pixel in the image.

Evaluation of the results was based on the one-to-one matching defined in [21], using the MatchScore table. Further details of performance evaluation can be found in the post-competition report [7].

To construct the MatchScore table (7), we start by defining I as the set of foreground pixels in the image, and R_i as the set of pixels recognized as belonging to the ith class, with G_j being the set of pixels in the jth class of the ground truth. The function $T(s)$ gives the number of elements in the set s. The MatchScore table assigns values in the range of [0, 1].

$$\text{MatchScore}(i, j) = \frac{T(R_i \cap G_j \cap I)}{T((R_i \cup G_j) \cap I)} \tag{7}$$

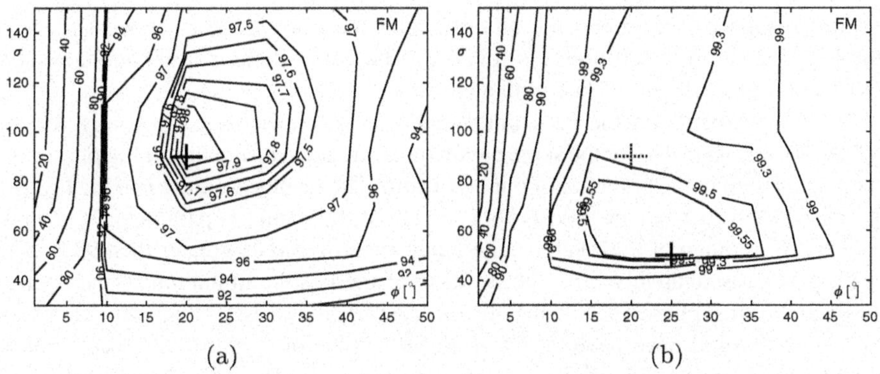

Fig. 5 FM metric for the threshold ρ equal 0.5 **a** training set, **b** benchmark data

A line i is considered a one-to-one match with ground truth line j only if MatchScore(i, j) is greater than the threshold $T_a = 0.95$, which was the accepted value during the ICDAR challenge. Let M be the number of recognized lines, N be the number of lines in the ground truth, and $o2o$ be the number of one-to-one matches. The metrics for detection rate (DR) and recognition accuracy (RA) are defined in (8), along with the value FM, which was used to rate applications during the competition.

$$DR = \frac{o2o}{N},\ RA = \frac{o2o}{M},$$
$$FM = \frac{2 \cdot DR \cdot RA}{DR + RA} \tag{8}$$

Similar competitions were also performed later, during the conferences ICFHR 2010 [8] and ICDAR 2013 [26], however they utilized data-sets that were harder to be segmentated. In the present work, we decided to stay with the simpler set to be able to compare our results with the other method which also incorporated Tensor Voting, presented in the paper [20] and with our previous work [4].

4.2 Experimental Results

The experiments were carried out on the previously described data-set of handwritten documents. We compared the detective performance of the algorithm described in the Sect. 3.2 for various parameters' values. Most of them are defined in (6). In the preliminary experiments, the threshold ρ was varied in the range $0 \div 0.9$, which means the range from accepting all saliency values to rejecting almost all but the largest ones. We found that our method is highly insensitive to this parameter, so the results for only one value are presented – $\rho = 0.5$. The voting scale σ was varied from 30 to 150, the angle ϕ – over the range $0 \div 50°$.

The results of the tuning the algorithm on the training data-set are shown in the Fig. 5a. The best solution was found for the parameters: $\sigma = 90$ and $\phi = 20°$. Location of this pair is marked with a cross.

The experiments on the benchmark data-set were performed not only for the selected values but also for the same parameters' range as during the tuning phase. We were interested in the potential of our algorithm. The results are shown in the Fig. 5b. It can be seen that our method is robust against the σ and ϕ parameters changes. For the wide range of their values, the quality metric *FM* is greater than 99%. The peak plateau is relatively big. The position of the best combination of parameters from the tuning phase is shown with the dashed line cross. This combination of parameters would have been used during the competition giving the result $FM = 99.43\%$. Comparing our results with those of the ICDAR 2009 competition, it can be seen that our algorithm would take the second place, losing only 0.1 percentage point to the winner ($FM = 99.53\%$).

Important observation is that many documents were segmented perfectly. Analysis of the 8 incorrectly segmented documents shown that the lines were detected quite correctly. Usually only some diacritics or touching characters in consecutive lines were misclassified. The restrictive rules of the competition require such cases to be rejected. Solely two lines (from $N = 4034$) were recognized completely wrongly.

5 Conclusions

In the paper, we present a new method of text lines segmentation in handwritten documents. The algorithm is based on a very fast variant of Tensor Voting using the idea of steerable filters. This version allows the substantial simplification of the preprocessing phase in our algorithm. Our procedure starts from the binary image, where each pixel directly becomes the tensor in the initial field to be voted.

Performed experiments showed that the proposed method gives very good results. This makes our method a reliable step in the full analysis of the document. The results are not very sensitive to the changes of the parameters, so the tuning process does not need to be accurate to obtain a satisfactory outcome. On the other hand, deeper investigation of the parameters' space shows that the results can be even better. It shows a need to refine the method of guessing the best parameters for a given document based on its features.

The data-set on which the algorithm was tested is rather simple. Texts are written with clear line spacing, the lines rarely touch each other. In the case of more complicated layouts, our method may behave worse. The labeling stage should be improved to deal with touching lines. Another direction of future development of the algorithm is the simplification of the line chains construction stage. Now, it is a rather complicated and time-consuming process.

References

1. Alaei, A., Nagabhushan, P., Pal, U.: Piece-wise painting technique for line segmentation of unconstrained handwritten text: a specific study with persian text documents. Pattern Anal. Appl. **14**(4), 381–394 (2011)
2. Arivazhagan, M., Srinivasan, H., Srihari, S.: A statistical approach to line segmentation in handwritten documents. In: Document Recognition and Retrieval XIV. vol. 6500, pp. 245–255. International Society for Optics and Photonics (2007)
3. Babczyński, T., Ptak, R.: Handwritten text lines segmentation using two column projection. In: Advances in Intelligent Systems and Computing, vol. 1173 AISC, pp. 11–20. Springer, Berlin (2020). https://doi.org/10.1007/978-3-030-48256-5_2
4. Babczyński, T., Ptak, R.: Line segmentation of handwritten text using histograms and tensor voting. Int. J. Appl. Math. Comput. Sci. **30**(3), 585–596 (2020). https://doi.org/10.34768/amcs-2020-0043
5. Baek, Y., Lee, B., Han, D., Yun, S., Lee, H.: Character region awareness for text detection. In: Proceedings of the IEEE/CVF Conference on Computer Vision and Pattern Recognition, pp. 9365–9374 (2019)
6. Franken, E., van Almsick, M., Rongen, P., Florack, L., ter Haar Romeny, B.: An efficient method for tensor voting using steerable filters. In: European Conference on Computer Vision, pp. 228–240. Springer, Berlin (2006)
7. Gatos, B., Stamatopoulos, N., Louloudis, G.: ICDAR2009 handwriting segmentation contest. Int. J. Doc. Anal. Recognit. (IJDAR) **14**(1), 25–33 (2011)
8. Gatos, B., Stamatopoulos, N., Louloudis, G.: Icfhr 2010 handwriting segmentation contest. In: 2010 12th International Conference on Frontiers in Handwriting Recognition, pp. 737–742. IEEE (2010)
9. Han, S., Lee, M.S., Medioni, G.: Non-uniform skew estimation by tensor voting. In: Workshop on Document Image Analysis, 1997. (DIA'97) Proceedings, pp. 1–4. IEEE (1997)
10. Kavallieratou, E., Dromazou, N., Fakotakis, N., Kokkinakis, G.: An integrated system for handwritten document image processing. Int. J. Pattern Recognit. Artifical Intell. **17**, 617–636 (2003)
11. Kennard, D.J., Barrett, W.A.: Separating lines of text in free-form handwritten historical documents. In: Second International Conference on Document Image Analysis for Libraries (DIAL'06), pp. 12–23 (2006). https://doi.org/10.1109/DIAL.2006.40
12. Lee, M.S., Medioni, G.: Inferred descriptions in terms of curves, regions and junctions from sparse, noisy binary data. In: Proceedings of International Symposium on Computer Vision, pp. 73–78 (1995)
13. Likforman-Sulem, L., Hanimyan, A., Faure, C.: A hough based algorithm for extracting text lines in handwritten documents. In: Proceedings of the Third International Conference on Document Analysis and Recognition, vol. 2, pp. 774–777. IEEE (1995)
14. Likforman-Sulem, L., Zahour, A., Taconet, B.: Text line segmentation of historical documents: a survey. Int. J. Doc. Anal. Recognit. (IJDAR) **9**(2), 123–138 (2007)
15. Louloudis, G., Gatos, B., Pratikakis, I., Halatsis, C.: Text line detection in handwritten documents. Pattern Recognit. **41**(12), 3758–3772 (2008)
16. Louloudis, G., Gatos, B., Pratikakis, I., Halatsis, C.: Text line and word segmentation of handwritten documents. Pattern Recognit. **42**(12), 3169–3183 (2009)
17. Medioni, G., Kang, S.B.: Emerging Topics in Computer Vision. Pearson Education, Upper Saddle River, N.J, Prentice Hall PTR, London (2004)
18. Mehta, N., Doshi, J.: Segmentation methods: A review. Int. J. Res. Appl. Sci. Eng. Technol. **8**, 536–540 (2020). https://doi.org/10.22214/ijraset.2020.31939
19. Mordohai, P., Medioni, G.: Tensor voting: A perceptual organization approach to computer vision and machine learning. Synth. Lect. Image, Video, Multimed. Process. **2**(1), 1–136 (2006)
20. Nguyen Dinh, T., Lee, G.S.: Text line segmentation in handwritten document images using tensor voting. IEICE Trans. Fundam. Electron., Commun. Comput. Sci. **E94.A**(11), 2434–2441 (2011)

21. Phillips, I.T., Chhabra, A.K.: Empirical performance evaluation of graphics recognition systems. IEEE Trans. Pattern Anal. Mach. Intell. **21**(9), 849–870 (1999)
22. Ptak, R., Żygadło, B., Unold, O.: Projection-based text line segmentation with a variable threshold. Int. J. Appl. Math. Comput. Sci. **27**(1), 195–206 (2017)
23. Pu, Y., Shi, Z.: A natural learning algorithm based on hough transform for text lines extraction in handwritten documents. Ser. Mach. Percept. Artif. Intell. **34**, 141–152 (2000)
24. Razak, Z., Zulkiflee, K., Idris, M.Y.I., Tamil, E.M., Noorzaily, M., Noor, M., Salleh, R., Yaakob, M., Yusof, Z.M., Yaacob, M.: Off-line handwriting text line segmentation: A review. Int. J. Comput. Sci. Netw. Secur. **8**(7), 12–20 (2008)
25. Shi, Z., Setlur, S., Govindaraju, V.: A steerable directional local profile technique for extraction of handwritten Arabic text lines. In: 10th International Conference on Document Analysis and Recognition, pp. 176–180 (2009). https://doi.org/10.1109/ICDAR.2009.79
26. Stamatopoulos, N., Gatos, B., Louloudis, G., Pal, U., Alaei, A.: Icdar 2013 handwriting segmentation contest. In: 2013 12th International Conference on Document Analysis and Recognition, pp. 1402–1406. IEEE (2013)
27. Vo, Q.N., Kim, S.H., Yang, H.J., Lee, G.S.: Text line segmentation using a fully convolutional network in handwritten document images. IET Image Process. **12**(3), 438–446 (2018)

Practical Approach to Introducing Parallelism in Sequential Programs

Denny B. Czejdo, Wiktor B. Daszczuk, and Wojciech Grześkowiak

1 Introduction

There is an apparent change of direction in processor development. Instead of looking for a method to speed up clocks continuously, multi-core architectures were created that revolutionized the personal computer market. Despite the change in computer architecture and the introduction of multiple computing units, many developers continue to develop purely sequential solutions for single-core machines. These programs cannot take advantage of the multi-core capabilities of the machines, so users feel no difference in application performance whether running on a 1-core or a 4-core processor. Creating parallel solutions requires deep knowledge, which is time-consuming for many companies to implement. Practical help is necessary to assist developers in mastering parallelism, identifying possible risks of parallel processing, and helping to address them. While some tools on the market support the development of applications for new platforms, there are no solutions that adapt existing applications to new environments, and there are no integrated solutions that focus on both methodology and tools. In addition, there is a need for educational tools that would serve both students and novice developers of concurrency.

Consequently, the main goal of this project was to create a methodology and tools to assist programmers with converting existing sequential into parallel programs. The other goal was to provide practical knowledge for educational and professional activities in this area. We have considered and experimented with two solutions. .NET

D. B. Czejdo
Department of Mathematics and Computer Science, Fayetteville State University, Fayetteville, NC 28301, USA
e-mail: bczejdo@uncfsu.edu

W. B. Daszczuk (✉) · W. Grześkowiak
Institute of Computer Science, Warsaw University of Technology, Nowowiejska Str. 15/19, 00-665 Warsaw, Poland
e-mail: wiktor.daszczuk@pw.edu.pl

W. Zamojski et al. (eds.), *Dependable Computer Systems and Networks*, Lecture Notes in Networks and Systems 737, https://doi.org/10.1007/978-3-031-37720-4_2

Framework (Parallel Extensions [1]) and a Native Approach. Our methodology and experimental tool are applicable to C# programs [2] However, the algorithms are applicable to a general class of high-level C-like languages including C, C++, C#, and VB. The contributions of this paper can be summarized as:

1. Algorithmic detection of sequential code fragments that can be "transformable" to parallel code.
2. Alternative algorithmic transformations of sequential code into parallel code.
3. A dynamic evaluation of transformed code that verifies the assumptions generated by the transformation algorithm.
4. Development of criteria to determine whether it is recommended to use the proposed parallelization based on its dynamic analysis.

The second section of the paper contains the literature review, including the current state and future of code parallelization using ML models. The third section describes the Parallel Extensions [1] as the parallelization tool. Section 4 describes the methodology and algorithms. Section 5 covers common synchronization considerations for proposed solutions. Section 6 concludes the paper.

2 Current and Future Work

To solve a problem in parallel and speed up its execution, it must generally be broken down into smaller subtasks that are as independent as possible. Dependent tasks require some effort for synchronization. The theoretical and practical solutions have been discussed in the literature for a long time. One of the first discussions of programs for parallel processing was by Bernstein in [3]. He established three basic conditions for two sequences of instructions to run independently and concurrently. Amdahl in [4] discussed the theoretical analysis of acceleration limits. Amdahl law was later generalized by Gustafson [5]. o cases of substantial parallelism. This fundamental work was followed by many researchers looking at the specific implementation and performance of massively parallel processors [6].

Another significant development was creating a framework for a parallel runtime environment, Parallel Extensions [1], as part of the .NET Framework. It introduces many helpful syntax elements, classes, and attributes that help creates highly scalable parallel applications, allowing researchers to develop new solutions based on this environment. An alternative framework widely used is OpenMP (Open Multi-Processing [7]), a collection of libraries, directives for the compiler, and environment variables that affect how the program runs and ensure the portability of applications.

Even with these frameworks, there is still a challenge in how to build an "optimal" parallel program. Discovering which code sections can be made to run in parallel is the step the programmers routinely struggle with [8]. There is a well-accepted need for tools to assist in creating parallel programs [8]. Extensive research is directed into the creation of tools to discover parallelism in sequential programs emphasizing static analysis [9], dynamic evaluation [10], or both [9]. The static analysis can use a control dependence graph to capture the maximal parallelism of tasks [11, 12], and dynamic evaluation can determine the effectiveness of the parallel programs [10]. Some research concentrates on addressing only specific aspects of the problem, e.g., uncovering hidden loop-level parallelism in sequential applications as described in [13]. More specific results are reported on loop parallelization with quadratic subscripts [14]. Some approaches concentrate on the widely spread parts of the sequential code that can run in parallel [15] and other emphasize fine-grained parallelism using a task-based approach [16]. Most tools are language dependent, often for C++ [12]. Understanding that most tools fail to auto-parallelize programs with a complex control and data flow, some research was intentionally restricted to provide profiling information of the sequential program, mainly to help the programmer to discover parallelism [17].

An alternative approach is to build a new compiler, based on the generated dependencies graph, to rewrite and organize the program in a task-oriented structure [18]. Yet another approach can be called speculative. The observation in this research is based on the fact that many data dependencies are unlikely during runtime and consecutive iterations of a sequential loop can be executed parallel [19]. Runtime parallelism is obtained when the speculation is correct [19]. A similar approach is used in [20], where the state of speculative parallel threads is maintained separately from the non-speculative computation state.

AI approaches described above can be referred to as algorithmic. In general, the corresponding algorithmic tools have limitations in analyzing complex programs [21]. There are also proposed new data-driven methods that can be applied to parallelism detection. These methods are ML-based (Machine Learning). One ML approach is to use the specialized deep learning model, e.g., implemented as a convolution neural network [21], which leverages the latest and artificial intelligence techniques. The model can accurately detect potential parallelism in sequential programs [21]. This is one of the first experiments with ML and therefore covers only a relatively narrow area of parallelization of sequential programs.

An alternative ML approach is to use a general, very large language model [22] that is pre-trained using many cross-lingual keywords used in a natural language and in programming languages. Recently, there is much discussion about natural language processing (NLP) models and their general capabilities [22]. The authors experimented with parallel program generation and conversion from sequential to parallel programs. The preliminary results of such experiments showed that direct use of such models is very limited and additional extensive training of such models is necessary. Another critical limitation of both general [22] and specific models [21] is that they do not explain the transformations. However, the continuous improvement of ML models explainability provides the potential to alleviate it in the future [23].

3 Parallel Extensions—The Framework Used for Parallelization

One of the important decisions converting sequential programs into parallel programs is to choose the translation target, i.e. a framework supporting parallel applications. We have considered and experimented with two solutions. .NET Framework (Parallel Extensions) and a Native Approach.

Parallel Extensions [1] is part of the .NET Framework runtime environment, developed and promoted by Microsoft. The extension introduces many helpful syntax elements, classes, and attributes that help create parallel applications. In addition, the environment itself is highly scalable, allowing for taking full advantage of different types of multi-core machines. Parallel Extensions can be used in languages supporting the .NET Framework version 4.0 or later (e.g., C++, C#, VB.NET).

The primary unit is a task, an independent functional part of the program that can be executed in parallel with other tasks. The environment manages resources (processors) and correct and optimal task allocation. For this purpose, a special planner (scheduler) was designed to optimize the allocation of tasks to the operating computing units and balance their load distribution. The task-stealing mechanism ensures that all available processors perform the task to be executed.

The environment starts a thread for every processor, with its own task queue. The requests allocate the tasks from the global queue, filled by the application, thread queues. Tasks can produce subsequent tasks, which are placed at the beginning of the thread queue. This practice is related to data caching. In turn, tasks at the end of a given queue can be stolen by other threads (processors), when the latter have no tasks to be performed in their queue or in the global queue.

Parallel Extensions introduce several special instructions that produce tasks, including parallel loops such as `Parallel.For`. An additional advantage of Parallel Extensions is portability—code compiled on a 4-core machine, for example, can use all the cores available on the machine on which it will be run (e.g., 8, 16 or 24 cores). This is not an advantage in applications where the number of threads is explicitly defined, e.g., a computational program using 4 permanently defined threads in the program code, will run on both 1, 4 and 8-core machines, but on the last one will not use all available cores.

For other languages, like C++ (not in .NET) or Fortran, an alternative framework OpenMP (Open Multi-Processing [7]), can be used. It is a collection of libraries, directives for the compiler, and environment variables that affect the program run, and ensure applications portability. It is a scalable system that can be used both on personal and supercomputer computers. OpenMP does not provide language extensions, the parallel constructs are built into the compiler and called using compiler directives. The programmer has deeper control over parallelism than in Parallel Extensions.

4 Conversion into Parallel Program

The essence of the work is to attempt to create algorithms for automatically finding locations in the code of the sequence program that can be parallelized and then transform the code into parallel computation. Such a modified program can be executed concurrently on a multi-processor (multi-core) machine.

In this paper we describe the three types of sequential code constructs that can be parallelized: function calls, instruction paths, and loops. We defined a set of necessary conditions for introducing parallelism. This leads to algorithms that check whether a particular occurrence of each of these constructions qualifies for parallelism. These are original algorithms containing explanation components to be used in both educational and professional environments.

The solutions are based on a static analysis of C# code and dynamic evaluation of performance difference between sequential and parallel approaches.

4.1 Asynchronous Function Call

This section describes the sequential calls to a function and the conditions that such a structure must meet in order to execute it in parallel with the code that is executed after the function is called.

Function calls. If a function is called, we have two sequences: in the function call and after the call. They can be dependent of each other if any Bernstein condition [3] is violated. Thus we can find a longest execution path after the function call that independent, concerning three properties: using a function result (if there is such), using variables passed by reference and using variables subject to the function side effects (of nested function calls side effects). If there is at least 1 statement between the function call and instruction with any of the above properties, it is called *deferred use*. The code from the function call statement to the first deferred use can be executed in parallel with the function body. If no such statement exists, the parallelism can be applied up to the end of the calling function.

Data structures. The algorithm of asynchronous call and the following algorithms require that each statement in the program be labeled with a label that uniquely identifies it. Data structures used in the algorithms are:

- *Call graph* is a directed graph in which the nodes represent functions in the source code being parsed and the edges represent function calls. The edge from node X to node Y means that function X calls function Y. Each function call produces an edge, so multiple edges are possible between a specific pair of nodes (so it is a multigraph in fact). To distinguish them (and for algorithmic purposes), the edges are labeled with the number of the function call.

- *Set of instructions that cause side effects* (CSE). The instruction produces side effects if it changes the value of a non-local variable, or it writes the value to output.
- *Set of instructions that rely on side effects* (DSE). For example, the instruction can read a non-local variable (that can be modified as a side effect).

Every entry in CSE/DSE contains a list of accessed non-local variables/written resources.

Algorithm for asynchronous calls. The algorithm identifies the feasibility of asynchronous call and transforms them in the source code.

Call graph. Initially, the call graph is constructed, based on the static analysis. For library functions, only thread-safe ones can be considered, on the basis of programming environment documentation. The graph construction is as follows: the terminal functions in a call graph form the initial set, or the level 0 set of the analysis. Subsequent levels are defined as follows: level i set contains all functions that directly call functions from level $i-1$ (and optionally, levels lower than $i-1$) and are not in any loop in the call graph (this excludes recursive calls).

In the call graph shown in Fig. 1, level 0 (red) contains H, I, F, K. Function D is not at level 0 because it is in the loop (D-B). Level 1 (blue) consists of E and J. G (orange) is level 2 and C (green) is level 3. A, B and D (grey) are not in any level because they either are in a loop or call functions of an unknown level.

If the function calls other functions (level > 0), then calling statements are included in CSE and/or DSE based on CSE/DSE of the called function. If the function calls library functions, the calling statements are included in CSE based on the documentation. If a library function in not thread-safe (for .NET, see [24]), the level of the calling function is unknown. The use of variables passed by reference go to CSE/DSE, although they are not subject to side effects. All CSE/DSE except access to

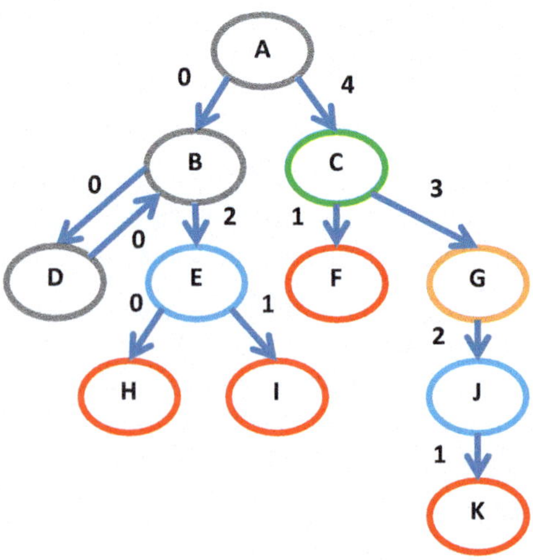

Fig. 1 Example call graph with analysis levels shown

local variables passed by reference are “pulled” to statements calling the function because it can be subject to further analysis.

Dynamic evaluation. *The parallelism up to the deferred use should be large enough* (based on the execution time of both parallel sequences). To analyze each function call, we need 3 timestamps: Z_1, Z_2, Z_3. The Z_1 timestamp should be taken at the beginning of the function called asynchronously, the Z_2 at the end of this function, and the Z_3 timestamp should be taken at the deferred use. We should take the average of timestamp differences from many possible program execution case: $Z = Z_1 - Z_2$, $S = Z_3 - Z_2$. Also, we need to measure the average time of starting a thread for the asynchronous call, *P*. The asynchronous call is reasonable when *Z* and *S* are comparable and both are substantially greater than *P*: $Z \approx S$, $Z \gg P$, $S \gg P$.

Introducing asynchronous call. The asynchronous function call must be inserted into the program in a transparent mode, equivalent to the sequential execution. Typically threads do not return a result; thus the function result must be stored in a local variable of the calling function (if the called function returns a result).

4.2 Parallel Statement Paths

The independent instructions of a function can be executed in parallel by grouping them in instruction paths. If the current state of the program is specified as *Z* and the next statement to execute is *I*, then the state of the program after the execution of the *I* statement, specified as $Z\{I\}$, can be defined as a function transformation $I(Z)$ such that $Z\{I\} = I(Z)$. The state of the system after the execution of two statements, successively I_1, I_2, is specified as: $Z\{I_1;\ I_2\} = I_1(I_2(Z))$.

For a program with a current *Y* state, and three statements I_0, I_1, waiting for the execution, I_1 is a statement that directly depends on I_0 if and only if $Y\{I_0;\ I_1\} \neq Y\{I_1;\ I_0\}$. Instruction I_k is indirectly dependent on I_0 if there is a sequence of instructions I_0, I_1, ..., I_{k-1}, I_k, where I_j directly depends on I_{j-1}, $j = 1, \ldots, k$. Independent statements can be executed in parallel.

An instruction path is a sequence on dependent instructions, ordered as in the source program. The two paths in a statement are independent of each other if they do not contain crosswise dependent statements. The maximum degree of concurrency for a given sequence is the number of independent paths that can be identified.

Functions that can be parallelized by introducing parallel instruction paths must have at most one `return` statement. If such a statement occurs, there must be the certainty that it always finishes the execution of the function (this information must be supplied externally if the statement is not the last one in a function). Otherwise, some instructions in parallel paths can be executed that would not be executed in sequential mode.

Algorithm for independent paths. The independence graph should be constructed, with nodes being instruction labels and used variables (local and non-local). The edges connect the instruction labels with the used variables. Additionally, the entire loop content is connected to the variables in the loop header, and the content of a conditional statement to variables in the condition. The independent path detection algorithm is based on code *slicing* methods [25]. If there are function calls, then their CSE and SDE must be calculated as in the asynchronous calls algorithm. They are the base of adding edges joining function call instructions with non-local variables and local variables passed by reference used in called functions. If there are variables that are only read, their edges are excluded from the graph (because of Bernstein conditions).

If the independence graph contains more than one separate component (with no edge connecting their nodes), it is subject to parallelization. Every component is a base of a separate path, and their ends must be synchronized at the function end.

If there are no separate components of the graph, a consistent part of the function, such as `if-else` statement or only `if`, or a loop content for a single run, can be the subject of parallelization.

Dynamic evaluation. Parallel execution of instruction paths involves creating a separate thread for each independent path, executing those paths in parallel, and joining the control at the end. The parallelization is reasonable when there are at least two paths with comparable execution times, both instantly greater than the average time of starting parallel threads, P. All paths whose execution time is substantially shorter than the two longest ones, especially those with time comparable to or smaller than P, should be incorporated into those longer paths.

The weak point is that the path execution times of the independent paths are hard to measure in a sequential program, as the instructions in the paths can interleave. Therefore, we need to apply the solution to a maximum number of paths, measure their execution times, and then withdraw those not meeting the timing requirements.

Introducing parallel paths. Parallel paths must start at a common beginning. Then for every parallel path a thread must be started, consisting of the instructions from the graph component, placed in the original order.

4.3 *Parallelized Loops*

The loop is typically identified as the most indicative and prone to parallelism. Because of its structure and behavior (repeatedly executed the same piece of code), if the conditions described below are met, the loop iterations can be parallelized.

Loops. In high-level languages, there are two main types of loops: `for` (`foreach` can be converted to `for`), and `while` (also covering `do-while`). The analysis of loops takes into account: the ability to predict the number of iterations, variables access (side effects, parameters, and iteration dependency), and loop nesting.

Number of iterations. If the loop contains premature termination of the loop itself, the containing function, or the entire program, it cannot be parallelized. The number of iterations of non-interruptible loop can be declared statically or dynamically.

Most often, `for` loops are used in cases where the number of iterations is predetermined as a constant, or dynamically calculated before the loop start. `for` loops are often used to iterate through elements of a collection, which size can be fixed, or settled before the loop starts. The `for` loop header contains the initializer, condition, and iterator. The initializer executes before the loop, the condition is checked before each iteration, and the iterator executes after each turn. If the condition is not true, the loop is terminated, i.e. the subsequent turns are not performed.

`while` loops are often used in cases where the number of iterations is unknown and depends on the continuation condition, It is common for a loop to change the values of the variables used in the condition or terminate the loop prematurely (for example, in `while(true)`). So it is extremely difficult to predict the number of turns. For these reasons, only `for` loops are analyzed.

Side effects, parameters, and iteration dependency. Each loop turn can produce side effects that can be exploited by other loop iterations. To the side effects, modification of function local variables is incorporated, including the loop counter. Additionally, the modification of variables used in the loop condition is prohibited. If a loop turn uses side effects caused by another iteration, they are dependent.

However, this definition of the loop iteration dependency is too strong as it results in the iterations using the modified counter in each loop cycle being dependent on each other.

Considering N iterations of the loop, and iterations counted from 1 regardless of actual iterator consecutive values:

- PI_n is a set of variables read in the loop header iterator after iteration n,
- PO_n is a set of variables modified in the loop header iterator after iteration n,
- CI_n is a set of variables that are read in the loop body in iteration n and in the condition of the loop,
- CO_n is the set of variables modified in the loop body in iteration n.

If the variable read in the body of the loop is modified after each iteration only in the loop iterator ($PO_n \cap CO_n = \varnothing \rightarrow PO_n \cap CI_n = P_n$), and its modification is not a function of the variables CO_n of any iteration n: ($\exists_{n=1...N} \exists_{m=1..N} PO_n \neq F(CO_m)$), then this variable value for each iteration n can be calculated before starting the loop turn.

If for $i, j \in (1, \dots, N)$ the following conditions are met: $CI_j \cap CO_i = \varnothing$, $CI_i \cap CO_j = \varnothing$, $CO_i \cap CO_j = \varnothing$ (modified Bernstein conditions [3]), and all variables modified in the loop iterator are loop parameters, then this `for` loop can be parallelized. This definition allows for the parallelization of `for` loops that use a counter iterated after each iteration.

Nested loops. Loop nesting can happen when iterating two-dimensional space or more, for example, rows and columns of an image or matrix multiplication. Four- or more-fold nests are very rare.

For nested loops, we try to parallelize the outermost loop, which provides the highest degree of parallelism. Considering a parallelized loop Z_1 nested inside loop Z_0, notice that at each iteration of Z_0, starts many Z_1 turns in parallel and waits for all iterations to complete. Only then the next Z_0 iteration can start. If Z_0 is parallelized, Z_1 turns from the different Z_0 iterations run parallel. Parallelization and waiting for parallel threads to complete occurs only once in this case.

Loop parallelism. The loop can be parallelized if the following 3 conditions hold:

- loop parameter values for each iteration must be known before the loop starts,
- all iterations should be independent of each other,
- the number of iterations should be known before starting a loop.

The conditions imposed are very restrictive and significantly limit the number of loops that can be parallelized.

Parallelized loop algorithm. Loop analysis concerns parameters, independence and iteration count. The data structures follow the previous techniques, namely call graph, CSE and DSE (the latter two for both non-local and local variables).

Loop parameter analysis. We define a loop parameter as a variable that is read in the loop body, but that is modified only in the loop header. Additionally, the modification of such a variable in an iterator must not be dependent on the variables being modified inside the loop.

If the intersection of the set of variables modified in the loop iterator and the set of variables modified inside the loop is not empty, then the loop cannot be parallelized, because the iterator value of this variable cannot be determined before the loop begins.

The next step is to analyze if the modification of the loop parameters is independent of any variables modified inside the loop. The sets of variables that are read in the iterator and the set of variables modified in the loop must be disjoint. If not, the loop cannot be parallelized since the necessary condition is violated.

Iteration independence analysis. The two sets of read variables and written variables must be calculated, with pulling the DSE and CSE of called functions. If there is a variable that is read and written, the race can occur between loop turns, which prohibits parallelization. This corresponds to the Bernstein condition between two iterations. In the following code, such a race occurs.

```
void count(int* data, int* result, int number)
{
   const int constant = 5;
   int lastResult = 0;
   for(int i=0; i<number; ++i)
   {
      int temp = Math.Pow(data[i], 3);
      temp += constant;
      lastResult = temp + data[i];
      result[i] = lastResult;
   }
}
```

A single iteration both reads and modifies the variable `lastResult`.

Iteration count analysis. If the condition and the iterator modification are in a restricted form and additional assumptions are met (described below), the number of iterations is known in advance.

We assume that the general condition is: <w_1> <op> <w_2>, where w_1 and w_2 are numeric constants or simple variable identifiers, op is an arithmetic operator, and:

- one of (w_1, w_2) is a *loop counter* identifier modified by the same value in every turn,
- the second (w_1, w_2) is a constant or a simple variable called *limit*,
- the *initial value* of the counter should be known before the loop starts,
- one of the following operators is used to modify the counter: + +, −−, +=, −=; the value added we call *modifier*, it is $\pm$ 1 for the former two cases, and it must be a constant or variable value $\pm p$ on the right to the operator in the latter two cases (if variable, it must be fixed during the loop execution),
- the op operator should be one of the following: = =, ! =, <, >, <=, >=.

If w_1 is the counter, w_{10} is the initial value of the counter, and w_2 is the limit, then the number of iterations can be computed depending on the operator used in the loop condition, counter modification operator and modifier value. The accurate analysis takes too much space; we can show for example: for $w_1 = w_2$, $p = 0$, op: +=infinitely many iterations; for $w_1 \neq w_2, w_2 < w_{1_0}$, modifier:++$\rightarrow$ unknown number of iterations; for $w_1 \neq w_2, w_2 > w_{1_0}, p > 0$, modifier $= ++ \rightarrow \frac{w_2 - w_{1_0}}{p}$ iterations. Only the latter case can be considered. We identified 55 cases, from which 18 cases give > 1 iterations. In the following code, the number of iterations can be identified (=`number`):

```
void countShortcuts(FILE* files, MD5* shortcuts, int number)
{
    for(int i=0; i<number; ++i)
    { shortcuts[i] = CountMD5(files[i]); }
}

MD5 CountMD5(FILE* file)
{ //long calculations }
```

Dynamic evaluation. For evaluation, a number of program runs are needed with different data, with measuring the time of every loop turn. There can be thousands of turns, thus a log must be output. If for the typical case at least two turns are of comparable time, substantially longer than the time needed for threads starting and final synchronization, then the parallelization can be reasonably applied.

Introducing loop parallelism. To apply the algorithm, every iteration must be started with its own variable representing the counter. The assigned value comes from the application of *n*-1 modifications (adding the modifier) to w_{1_0}. The condition is unnecessary since it is consumed by parallelism.

In the case of some high-level languages (such as C#), a parallel version of the loop can be used instead of threads, as long as the language or library supports this functionality. In C#, the `for` loop can be replaced by a parallel one—the `Parallel.For` loop available in the Parallel Extensions library.

5 Common Considerations

The asynchronous call can be applied as is, but the number of parallel threads started in parallel paths or parallelized loop can be biggish. It can exceed the physical number of processors/cores or the number of parallel threads assigned to the application by the operating system. If the difference over the physical or logical limit is substantial, threads should be started in groups comparable to the limit.

Synchronization is an important issue in parallelization. An environment like Parallel Extensions offers synchronization points, in which the thread ends are collected before the next instruction starts. Such a synchronization point must be applied before a deferred use occurs in an asynchronous call, function (or its part) end in parallel paths, and the loop end if it is parallelized. If such a mechanism is not available, an available synchronization mechanism must be applied. For example, the instruction after the required synchronization point waits on a lowered semaphore. A thread counter is initialized before the threads branching, and every finished thread decrements the counter in a critical region. The semaphore is raised then the counter reaches 0. In C#, the event mechanism can be used.

There were some limitations of our algorithms, and they include: nesting function calls, such as $f1(f2())$, using exceptions, and using virtual function. The virtual function mechanism was denied because static code analysis cannot predict which type of object may be under a given variable, so it is not known which method, whether the base or inheriting class, will be called.

A side effect is a change in the state of an object (resource) that is not a parameter of the call or interaction with the surrounding environment. Statements that have side effects on the same resource cannot be executed in parallel (without the additional effort of synchronization), e.g., a screen. Resource operations that can be executed in parallel should be determined by the developer.

6 Summary

This paper focuses on the practical methodology for converting sequential programs in C-like languages into parallel programs. Specifically, we provided detailed examination of three fundamental sequential code constructs i.e., function calls, instruction paths, and loops. Careful analysis of these structures has revealed a set of conditions necessary to introduce parallelism in these structures and resulted in algorithms that search for and parallelize these constructs.

We developed an experimental tool that implements the loop parallelization. The tests have been performed on several different programs: multiplication of two-dimensional matrices, searching for the neighborhood of points in N-dimensional space, image recognition, and computing the entropy of information.

The most interesting observation was that most of the loops could not be translated into constructs that could be parallelized. In each such case, the iterations used an external resource, such as an image or a collection of data analyzed by the application. The similar but not that extreme was the result of manual analysis for other 2 sequential constructs discussed in the paper (asynchronous calls and parallel paths).

In cases of matrix multiplication and neighborhood search, 8 loops out of 16 available could be parallelized. Only in the former case the introduction of parallelism resulted in an increase in the efficiency of a given piece of code of about 40%. In other cases, the loop iterations were too short and the introduction of the parallelizing mechanism increased the execution time of the loop.

In our approach the analysis of all structures involved test for independence, allowing for parallelism without the need for additional resource synchronization mechanisms. Future work will may involve examining the dependency and using the information to automatically introduce parallelism with resource synchronization. Such an approach would allow increasing the number structures that could be parallelized.

An alternative discussed in this paper was to use a general purpose ML model to convert sequential programs into parallel. The ML model limitations are described in the paper. They can be alleviated by adding more example data to a ML model.

In summary, the automatic introduction of parallelism has proved to be a very broad subject and one that requires a great many factors to be analyzed. Future alternatives may possibly bypass this necessity by data-driven ML approaches.

References

1. Campbell, C., Johnson, R., Miller, A., Toub, S.: Parallel programming with microsoft .NET: Design Patterns for Decomposition and Coordination on Multicore Architectures. Microsoft Press, Redmond, WA, 224p. (2010). ISBN: 978-0-7356-5159-3
2. ECMA_International: C# language specification. https://www.ecma-international.org/publications-and-standards/standards/ecma-334/. Accessed 26 Jan 2023
3. Bernstein, A.J.: Analysis of programs for parallel processing. IEEE Trans. Electron. Comput. **EC-15**, 757–763 (1966). https://doi.org/10.1109/PGEC.1966.264565
4. Amdahl, G.M.: Validity of the single processor approach to achieving large scale computing capabilities. In: Spring Joint Computer Conference on—AFIPS '67 Atlantic City, NJ, April 18–20, 1967, p. 483. ACM Press, New York, NY (1967). https://doi.org/10.1145/1465482.1465560
5. Gustafson, J.L.: Reevaluating Amdahl's law. Commun. ACM. **31**, 532–533 (1988). https://doi.org/10.1145/42411.42415
6. Danelutto, M., Garcia, J.D., Sanchez, L.M., Sotomayor, R., Torquati, M.: Introducing parallelism by using REPARA C++11 attributes. In: 24th Euromicro International Conference on Parallel, Distributed, and Network-Based Processing (PDP), Heraklion, Greece, 17–19 Feb 2016, pp. 354–358. IEEE (2016). https://doi.org/10.1109/PDP.2016.115

7. Chapman, B., Jost, G., van der Pas, R.: Using OpenMP—Portable Shared Memory Parallel Programming. MIT Press, Cambridge, MA, 384p. (2007). ISBN: 978-0-262-53302-7
8. Atre, R., Jannesari, A., Wolf, F.: Brief announcement: meeting the challenges of parallelizing sequential programs. In: 29th ACM Symposium on Parallelism in Algorithms and Architectures, Washington, DC, 24–26 July 2017, pp. 363–365. ACM, New York, NY (2017). https://doi.org/10.1145/3087556.3087592
9. Dig, D.: A refactoring approach to parallelism. IEEE Softw. **28**, 17–22 (2011). https://doi.org/10.1109/MS.2011.1
10. Li, Z.: Discovery of Potential Parallelism in Sequential Programs, Ph.D. thesis, Technische Universität Darmstadt, Department of Computer Science Parallel Programming, 135p. (2016). https://tuprints.ulb.tu-darmstadt.de/5741/7/thesis.pdf/. Accessed 26 Jan 2023
11. Allen, F., Burke, M., Cytron, R., Ferrante, J., Hsieh, W.: A framework for determining useful parallelism. In: Lenfant, J. (Ed.) 2nd International Conference on Supercomputing—ICS '88, St. Malo, France, 1 June 1988, pp. 207–215. ACM Press, New York, NY (1988). https://doi.org/10.1145/55364.55385
12. Huang, T.-W., Lin, C.-X., Guo, G., Wong, M.: Cpp-Taskflow: Fast task-based parallel programming using modern C++. In: 2019 IEEE International Parallel and Distributed Processing Symposium (IPDPS), Rio de Janeiro, Brazil, 20–24 May 2019, pp. 974–983. IEEE (2019). https://doi.org/10.1109/IPDPS.2019.00105
13. Zhong, H., Mehrara, M., Lieberman, S., Mahlke, S.: Uncovering hidden loop level parallelism in sequential applications. In: 2008 IEEE 14th International Symposium on High Performance Computer Architecture, Salt Lake City, UT, 16–20 February 2008, pp. 290–301. IEEE (2008). https://doi.org/10.1109/HPCA.2008.4658647
14. Lim, A.W., Lam, M.S.: Maximizing parallelism and minimizing synchronization with affine partitions. Parallel Comput. **24**, 445–475 (1998). https://doi.org/10.1016/S0167-8191(98)00021-0
15. Li, Z., Atre, R., Huda, Z., Jannesari, A., Wolf, F.: Unveiling parallelization opportunities in sequential programs. J. Syst. Softw. **117**, 282–295 (2016). https://doi.org/10.1016/j.jss.2016.03.045
16. Tagliavini, G., Cesarini, D., Marongiu, A.: Unleashing fine-grained parallelism on embedded many-core accelerators with lightweight OpenMP tasking. IEEE Trans. Parallel Distrib. Syst. **29**, 2150–2163 (2018). https://doi.org/10.1109/TPDS.2018.2814602
17. Rul, S., Vandierendonck, H., De Bosschere, K.: A profile-based tool for finding pipeline parallelism in sequential programs. Parallel Comput. **36**, 531–551 (2010). https://doi.org/10.1016/j.parco.2010.05.006
18. Fonseca, A., Cabral, B., Rafael, J., Correia, I.: Automatic parallelization: executing sequential programs on a task-based parallel runtime. Int. J. Parallel Program. **44**, 1337–1358 (2016). https://doi.org/10.1007/s10766-016-0426-5
19. Du, Z.-H., Lim, C.-C., Li, X.-F., Yang, C., Zhao, Q., Ngai, T.-F.: A cost-driven compilation framework for speculative parallelization of sequential programs. ACM SIGPLAN Not. **39**, 71–81 (2004). https://doi.org/10.1145/996893.996852
20. Tian, C., Feng, M., Nagarajan, V., Gupta, R.: Copy or Discard execution model for speculative parallelization on multicores. In: 41st IEEE/ACM International Symposium on Microarchitecture, Como, Italy, 08–12 November 2008, pp. 330–341. IEEE (2008). https://doi.org/10.1109/MICRO.2008.4771802
21. Shen, Y., Peng, M., Wang, S., Wu, Q.: Towards parallelism detection of sequential programs with graph neural network. Futur. Gener. Comput. Syst. **125**, 515–525 (2021). https://doi.org/10.1016/j.future.2021.07.001
22. OpenAI: ChatGPT: Optimizing Language Models for Dialogue. https://openai.com/blog/chatgpt/. Accessed 26 Jan 2023
23. Guidotti, R., Monreale, A., Ruggieri, S., Turini, F., Giannotti, F., Pedreschi, D.: A survey of methods for explaining black box models. ACM Comput. Surv. **51**, 1–42 (2019). https://doi.org/10.1145/3236009

24. Microsoft: Thread-Safe Collections. https://learn.microsoft.com/en-us/dotnet/standard/collections/thread-safe/. Accessed 26 Jan 2023
25. Harman, M., Hierons, R.: An overview of program slicing. Softw. Focus. **2**, 85–92 (2001). https://doi.org/10.1002/swf.41

The Digital Twin to Train a Neural Network Detecting Headlamps Failure of Motor Vehicles

Aleksander Dawid, Paweł Buchwald, and Bartłomiej Pawlak

1 Introduction

The lighting system of a motor vehicle plays a critical role in ensuring the safety of the driver and other road users. The lighting system is responsible for illuminating the road ahead, signaling the vehicle's intentions to other drivers, and providing visibility in low-light conditions. The headlamps are the crucial part of a motor vehicle's lighting system. Headlamps are located at the front of the vehicle and are used to illuminate the road ahead. They are typically equipped with high and low beams, which are used for different lighting conditions. Nowadays, when most streets are illuminated by street lamps, drivers may not be aware of the fact that they drive without headlamps switched on. It can lead to car or pedestrian accidents. Modern automatic headlamps are equipped with sensors that automatically turn the headlamps on and off depending on the lighting conditions. Besides this, there are regulations in many countries that headlights should be switched on when the vehicle is on the move. Continuous use of headlights can lead to their failure. According to National Highway Traffic Safety Administration (NHTSA) [1] in the US, Inadequate lighting is a contributing factor in approximately 2% of all motor vehicle crashes. This highlights the importance of proper lighting systems in vehicles. Furthermore, The American Automobile Association (AAA) [2] states that more than half of all pedestrian fatalities occur in low-light conditions, between 6 p.m. and midnight. Therefore, it is important to regularly check and maintain the lighting system to

A. Dawid (✉) · P. Buchwald · B. Pawlak
Department of Transport and Computer Science, WSB University, 1c Cieplaka St., 41-300 Dąbrowa Górnicza, Poland
e-mail: adawid@wsb.edu.pl

P. Buchwald
e-mail: pbuchwald@wsb.edu.pl

B. Pawlak
e-mail: bartlomiej.pawlak@wsb.edu.pl

W. Zamojski et al. (eds.), *Dependable Computer Systems and Networks*, Lecture Notes in Networks and Systems 737, https://doi.org/10.1007/978-3-031-37720-4_3

ensure that it is functioning properly and to ensure maximum visibility while driving. There are some situations on the road in which other road users tries to inform driver to switch on the vehicle lighting system. Usually, by the short blink of long lights. In modern road infrastructure systems, we can find a lot of cameras. They are mostly used to monitor road traffic. There are a lot of scientific works about automatic vehicles detection techniques for collision avoidance system [3, 4] that uses cameras as a main tool. Vehicle detection is far from perfect due to the huge variability of the vehicle shape, cluttered environment and varying lighting conditions. To get a more robust detection system machine learning (ML) techniques are often used [5–7]. Especially, the convolutional neural network (CNN) significantly improved vehicle detection systems. These systems are mostly applied in autonomous vehicles and road safety. In this work, we will use ML methods to detect the proper operation of the vehicle's headlights. As the training set, we will use the images of the vehicle's 3D models equipped with headlamps. Finally, with the help of CNN, we try to detect if the vehicle's headlights are switched on or off using a camera.

2 Parameters of Proper Car Lighting

The headlights of a motor vehicle should be white (or selective white). In the case of xenon lamps, colors from light blue to almost violet may also occur, provided that their brightness is within the acceptable limits. In addition, recent changes in the regulations have also resulted in the addition of separate daytime running lights. More and more often we can meet here with LED lighting, which is characterized not only by increased durability, but also by much lower energy consumption. Rude or simply old headlights can dull and turn yellow over time, interfering with the process of proper road illumination. Another thing we should take care of is that the lights in our vehicle have not only the correct, but above all the same color. This problem basically does not occur in the case of traditional and halogen bulbs. In the event of a burnout, we only replace one piece. We have to deal with xenon lamps a bit differently. Their filaments wear out over time and it is extremely difficult to match a new copy to the brightness of its predecessor. The general technical conditions for the operation of motor vehicles headlights are described in regulation of the Economic Commission for Europe of the United Nation [8]. Collecting the right number of photos of motor vehicles that reflect samples of correct and incorrect lighting is a difficult task. Moreover, the preparation of a sufficiently large set of data requires a long time with the simultaneous risk of not being able to register a vehicle with certain parameters of lighting irregularities. In addition, collecting a database of photos of motor vehicles, which may contain sensitive data from the point of view of the Personal Data Protection Act, is associated with additional difficulties. For this reason, to obtain the appropriate images of motor vehicles with different lighting parameters, it was decided to design a digital simulator following the digital twin concept.

3 The Digital Twin Concept

The "digital twin" approach in computer simulation means creating a digital representation of a real object or system to simulate its operation and test different variants of operation or solutions. The digital twin can be used to simulate different operating conditions of real systems, test new solutions, or predict future system states. This is especially useful for complex systems or facilities that are difficult or impossible to build or study in real life. The digital twin approach is also applicable when real-world testing would be costly or dangerous. In the experiment, the Unity 3D environment was used to build a simulator that allows setting the parameters of the lighting of a car vehicle and the parameters of environmental lighting. The simulator also allows you to personalize the camera settings of a specific scene. Thanks to this, it was possible to generate a large number of photos of motor vehicles with different lighting parameters of the car itself and external conditions. To carry out the simulation, digitized models were used to reflect the external environment and the vehicle. Using the camera component, graphic files showing examples of a vehicle with correct and incorrect lighting were obtained. The use of a digital twin that maps the operation of the car's light system allows you to simulate the switching on or off of car lighting, and to set its parameters. The simulator allows you to automate the generation of car images at different distances and different angles. In addition, environmental lighting setting functions are available to achieve conditions such as driving against daylight, day or night driving conditions. It is also possible to introduce various failures and irregularities into the simulation, such as other parameters of the right and left lights, or values of the lighting emission angles that are unacceptable for road conditions.

4 Unity 3D Environment

The scene model for the simulation of car lighting has been prepared in the Unity 3D environment. To prepare the simulation, standard asset, such as Terrain, Skymaps, were used. We used car models, which were obtained using scanning and photogrammetric methods [9], created on the basis of BluePrint images and obtained from free FBX files. The most important element of the scene for generating target car photos was lighting modeling. In the Unity 3D environment, it is possible to use the following types of Light object (Fig. 1), which is used to simulate lighting in the designed scene.

- **Point Light**—a light source that is located at a single point and emits light in all directions equally
- **Spot Light**—a light source that has an emission point and emits light in a specific direction defined by a light emission cone. It is used to reflect artificial light sources such as flashlights, lamps or vehicle lights on the stage.
- **Directional Light**—a light that is infinitely far away and only emits light in one direction

Fig. 1 Unity 3D scene with light objects

- **Area Light**—a light defined by a rectangle or disk in the scene and emitting light in all directions evenly over its entire area, but only on one side of the rectangle or disk

The parameters of the Light object of the Unity3d environment that allow you to personalize the applied light source are presented in Table 1.

The appearance of the object presented on the scene in the generated image depends not only on the parameters of the external lighting but also on the environmental lighting settings selected in the Windows- > Rendering- > Lighting menu. Unity 3D also allows you to set the environment material (skybox) constructed from 6 photos reflecting the environment of the scene. Environmental parameters also allow you to select the color of the ambient light, which affects the appearance of all illuminated objects. The appearance of the generated images also depends on the settings of the camera object of the Unity 3D environment, which can map the physical parameters of the camera. By manipulating the position properties of the camera object and the car model, it is possible to obtain the movement of the car

Table 1 Parameters of the object simulating lighting in the unity 3D environment

Property	Description
Type	Light source type
Range	It defines how far the light emitted from the centre of the object travels
Spot Angle	Specifies the angle covered by the emitted light beam
Color	Allows you to specify the color of the emitted lighting
Mode	It allows you to change the lighting mode, this setting is related to computational efficiency
Intensity	Specifies the illumination intensity for light sources

and take virtual photos from a set distance. The enabled property of the light class object that was used to reflect the car lights allows you to turn the car lights on or off. The capabilities of the Unity 3D environment allow you to generate a satisfactory number of sample images without the need for long-term observation and waiting for the appearance of a car with the searched parameters of the lights.

5 The Detection Algorithm

In our work, we have used YOLO (You Only Look Once) for detecting proper car lighting. It is a real-time object detection algorithm that is widely used in the field of computer vision. The main idea behind YOLO is to perform object detection in a single pass, without the need for a region proposal step, which is typically required in other object detection algorithms such as R-CNN, Fast R-CNN and Faster R-CNN. This makes YOLO much faster and more efficient than other object detection algorithms. The YOLO algorithm works by dividing the input image into a grid of cells, where each cell is responsible for detecting objects within its corresponding region. The algorithm then uses a convolutional neural network (CNN) to predict the probability of an object being present in each cell, as well as the bounding box coordinates of the object. The YOLO algorithm also includes a Non-Maxima Suppression (NMS) step which is used to remove multiple bounding boxes that correspond to the same object. This step is important because the same object may be detected multiple times in different cells. YOLOv7 is the latest version of the YOLO object detection algorithm [10].

6 Learning Model

Our learning model consists of two classes. We have named them "lights on" and "lights off", which signal the appropriate state of motor vehicle headlamps. The YOLO network was trained to detect these two states using 872 images per class. Our dataset then was divided into a training dataset (748 images) and a validation dataset (124 images), which in turn gives 86% images for training and 14% images for validation. We keep the same format for each image. The resolution was set to 640 $\times$ 640 pixels, color model to RGB, and the file format to JPEG. In Table 2 and Fig. 2, we can see the examples of images taken for our training set. All images have been taken from our digital twin simulation in the Unity3D engine. The photos of vehicles were made according to the rules described in Table 2. We have applied these rules to vehicles with headlights switched on and off. The snapshots of the digital twin simulation were taken in three different lighting conditions named; night, dawn, and midday. The snapshot of these three conditions is shown in Fig. 2. We have not made any further image enhancements in an image editor. The YOLO network required the images to be labelled.

Table 2 The position of the unity 3D camera object relative to the car model

Distance from vehicle	Snapshot's angle names			
	Left	Right	Center	Top
Near	X: -2 Y:1,5 Z: 3	X: 2 Y: 1,5 Z: 3	X: 0 Y: 1,5 Z: 3	X: 0 Y: 3 Z: 3
Middle	X: -2 Y: 1,5 Z: 15	X: 2 Y: 1,5 Z: 15	X: 0 Y: 1,5 Z: 15	X: 0 Y: 3 Z: 15
Far	X: -2 Y:1,5 Z: 30	X: 2 Y:1,5 Z: 30	X: 0 Y: 1,5 Z: 30	X: 0 Y: 3 Z: 30

Fig. 2 Snapshots of vehicle model at three different lighting conditions; **a** night, **b** dawn, **c** midday

For this purpose we have used our labelling program written in Python language with OpenCV library (Fig. 3). In our training set we have used a standard tiny YOLO model suitable for edge IoT computing devices like for example, Jetson Nano. We have performed training procedures using GPU NVIDIA GeForce GTX 960 M with CUDA 5.0 capability.

Fig. 3 Labelling program

7 Results

In our experiment, we used 872 snapshots from Unity 3D game engine as the training set for the YOLO version 7 network. We have trained our network for 500 epochs. The quality of the inference for our model can be measured by the confusion matrix (Fig. 4a). The matrix is made of actual class columns and predicted class rows. The quality of the confusion matrix depends on its diagonal elements. The higher the diagonal values, the better. Looking at the matrix, we can tell that our network can recognize 3D-modeled cars with switched-on and off headlamps with 98% and 97% certainty, respectively. Background FP in this figure refers to background objects that do not belong to either of the classes but are detected as one of them. Our network works fine within computer-generated 3D graphics, but what about real-life pictures of cars? Is this level of graphical detail enough to train the YOLO network to recognize cars with switch on and off headlamps? To test this, we have prepared 110 photos of real cars with switched-on and off headlamps. Using the testing tool included in YOLO software we were able to validate our trained best weights against the real pictures. In the real condition, the confusion matrix looks slightly different (Fig. 4b). The quality of inference, for both classes, is equal to 80%. Besides less precise recognition of classes, the YOLO network also recognizes not present objects. Background FN in the confusion matrix refers to Trash or Non-trash objects missed by the detector and considered as some other background objects. An example of the predicted batch is shown in Fig. 5. We see there are 16 pictures with bounding boxes labelled with the names of classes and the probabilities of their occurrence. Two of them can be considered as mistaken and on another two, the classes were not recognized at all. The lighting in real pictures differs from the simple lighting model of computer graphics. That's why we can observe inference of non-existence classes in the example batch. The network recognizes switched-off headlamps as a part of the bumpy road. The precision (P) of recognition is expressed by several true positive (TP) and false positive (FP) cases expressed by the formula $P = TP/(TP + FP)$. The precision is usually measured at the given confidence. The question is at what confidence the value of the precision is equal to 100%? In Fig. 6a we see the plot of the precision-confidence curve. For all classes, the precision is equal to 100% at the confidence is equal to 86.4%. Another significant parameter describing the quality of machine learning is recall known also as true positive rate or sensitivity. The recall (R) is calculated as the value of $TP/(TP + FN)$. FN is the false negative case. In our batch pictures (Fig. 5) we can see one case of FN. The class was recognized but it is a false class. Recall can be thought of as the fraction of positive predictions out of all positive instances in the data set. Similar to precision, recall is considered in the view of confidence. The plot of recall-confidence is shown in Fig. 6b. If we set the threshold at 0% confidence we can obtain all predictions from our ML model no matter if it is correct or not. At the 50% confidence threshold, we still have 78.6% predictions. We see that the recall curve decays to zero at the 90% confidence threshold. It means that if we set conf = 0.9 in our trained model for the YOLO network we will obtain no prediction. For the perfect classifier, the recall curve should be then the constant

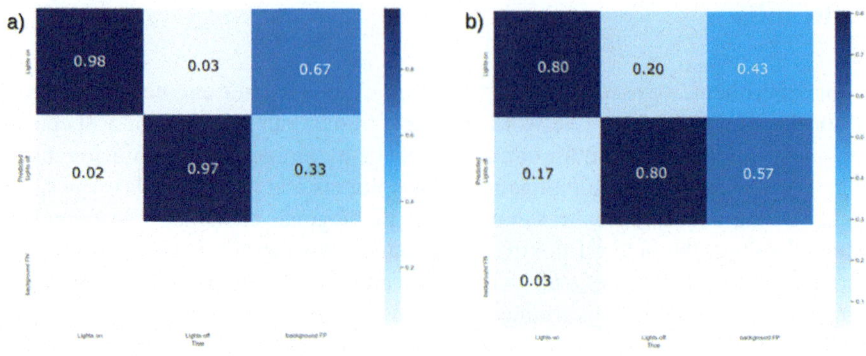

Fig. 4 The confusion matrix for a) the training set and b) the images of the real-life vehicles

Fig. 5 An example of the predicted batch

function of confidence equal to 1.0. It is desired that the algorithm should have both high precision, and high recall. The quality of the classifier is usually measured by the precision-recall (PR) curve (Fig. 7a). The measure of a decent classifier is a high value of the area under the PR curve. Looking at this plot we can tell that our classifier recognizes more "switch on" than "switch off" classes.

For all classes, the best value of confidence that balances precision and recall is 83.3%. If we consider using our classifier in for example road lights sensor we have to know the best confidence threshold. To achieve this goal we generate the plot of the F_1 score function. It is a special case of the more general function F_β [11].

$$F_1 = 2 \cdot \frac{P \cdot R}{P + R}$$

Using the F_1 score curve (Fig. 7b), the balance between precision and recall can be visualized. The best confidence threshold for our classifier is 0.417. At this point, the F_1 score function has its maximum value. The robustness of the network can also be improved by generating additional spatial distorted images that can be included into the training set. There are a lot of examples that this method can improve the quality of trained ML model [12, 13]. This work shows that even a simple model of lighting in the 3D scene can lead to good results in the recognition of vehicles'

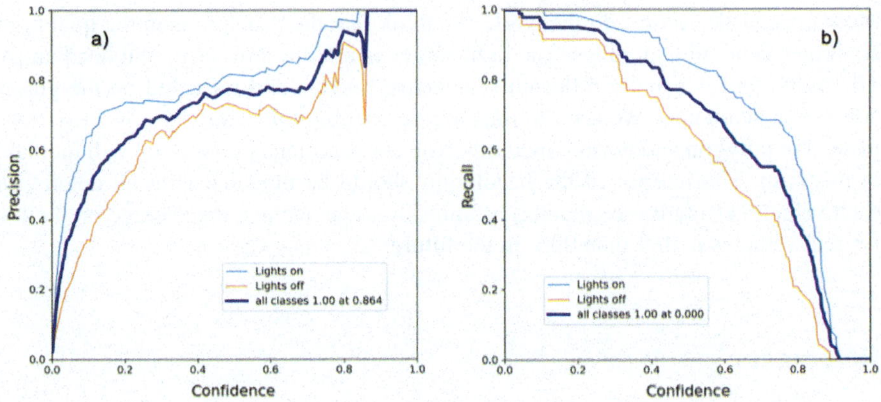

Fig. 6 Confidence dependent **a** precision and **b** recall curve

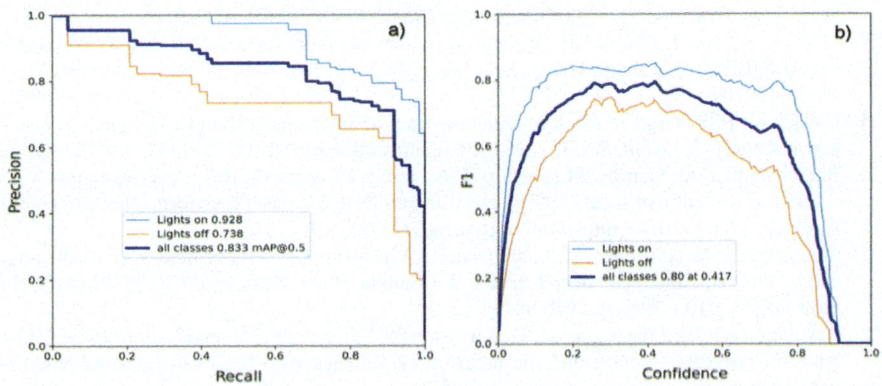

Fig. 7 The plot of **a** precision-recall curve and **b** F1 score function

headlight status. Our research proves the usefulness of digital twins in the machine learning training process. The costs of training the neural network using digital twins are much lower than using real-life images. The process of learning can also be more automatized in the case of digital twin.

8 Conclusions

In conclusion, our work proves that simple digital twins made of polygons can be used as a training set for the machine learning algorithm. Trained YOLO v7 network was able to recognize vehicles headlamps status with average probability equal to 80% in 110 test real vehicles images. We have tried to train the network using more generated images but no significant improvement in inference was achieved.

More epochs also didn't do the work. We think that the problem is in the quality of computer-generated graphics. The light source is not the same as in real headlamps. The light there is usually reflected in concave mirrors and refracted on the glassy cover of a headlamp. We see the improvement in digital twins if we use ray-trace instead of polygon and normal methods. The cost of image preparation will increase in this case. The interior of the headlamps should be modeled with all details for each vehicle. Using these enhanced graphics we can achieve the average prediction for real vehicles higher than 90% in the future.

References

1. Home|NHTSA. https://www.nhtsa.gov/. Accessed 17 Jan 2023
2. AAA|American Automobile Association. https://www.ace.aaa.com/. Accessed 17 Jan 2023
3. Sun, Z., Bebis, G., Miller, R.: On-road vehicle detection: a review. IEEE Trans. Pattern Anal. Mach. Intell. **28**, 694–711 (2006). https://doi.org/10.1109/TPAMI.2006.104
4. Mukhtar, A., Xia, L., Tang, T.B.: Vehicle detection techniques for collision avoidance systems: a review. IEEE Trans. Intell. Transp. Syst. **16**, 2318–2338 (2015). https://doi.org/10.1109/TITS.2015.2409109
5. Dawid, A.: PSR-based research of feature extraction from one-second EEG signals: a neural network study. SN Appl. Sci. **1**, 1536 (2019). https://doi.org/10.1007/s42452-019-1579-9
6. Bhattacharyya, A., Sharma, M., Pachori, R.B., Sircar, P., Acharya, U.R.: A novel approach for automated detection of focal EEG signals using empirical wavelet transform. Neural Comput. Appl. **29**, 47–57 (2018). https://doi.org/10.1007/s00521-016-2646-4
7. Madhan, E.S., Neelakandan, S., Annamalai, R.: A novel approach for vehicle type classification and speed prediction using deep learning. J. Comput. Theor. Nanosci. **17**, 2237–2242 (2020). https://doi.org/10.1166/jctn.2020.8877
8. Regulation No 48 of the Economic Commission for Europe of the United Nations (UNECE)—Uniform provisions concerning the approval of vehicles with regard to the installation of lighting and light-signaling devices [2019/ 57].
9. Palka, D., Sobota, M., Buchwald, P.: 3D object digitization devices in manufacturing engineering applications and services. Multidiscip. Asp. Prod. Eng. **3**, 450–463 (2020). https://doi.org/10.2478/mape-2020-0038
10. Wang, C.-Y., Bochkovskiy, A., Liao, H.-Y.M.: YOLOv7: Trainable bag-of-freebies sets new state-of-the-art for real-time object detectors, http://arxiv.org/abs/2207.02696 (2022). https://doi.org/10.48550/arXiv.2207.02696
11. Goutte, C., Gaussier, E.: A probabilistic interpretation of precision, recall and F-score, with implication for evaluation. In: Losada, D.E., Fernández-Luna, J.M. (Eds.) Advances in Information Retrieval, pp. 345–359. Springer, Berlin, Heidelberg (2005). https://doi.org/10.1007/978-3-540-31865-1_25
12. Szyc, K.: An impact of data augmentation techniques on the robustness of CNNs. In: Zamojski, W., Mazurkiewicz, J., Sugier, J., Walkowiak, T., and Kacprzyk, J. (Eds.) New Advances in Dependability of Networks and Systems, pp. 331–339. Springer International Publishing, Cham (2022). https://doi.org/10.1007/978-3-031-06746-4_32
13. Shorten, C., Khoshgoftaar, T.M.: A survey on image data augmentation for deep learning. J. Big Data. **6**, 60 (2019). https://doi.org/10.1186/s40537-019-0197-0

Dynamic Change of Tasks in Multiprocessor Scheduling

Dariusz Dorota

1 Introduction

Scheduling issues model real systems dedicated to various applications. Embedded IoT (*Internet of Things*), IoV (*Internet of Vehicles*) systems with an enlarged level of security, lead to a special class of problems called multiprocessor task scheduling. Taking into account the trends in the uniformization and modularization of hardware, increased reliability is achieved by introducing hardware redundancy, which allows for software redundancy. Redundancy means to make automatically the right decisions by "voting" among duplicated, identical devices, performing the same functions and executing the same programs for identical data. Unfortunately, if the two devices provide contradictory results, we need to re-calculate the task by three devices or call an arbiter. Systems of this type are already widely used, among others, in aviation, space vehicles, military equipment, rockets, means of transporting hazardous materials, nuclear installations, chemical installations, mining, drones. Ease of implementation extends such solutions to other areas of human activity.

2 Problem Formulation

We start from the formulation of the problem in deterministic case, then we extend this problem in the Sect. 4 to the dynamic case in an unpredictable environment. The initial optimization problem is defined as follows. Set of task $T = \{T_1, T_2, T_3, \ldots, T_n\}$ has to be processed on the set of identical machines (processors) $M = \{M_1, M_2, M_3, \ldots, M_m\}$. Each task T_i requires for processing simultaneous access to $a_i \geq 1$ processors and needs p_i time for processing. Tasks have

D. Dorota (✉)
Cracow University of Technology, 31-155 Cracow, Poland
e-mail: ddorota@pk.edu.pl

W. Zamojski et al. (eds.), *Dependable Computer Systems and Networks*, Lecture Notes in Networks and Systems 737, https://doi.org/10.1007/978-3-031-37720-4_4

to be processed to fulfill some partial precedence constraints given by the graph $TG = (T, E)$, $E \subset T \times T$. Path in the graph ending with the task T_i is denoted by A_i. Our aim to find the schedule of tasks (allocation plus time events) which minimizes the makespan, i.e. length of the schedule. The considered case is known in the literature as scheduling multiprocessor tasks.

The basic assumption is that the task T_i contains one indivisible operation performed simultaneously and synchronously on the predefined number of processors $a_i \geq 1$ in the time p_i. If task interruptions are allowed, task $T_i \in T$ is assumed to consist of a sequence of some x_i operations $O_i = (o_{i,1}, o_{i,2}, o_{i,3}, \ldots, o_{i,x_i})$, executed in this order, each requiring synchronous use of a_i processors. Of course, the below requirement must meet $\sum_{k=1}^{x_i} |o_{i,k}| = p_i$, where $|o_{i,k}|$ is the duration of the operation $o_{i,k}$. In this case, the division of the sentence into operations is not fixed and subject to choice.

In the practice of embedded systems, we most often meet $a_i \leq 3$—for multiprocessor tasks, depending on the required level of reliability or known notion *true parallelism*. Denotation w_i refers to the priority of task T_i, so that $w_i = a_i \cdot p_i$, whereas $w_{i,CT}$ denotes the highest priority of task T_i per time unit.

3 The Current State of Knowledge

One of the first papers on the synchronous use of at least two identical processors to perform tasks in the deterministic case is [3]. Algorithms of polynomial complexity are presented there, however, for a very special case of task execution times and the number of requested processors $1 \ldots a_i \leq k$, where $k \leq m$ is a fixed number. Moreover, the problem has been shown to be NP-hard even for the mentioned run times and arbitrary a_i. A polynomial algorithm is also given for the case with arbitrary run times and 1 or k processor requests. In a special case, the optimization problem was transformed into a linear programming problem. Another paper examining the problem of scheduling multiprocessor tasks is [6], where the case of a four-processor system was analyzed. It has been shown that the scheduling problem without preemption is NP-complete already for $m = 2$. The authors then consider scheduling without preemption of a specific task on any processor in the set of executives for a four-processor architecture. An algorithm was proposed that solves the problem in pseudo-polynomial time. The authors of [2, 7], trying to generalize the problem considered in [3], using single and multilayered neural networks, determine a set of N indivisible, independent, multi-variant and multiprocessor tasks on many identical and parallel processors so that all tasks without violating the time constraints meeting the imposed criterion. At the same time, they assume the occurrence of many variants of one task that may require any number of processors to perform in the range of $1 \ldots M$. The article [10] considers m machines (M) where each machine has a certain number of processors running in parallel. Each M machine can run n tasks $(1 \ldots n)$. Unit processing times were considered, assuming that a single task requires the use of 1 or k of processors. In such an environment, a linear time-optimal algorithm was proposed for a special case of the considered problem. The problem of handling dynamic redundancy on demand in the area of critical embedded systems was considered in the work [4]. Designing systems of this type is aimed at maintaining an appropriate level of reliability, especially in the case of transient failures. The work generated a set of tasks for the study of various redundancy strategies on demand. The tests showed that in the case under consideration, the Quality of Service (QoS) was improved on average by 29% compared to the static redundancy approach. Another argument justifying the use of multiprocessor tasks is the so-called true parallelism, which defines a construction in which functional units are duplicated and more than one program instruction is executed at any time [8]. The paper mentions the following types of synchronous parallel calculations: (A) basic; (B) superscalar; (C) parallel vector; (D) multiprocessor. Another work [11] presents the problem of reliability in flow scheduling problem with reduced resources. The lower bound of the scheduling reliability level is analyzed when managing a greedy redundancy minimization algorithm. The paper proposes three algorithms to meet the given constraints: (A) RR, (B) DRR, (C) dynamic algorithm for long-term failure prediction. The conducted research shows that the proposed approach significantly reduces the level of use of computing and communication resources. On the other hand, authors of [9] concentrate on the problem of scheduling some HPC (high performance computing) applications with the use of parallel processing environments. It is assumed that an application uses multiple type resources, which are available partially or temporary. The impact of various parameters of the problem and solution algorithms on the solution quality were considered. The goal was to minimize the overall execution time or makespan. There were considered a set of moldable tasks under multiple-resource constraints. Authors of [9] compared basically two approaches, namely, list scheduling and pack scheduling.

4 Dynamic Change of the Set of Tasks During Scheduling

We assumed in this paper that each task requires a number of processors to operate organized in the NoC network architecture with $a_i \leq 3$. Let us consider implications of this assumption in details. If for the task T_i we have $a_i = 1$ then the result is unquestioned, since there are no formal objections to this decision. If $a_i = 3$ then the result is usually set by "voting" among three answers, where majority of answers are deciding. So formally, we do not need to re-run this task once again. The most controversial is the case $a_i = 2$. If this task ends with the ambiguous decision, we have to call on-line an additional three-processor task or single-processor arbitration task in order to resolve conflicting decision. This means that we have to change (enhance) dynamically the set of tasks to perform, in the in-predictive, not forecasting way.

We are trying to formulate our approach starting from a simplified model. Let us assume in the sequel that each task T_i with $a_i = 2$ provides acceptable results with probability $(1 - p)$ (for some fixed $p < 1$) and with probability p forces to call additional task with $a_i = 3$ and the same as previously processing time p_i. Taking into account these three cases we can classify scheduling problems as follows: (a) there are no tasks with $a_i = 2$, (b) all task have $a_i = 2$, (c) we have tasks with mixed $a_i \leq 3$. The area of interests of the this paper are cases (b) and (c). More precisely, we consider next only case (c) as the most general and clearly containing case (b).

Scheduling using the dynamic change of the set of tasks consists in *conditional execution* of tasks with $a_i = 2$ in accordance with the algorithm described in Chap. 6. Conditional means that if the responses on different processors are detected to be different, then a three-processor task is added to the task set, planned and performed after the execution of the two-processor task. This case we will call sometimes as the $fault$. This, of course, will increase $C_{\max}$ as well as may causes re-allocation of tasks on processors. Thus, assuming the existence of a probability correlated with the task describing respectively: (A) probability of completing the task, (B) probability of incorrect performance of the task. It may be assumed that (B) implies (A).

We start from the originally defined set of tasks T. Then, we denote by

$$Q = \{T_i \in T : \ a_i = 2\} \tag{1}$$

the set of tasks which require two processors. Without losing generality we can assume that tasks T_i are indexed so that $Q = \{T_1, T_2, \ldots, T_q\}$, $q = |Q|$, $q \leq n$. Let us define the vector of random variables $X = (x_1, x_2, \ldots, x_q)$ where random variable x_i denotes the final result of T_i run, $x_i = 0$ if both processors provided the same result, and $x_i = 1$ if both processors provided different results. In the former case, task T_i is not repeated, in the latter case we have to extend task set T by adding the new three-processor task T_i' with the processing time equal p_i. It is clear that X takes 2^q different values. Thus this describes the set of events for X. Since x_i takes value 0 or 1 independently, then the probability that X takes a particular value $x = (x_1, \ldots x_q)$ is

$$P(X = x) = \prod_{T_i \in Q; x_i = 1} p \prod_{T_i \in Q; x_i = 0} (1 - p) = p^{q-r}(1 - p)^r, \tag{2}$$

where $r = |\{T_i \in Q : \ x_i = 0\}|$. Note that x influences on the schedule as well as on $C_{\max}$ value in an in-predictive way, thus we employ hereinafter denotation $C_{\max}(x)$ for a realisation of vector X.

5 Multiprocessor Tasks

A two-processor task can be treated as a mutual verification of the result of synchronous calculations, as shown in the Fig. 2. If the answers are identical, the task is considered correctly processed and the result is beyond doubt. Otherwise, due to divergent results, an independent arbitrator is needed determining the correctness of the result. This can be done as: (1) executing the computational process a priory as a

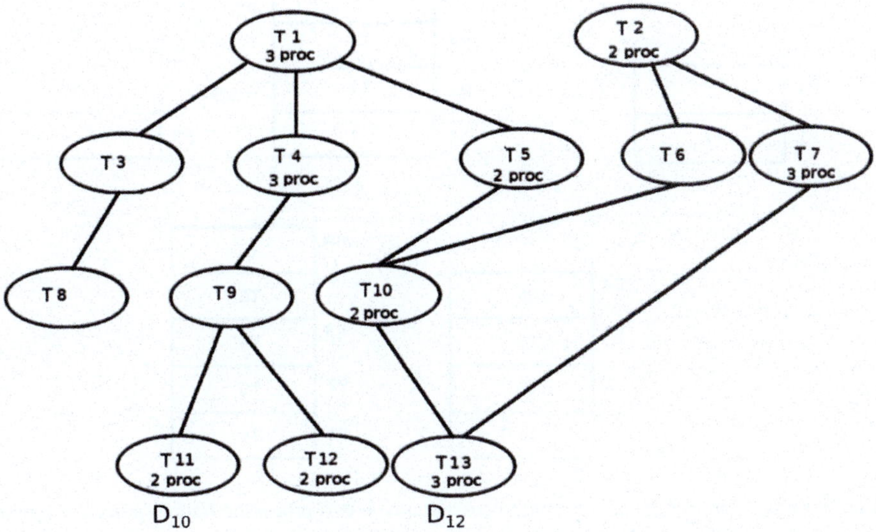

Fig. 1 An instance of depending tasks, containing 1-, 2- and 3-processor tasks

three-processor task (majority vote), (2) executing an additional three-processor task launched on-line on-demand after the failure of the two-processor task, (3) executing on-line an "additional" single-processor arbitration task, launched on demand after a negative result of a two-processor task. Of course, variants (2) and (3) are conditional only when the two-processor task did not give an unambiguous solution.

Thus, the most general concept assumes the simultaneous occurrence of one-processor, two-processor and three-processor tasks, where single-processor tasks may occur always (no computational redundancy is required) or on demand. In order not to unnecessarily increase the size of the entire system, redundancy is implemented only for critical tasks (located on the critical path). These activities ultimately result in increasing the degree of system reliability, which has been demonstrated, among others, by work at [5]. A natural extension of the multiprocessor concept to tasks requiring $a_i > 3$ processors are advanced robotic systems, parallel communication channels, and the so-called true parallelism. The problem specification is usually presented in a task graph, see Fig. 1 for an example. The convention of describing tasks through graph nodes was used there.[1] The actual reliability level can be calculated by the formula

$$D_x = \frac{\sum_{i=x}^{n}(w_i)}{\sum_{i=x}^{n}(p_i)} \quad (3)$$

where numerator is the sum of task priorities (see the algorithm below), and denominator is the sum of processing times.

[1] No a_i entry means there $a_i = 1$.

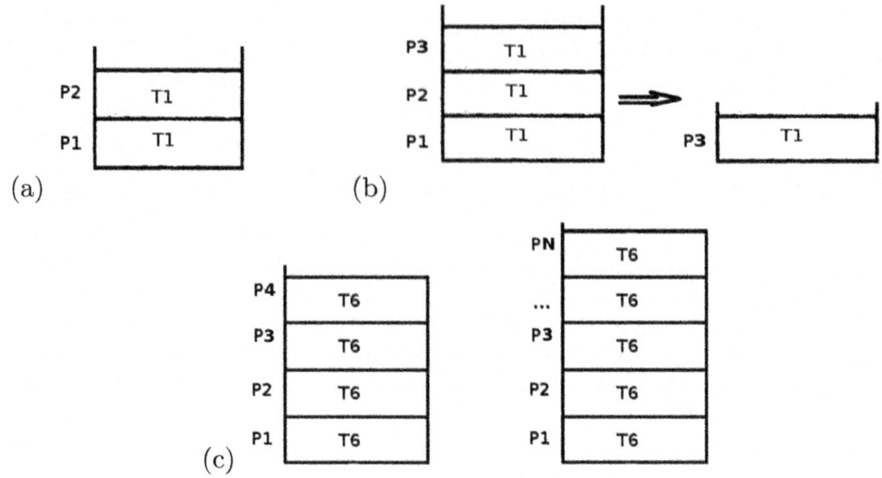

Fig. 2 Suggestions for the use of multiprocessor tasks: **a** two-processor, **b** three-processor (possibility of performing one task on a processor other than the current one), **c** representation of quad and N-processor task

6 Algorithms

In fact, we do not know in advance the current realisation x of random variable X, it has an dynamic character. The values x will be revealed during the scheduling process. Therefore we propose an algorithm that has built in the mechanism of evaluating and reacting on the current x.

Algorithm 1 *Dynamic scheduling of multiprocessor tasks*

1. $t = 0$; $TG_{struct} = read(TG)$;
2. `for` $T_i \in T^Z$ `do` $TG_{level} = calculate\_levels(TG_{struct})$; /* *calculation of task priorities* */
3. $MP = TG_{struct}.p$; /* MP is a set of priorities */
4. $MMP = \max\{w_i : T_i \in MP\}$; /* setting maximum priorities in the MP set */
5. $HP = \{T_j \in T^Z : w_j = MMP\}$ /* find all tasks from the MMP set */
6. `for` $T_i \in HP$ `do`
7. `if` $(w_i > 1)$ /* analyze tasks with non-zero priorities */
8. `if` /* If H_i is identical for more than one task */
9. $LHP = \max(TG_{struct}.high)$; /* LHP is a set of jobs with the same H_i from set HP */
10. `if`
11. $w_i = \max\{w_j : T_j \in LHP\}$ /* select the task with the highest index from the LHP */
12. `else` $w_i = \max\{w_j : T_j \in HP\}$
13. `else` $w_i = MMP$;

14. `end for;`
15. `if` $T_j^{rt} == t$
16. `if` $(p_j - [p_j] == 0)$
17. $x = 1$;
18. remove w_j from the set MP;
19. `if` $(a_j == 2)$ `and` $(x_j == 1)$
20. addTo the set T of tasks the additional task (T_j' with $a_j = 3$), Modify TG_{struct}
21. goto step 2
22. `else` $x = p_j - [p_j]$;
23. `end for;`
24. `for` $T_j \in TG_{level}$ `do`
25. rank T_j for the time unit x according to Mc'Naughton's algorithm;
26. `if`
27. `go to` step 8 /* The $P(t)$ function checks if there is an unoccupied processor/processors in the time unit x, in this case the processor is the same as the service station */
28. `else`
29. $p_j = p_j - x, t = t + x$;
30. `end for;`

Algorithm 2 *Function* $calculate\_levels(TG_{struct})$ *for task leveling:*

1. `for` $T_j \in TG_{path}$ `do`
2. `if` $(a_j == 2)$ *and* $(x_j = 1)$
3. *find*(x_i) /* *recalculate* x_i *for* (T_j' *with* $a_j = 3$) */
4. $w_j = p_j \cdot a_j$;
5. $TG_{level}.add(w_j, p_j)$;
6. `end for;`
7. $w_j = \sum_{k \in A_j} p_k \cdot a_k$;
8. $TG_{level}.add(w_j, p_j)$;
9. `end for;`
10. `for` $w_i \in TG_{level}$ `do`
11. `if` $\exists(\max(TG_{level}.time) \leq D_i)$
12. $w_i = \max(TG_{level}(w_j))$
13. `else`
14. *There is no timetable*
15. `end for;`

A brief description of the algorithm activity is as follows. The data for the algorithm contains $p_j, a_j, j = 1, \ldots, n$, and the precedence of tasks in the form of graph TG, see for example Fig. 1, where requirements for multiprocessor tasks have been marked. If TG has no arcs, we say that tasks are independent. For any 2-processor task, there is a 3-processor equivalent which will be run if the 2-processor task finishes with ambiguous decision. Thus, Algorithm 1 is run for the tasks from initial set

Table 1 Data for the instance $n = 13$. Independent tasks

task	p_j	a_j	w_j
T1	30	3	90
T2	10	2	20
T3	5	1	5
T4	10	3	30
T5	15	1	15
T6	20	1	20
T7	25	3	75
T8	25	1	25
T9	10	1	10
T10	20	2	40
T11	20	2	40
T12	10	2	20
T13	20	3	60

T dynamically extended by additional tasks from the set $\{T_j : j \in Q, x_j = 1\}$, see step 19. The algorithm generates the schedule with the makespan $C_{\max}(x)$ depending on the distribution of detected faults.

The main idea of the algorithm consists in the analysis of task priorities $w_j = p_j \cdot a_j$, taking into account the position of each task on the path (if any exists). Task priorities, called next $levels$, are calculated by auxiliary $calculate\_levels(.)$ function, see Algorithm 2. Algorithm 2 analyse all paths A_j in TG ending in the node T_j, so that T_j is available for processing from the graph of precedence point of view and taking into account constraint on D_i. This algorithm returns suitable priorities. Coming back to Algorithm 1, the subset HP of tasks, which are ready for execution at the fixed time moment $ready\ time$ is considered. Tasks with the highest value of priority MMP, selected among the set of priorities MP, are searched for possible scheduling if only there are idle processors. Allocation is done according to the McNaughton algorithm, see for example [12], for a given time unit. If there are no idle processors in this time unit, or there is no way to assign any tasks to unused processors, then the time will be shifted to the next time unit, and the algorithm is repeated until it schedules all tasks.

7 Experiments and Results

For the approach discussed in Sect. 4 and the Algorithm 1 described above, simulation experiments were performed to check the quality of the idea. We consider several instances with various properties. The main instance has $n = 13$ and appears in two versions, with independent and with dependent tasks. Data for this instance are

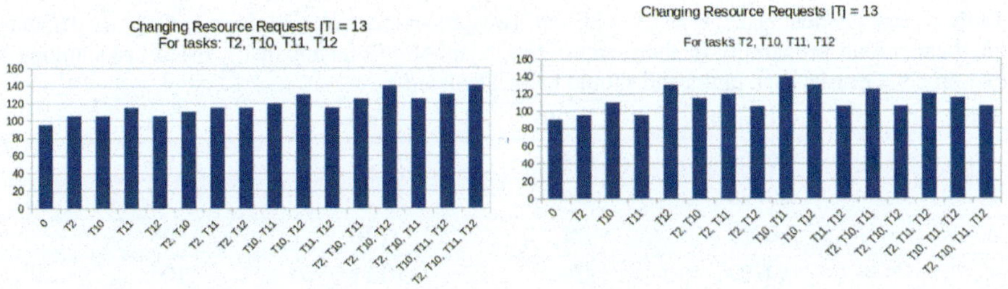

Fig. 3 Histogram for task scheduling with dynamic change of resources, $n = 13$; dependent tasks (left), independent tasks (right)

given in Table 1 and these data are used for the case of independent tasks, therefore priorities w_j in the table have been calculated for independent version. For dependent version we get data from the Table 1 and precedence graph given in the Fig. 1. The instance has four tasks with $a_i = 2$, namely T_2, T_{10}, T_{11}, T_{12}. Therefore we have $Q = \{T_2, T_{10}, T_{11}, T_{12}\}$, $r = 4$ and $|X| = 2^4 = 16$ called next as *cases*. For each realisation x of the random variable X we calculate the probability of occurrence the event and the makespan. We set in this experiment probability of correct answer for two-processor tasks equals $1 - p = 0.9$, so $p = 0.1$. Results are shown in Table 2 for independent tasks and also in Fig. 3 for dependent tasks (left) and independent tasks (right). Symbol a/b means that a listed tasks ends with *faults* and need to be recalculated.We observe the decreasing probability with the increasing number of faults. Indeed, zero faults appears with probability 0,65, whereas four faults with probability 0,0001. From the table one can calculate the mean value of the makespan $E[C_{\max}(x)] \approx 94.1$. More detailed analysis reveals that basic case (zero faults) has participation in the mean on the level $\approx 63\%$, and those with at most two faults takes a part $\approx 1\%$ each. The influence of the remain cases (more than two faults) on $E(.)$ is insignificant. Indeed, calculating the mean value for cases with at most two faults we get 93.6, which is a quite good approximation. These means that we can reduce in this instance the number of various cases to a few primal configurations (assuming a small number of faults).

To confirm the suggested hypotheses we carried out additional experiments with $n = 9$ and $n = 15$, see Figs. 4 and 5. Both instances are obtained through some modification of mentioned already instance $n = 13$, see own paper [5] for details of the data. Instance $n = 9$ allows at most 2 faults among 5 tasks. Probabilities of successive cases are from the left: 0.81, 0.09, 0.09 and 0.01. The "potential"participation of wrong answers is $2/9 \approx 22\%$ and is rather low. Instance $n = 15$ has this participation $5/15 \approx 33\%$. Correct answer is provided with probability 0.59, whereas 5 faults appear with probability 10^{-5}, thus are insignificant. In both instances the differences between makespans are small.

For each task specification, two additional cases have been considered as the reference solutions: (A) the original system specification without changing the resource requests; (B) replacing a prior all 2-processor tasks with 3-processor tasks. The

Table 2 Independent tasks, $n = 13$, with four two-processors tasks. Makespan C_{max} for 16 identified cases with probability of their occurrence. Symbol a/b means that a listed tasks among b two-processor tasks have faults and should be recalculated

Case	Probability	Faults	Tasks	C_{max}
1	0.6561	0/4		90
2	0,0729	1/4	T2	95
3	0,0729	1/4	T10	95
4	0,0729	1/4	T11	105
5	0,0729	1/4	T12	105
6	0,0081	2/4	T2, T10	110
7	0,0081	2/4	T2, T11	115
8	0,0081	2/4	T2, T12	105
9	0,0081	2/4	T10, T11	120
10	0,0081	2/4	T10, T12	105
11	0,0081	2/4	T11, T12	115
12	0,0009	3/4	T2, T10, T11	130
13	0,0009	3/4	T2, T10, T12	120
14	0,0009	3/4	T2, T11, T12	125
15	0,0009	3/4	T10, T11, T12	130
16	0,0001	4/4	T2, T10, T11, T12	140

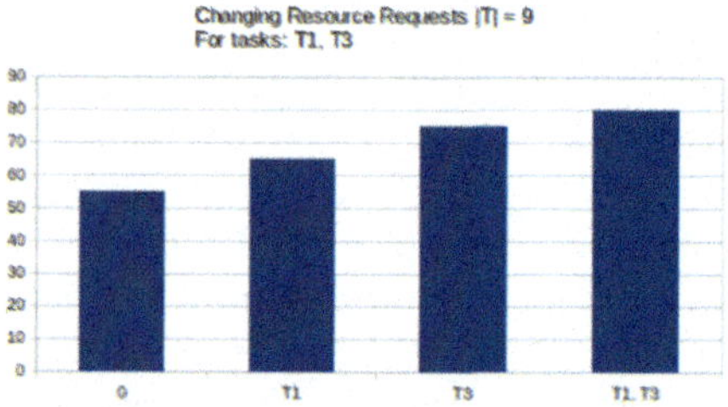

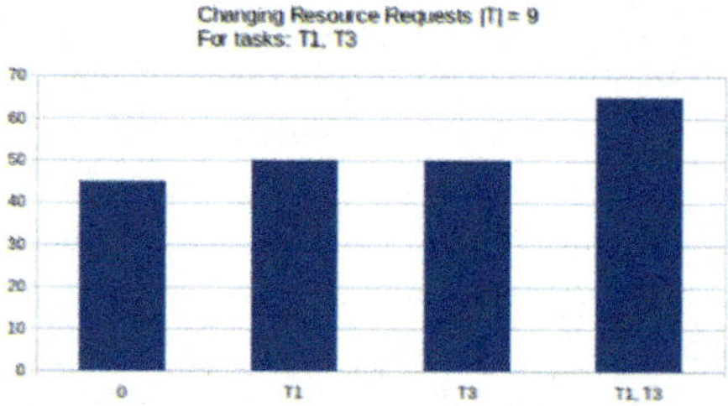

Fig. 4 Histogram for task scheduling with dynamic change of resources, $n = 9$; independent tasks (left), dependent tasks (right)

experiments were carried out on an architecture consisting of 5 processors organized in a NOC network. That approach can be applied in embedded system organized in NoC topology. When that kinds of system can allows to reconfigure, the idea of apply dynamic change of tasks makes more sense. So ideally architecture for that approach may by SoC-FPGA [1].

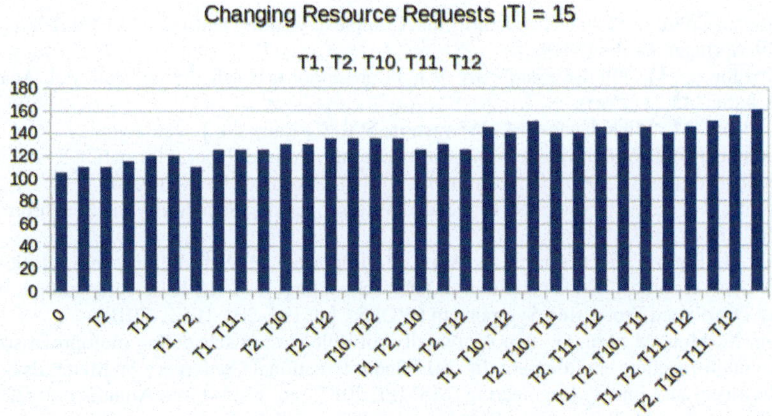

Fig. 5 Histogram for task scheduling with dynamic change of resources, $n = 15$, dependent tasks

8 Summary

One can observe a direct increase in the length of the schedule to the increasing number of changes in resource requests. This parameter increases for both dependent and independent tasks, and is in most cases (the maximum possible number of processor replacement requests) the same as for replacing all 2-processor tasks with 3-processor tasks. In a few cases, the length of the scheduling is shorter than when changing tasks to 3-processor, but this applies only to cases with dependent tasks. It can therefore be concluded that in systems with a high probability of incorrect results when performing tasks on two processors for all tasks, it is more advantageous to change them to 3-processor tasks. In other cases (other than the maximum possible number of requests for resource changes), it is beneficial to use the change of resource requests. Considering the number of events 2^q, it seems reasonable to narrow down the number of experiments, ignoring high probabilities of correct task completion.

References

1. Alkhafaji, F.S., Hasan, W.Z., Isa, M., Sulaiman, N.: Robotic controller: Asic versus fpga-a review. J. Comput. Theor. Nanosci **15**(1), 1–25 (2018)
2. Błądek, I., Drozdowski, M., Guinand, F., Schepler, X.: On contiguous and non-contiguous parallel task scheduling. J. Sched. **18**, 487–495 (2015)
3. Blazewicz, J., Ecker, K.H., Schmidt, G., Weglarz, J.: Scheduling in Computer and Manufacturing Systems. Springer Science & Business Media (2012)
4. Caplan, J., Al-Bayati, Z., Zeng, H., Meyer, B.H.: Mapping and scheduling mixed-criticality systems with on-demand redundancy. IEEE Trans. Comput. **67**(4), 582–588 (2017)
5. Dorota, D.: Dual-processor tasks scheduling using modified muntz-coffman algorithm. In: Contemporary Complex Systems and Their Dependability: Proceedings of the Thirteenth Inter-

national Conference on Dependability and Complex Systems DepCoS-RELCOMEX, pp. 151–159. Springer, Berlin (2019)
6. Drozdowski, M.: On the complexity of multiprocessor task scheduling. Bull. Pol. Acad. Sci. Tech. Sci. **43**(3) (1995)
7. Kłopotek, M., Michalewicz, M., Wierzchoń, S.T., Czarnowski, I., Jędrzejowicz, P.: Artificial neural network for multiprocessor tasks scheduling. In: Intelligent Information Systems: Proceedings of the IIS'2000 Symposium, pp. 207–216. Springer, Berlin (2000)
8. Sriram, S., Bhattacharyya, S.S.: Embedded Multiprocessors: Scheduling and Synchronization. CRC Press (2018)
9. Sun, H., Elghazi, R., Gainaru, A., Aupy, G., Raghavan, P.: Scheduling parallel tasks under multiple resources: List scheduling vs. pack scheduling. In: 2018 IEEE International Parallel and Distributed Processing Symposium (IPDPS), pp. 194–203. IEEE (2018)
10. Xin, X., Mou, M., Mu, G.: A polynomially solvable case of scheduling multiprocessor tasks in a multi-machine environment. In: 2017 2nd International Conference on Materials Science, Machinery and Energy Engineering (MSMEE 2017), pp. 1746–1749. Atlantis Press (2017)
11. Zhao, L., Ren, Y., Sakurai, K.: Reliable workflow scheduling with less resource redundancy. Parallel Comput. **39**(10), 567–585 (2013)
12. Żurowski, M., et al.: Podzielne szeregowanie zadań z pozycyjno-zależnymi czasami wykonywania na dwóch równoległych identycznych maszynach. Ph.D. thesis, Adam Mickiewicz University, Poznań (2019)

Regression Models Evaluation of Short-Term Traffic Flow Prediction

Paweł Dymora, Mirosław Mazurek, and Maksymilian Jucha

1 Introduction

In most cities around the world, the problem of traffic flow is a challenge for city traffic managers. The rapid increase in the number of vehicles causes an increase in traffic volumes for which the urban street network is not prepared. This phenomenon consequently causes difficulties for individual and public transport, decreases travel time and vehicle regularity; vehicle operating costs increase. In traffic control and management algorithms, elements of artificial intelligence can be applied, attempts can be made to make predictions on the basis of historical data of the traffic situation, and signaling can be controlled to improve the crossing of the city. In optimizing traffic control and management, the parameters responsible for traffic flow and capacity are important. With this in mind, the issue of traffic analysis for finding certain patterns or mathematical models with the right fit seems worth investigating. Such a solution could support the design of road infrastructure, especially in urban areas. It could also be useful when analyzing changes in traffic volumes in particular sectors of the city.

This paper aims to model, based on an analysis of available historical data, certain traffic characteristics and to create, using these, a model capable of determining predictable traffic volume values. The research was aimed at analyzing the possibility of using and the fit degree of different variants of the regression model: polynomial, trigonometric, and polynomial-trigonometric, based on the Random Forest algorithm

P. Dymora (✉) · M. Mazurek · M. Jucha
Rzeszów University of Technology, 35-959 Rzeszów, Al. Powstańców, Warszawy 12, Warszawy, Poland
e-mail: pawel.dymora@prz.edu.pl

M. Mazurek
e-mail: mirekmaz@prz.edu.pl

M. Jucha
e-mail: 1641161@stud.prz.edu.pl

W. Zamojski et al. (eds.), *Dependable Computer Systems and Networks*, Lecture Notes in Networks and Systems 737, https://doi.org/10.1007/978-3-031-37720-4_5

and linear regression in MS Excel. The results are presented in the form of graphs and charts.

2 Selected Methods of Determining Regression Models

Traffic modeling and load forecasting are important research problems. The problem of short-term traffic flow analysis is a relatively complex issue. This is primarily due to the fact that the behavior and decisions of drivers on the road are influenced by a great many factors, which are often difficult to define precisely and even more difficult to measure accurately. Many studies have been carried out over the years. Over the years, a large number of studies have been produced on the subject, and the spectrum of aspects covered in these. The spectrum of aspects covered is very diverse. To model the road network, the concept of a graph has been used. A graph is nothing more than a collection of points, called vertices or nodes, between which there are connections, called edges [1, 2]. In the context of traffic, it seems most obvious to represent intersections as vertices and streets as the edges connecting them. However, due to the nature of the survey data, this is not possible. For the study, a model was adopted that took the form of a directed graph with streets as vertices and the edges as the connections between these streets. The weight of each edge was the current traffic volume on it, described by a mathematical function. These functions were determined by training a regression model on real measurement data. A directed graph is a type of graph whose edges are 'unidirectional', i.e. movement on them is only possible in one direction. Using this type of graph allows separate weights to be assigned to a pair of edges representing bidirectional movement between two vertices.

Each edge of the graph is assigned a regression model that determines its importance at a given point in time. A regression model defines a numerical relationship between a set of explanatory (input) variables and an explanatory (input) variable [3, 4]. The difference between the actual value and the value predicted by the regression model is called the residual of the model. In this paper, different variants of the regression model were tested to achieve the highest possible precision in approximating the actual measurement data. The variants tested included: the polynomial model, trigonometric model, polynomial-trigonometric model, and model based on the Random Forest algorithm.

2.1 Polynomial Model

The classic type of regression model takes time to raise to different powers as the describing variables. The model can be written in the general form [5]:

$$f(t) = \beta_0 + \beta_1 \cdot t + \beta_2 \cdot t^2 + \beta_3 \cdot t^3 + \beta_4 \cdot t^4 + \ldots,$$

where: $\beta_0, \beta_1, \beta_2, \ldots$ are the successive coefficients of the model and t is the time given in hours.

2.2 Trigonometric Model

This type of model uses a set of transformed sinusoids and cosinusoids as explanatory variables. It works very similarly to the Fourier series [6]. The model can be written in general form:

$$f(t) = \beta_0 + \beta_1 \cdot \sin\frac{2t\pi}{24} + \beta_2 \cdot \cos\frac{2t\pi}{24} + \beta_3 \cdot \sin\frac{4t\pi}{24} + \beta_4 \cdot \cos\frac{4t\pi}{24} + \ldots,$$

where: $\beta_0, \beta_1, \beta_2, \ldots$ are the successive coefficients of the model and t is the time given in hours.

2.3 Polynomial-Trigonometric Model

It is a model that combines some features of the polynomial and trigonometric models. It takes both times raised to different powers and transformed sine and cosine as explanatory variables. The model can be written in general form:

$$f(t) = \beta_0 + \beta_1 \cdot t + \beta_2 \cdot t^2 + \beta_3 \cdot \sin\frac{2t\pi}{24} + \beta_4 \cdot \cos\frac{2t\pi}{24} + \ldots,$$

where: $\beta_0, \beta_1, \beta_2, \ldots$ are the successive coefficients of the model and t is the time given in hours.

2.4 Model Based on Random Forest Algorithm

The Random Forest algorithm uses a random set (forest) of decision trees to determine the relationship between the explanatory variables and the explanatory variable. This means that the results obtained in this way are less stable than the previously mentioned methods, which are based on linear regression [7]. The number of decision trees in the forest has a huge impact on the precision of the fit of this model.

3 Traffic Flow Dataset and Rzeszow City Road Network Model

The dataset contains information about traffic volume measured on individual connections within an intersection with hourly intervals. The measurements were taken in Rzeszów on July 12th, 2022. The goal is to create a graph with edge weights corresponding to the current traffic volume measured in vehicles per minute. In order to transform distinct measuring points into continuous functions, we prepared a regression model for each edge of the graph. The most effective way to do so is to use one universal model with a fixed set of explaining variables and exchangeable sets of coefficients, one set for each edge. To test the model, the dataset was limited to the selected intersection. This intersection has 10 monitored connections:

1. Dąbrowskiego -> Podkarpacka (*Db_Pk*)
2. Powstańców Warszawy -> Batalionów Chłopskich (*PW_BCh*)
3. Podkarpacka -> Dąbrowskiego (*Pk_Db*)
4. Batalionów Chłopskich -> Powstańców Warszawy (*BCh_PW*)
5. Dąbrowskiego -> Powstańców Warszawy (*Db_PW*)
6. Dąbrowskiego -> Batalionów Chłopskich (*Db_BCh*)
7. Powstańców Warszawy -> Dąbrowskiego (*PW_Db*)
8. Podkarpacka -> Powstańców Warszawy (*Pk_PW*)
9. Batalionów Chłopskich -> Dąbrowskiego (*BCh_Db*)
10. Batalionów Chłopskich -> Podkarpacka (*BCh_Pk*)

A road network can be represented in the form of a graph. There are at least two possible approaches to modeling a road network as such. The first idea is more suitable for models that are broader in scope. It focuses on the traffic between intersections. It can also include inbound and outbound external traffic. The second idea is a better fit for more detailed models. Its main focus is on the traffic between individual streets or street sections. Due to the nature of the dataset, the second approach was used (Fig. 1).

3.1 *Model Development*

The next step is developing the model. It took 13 generations of different models to come up with the definitive solution (Fig. 2).

Each generation features a slightly different approach and learns from the mistakes of the earlier generations. Some of them only apply minor tweaks to their predecessors, and some start entirely from scratch. All models were trained and tested on connection *BCh_Pk*. Figure 3 shows selected generations obtained in the study. The final generation (no. 13 in Fig. 3) is a multiple linear regression model described by the following formula:

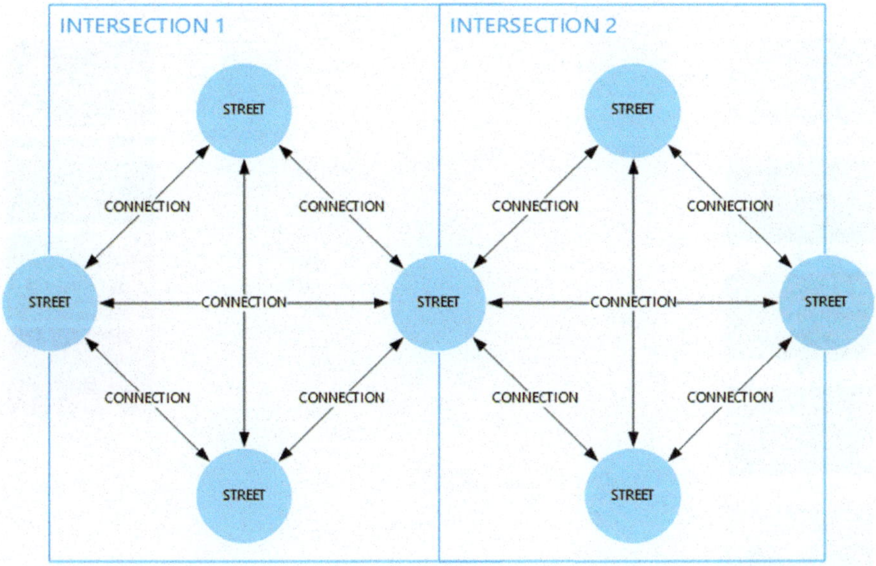

Fig. 1 City road network model as a graph: streets as vertices, individual connections as edges

$$f(t) = \beta_0 + \sum_{i=1}^{6} (\beta_{2i-1} \cdot \sin\frac{2it\pi}{24} + \beta_{2i} \cdot \cos\frac{2it\pi}{24})$$

The coefficients of the generated model for generation 13 took on the values:

$$\begin{aligned} f(t) = {} & 158,083307988767 - 47,9039636409462 \cdot \sin\frac{t\pi}{12} \\ & - 121,274012157452 \cdot \cos\frac{t\pi}{12} - 27,4628917006844 \cdot \sin\frac{t\pi}{6} \\ & - 16,8789398030092 \cdot \cos\frac{t\pi}{6} - 7,25741639975511 \cdot \sin\frac{t\pi}{4} \\ & + 24,5283775663424 \cdot \cos\frac{t\pi}{4} + 17,43583838101 \cdot \sin\frac{t\pi}{3} \\ & - 6,30189719870473 \cdot \cos\frac{t\pi}{3} - 0,10422614995114 \cdot \sin\frac{5t\pi}{12} \\ & - 12,6520638594728 \cdot \cos\frac{5t\pi}{12} - 19,9622027243579 \cdot \sin\frac{t\pi}{2} \\ & + 3,531723639023 \cdot \cos\frac{t\pi}{2} \end{aligned}$$

Fig. 2 Model development process

3.2 *Testing the Model on Other Connections*

MS Excel was used to calculate model coefficients (β_0 − *intercept*, β_1 − *sin*(t), ..., β_{12} − *cos*($6t$)) for each of the monitored connections. The coefficients are stored in a single data frame for easy use. The model can be used for volume traffic prediction

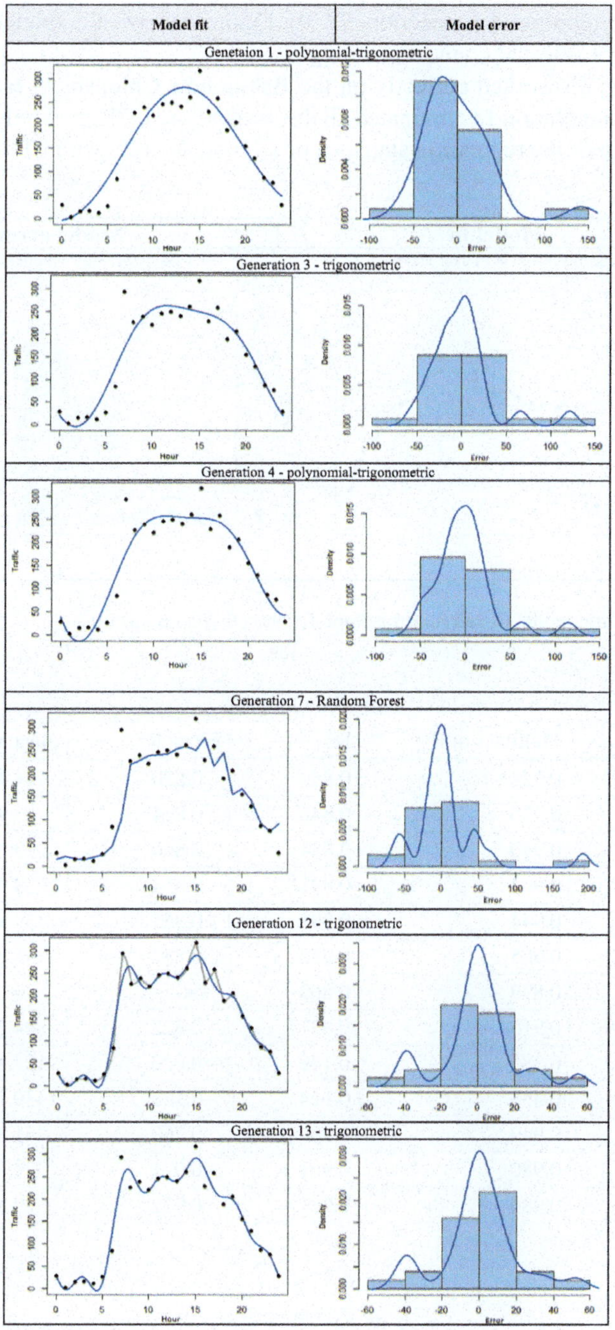

Fig. 3 Visualization of the most interesting model generations

on any of the monitored connections [8, 9]. Figure 4 shows the model fit and error distribution for selected connections only.

Generation 13 worked perfectly on the Batalionów Chłopskich -> Podkarpacka connection. Generation 13 eliminated all the weaknesses of the previous generations. A comparison of the regression statistics of all models is shown in Table 1.

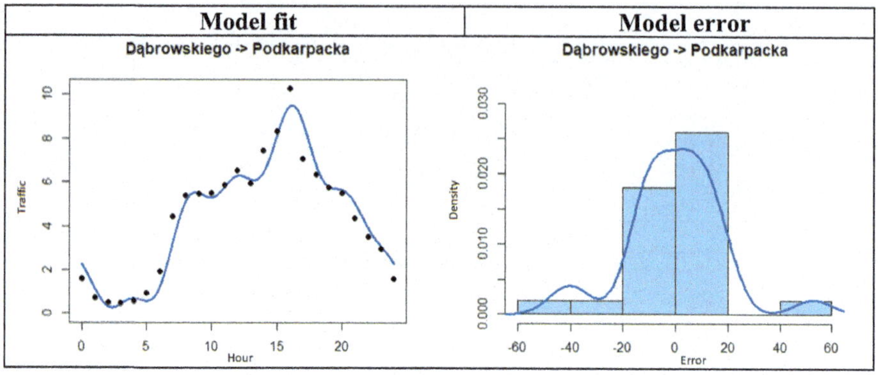

Fig. 4 Generation model fit and fit error graph for the *Db_Pk* connection

Table 1 Comparison of regression statistics of all models

Generation	Multiples of R	R^2	Fitted R^2	Standard error
1	0.922	0.850	0.820	44.408
2	0.920	0.846	0.824	43.928
3	0.940	0.884	0.860	39.054
4	0.944	0.891	0.854	39.935
5	0.944	0.891	0.855	39.901
6	0.965	0.932	0.883	35.861
7	0.899	0.809	–	–
8	0.915	0.837	0.322	44.205
9	0.385	0.148	−0.022	187.821
10	0.353	0.124	−0.168	1430.158
11	0.949	0.900	0.760	51.344
12	0.980	0.961	0.921	29.378
13	0.980	0.961	0.922	29.320

3.3 Model Normalization

In its current form, the model fits the original data but its output isn't quite descriptive as a measure of traffic volume. To remedy this the model needs to be normalized.

The first step is to move the curve forward by 30 min. That's because, in the original dataset, measurement points are always placed at the start of the measurement interval, which is 1 h long. By moving the curve forward, the model acts as if the measurement points were placed in the middle of the measurement interval. The second step is to divide the model output by 60, thus making the model predict the current traffic volume in vehicles per minute, which is more convenient.

MS Excel makes it easy to prepare a model based on linear regression. The *Data Analysis* tool is used for this. To generate the model, the input parameters (explanatory variables and the explanatory variable) need to be prepared [8]. The results were verified in the R environment. R is an excellent environment to test the model in practice [10–12]. The universal model uses the same parameters for all links, but each link requires its own set of coefficients. The *model_coeff* data frame was used to store these. The *predict_traffic* function was then written, which returns the universal model's predicted traffic volume on the link indicated by the user and at the indicated time. From an analytical point of view, the flow of traffic flowing in and out of each node in the graph, as well as the balance of this traffic, can be of interest. Their observation can provide key information on the traffic characteristics of the study area, as shown in Fig. 5.

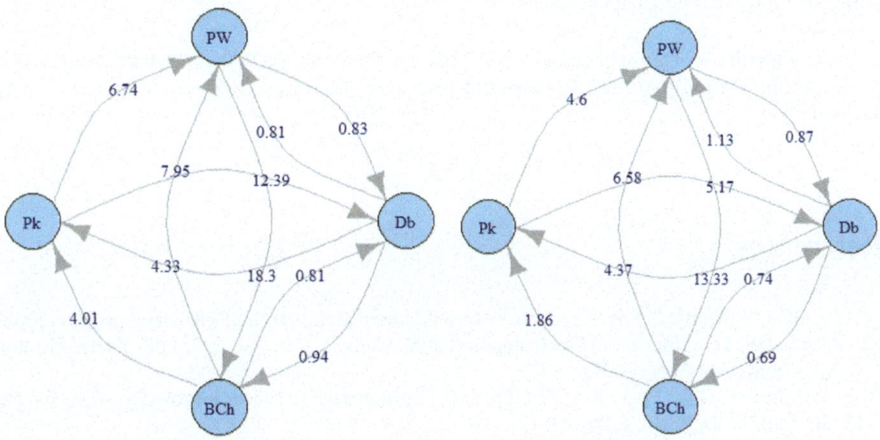

Fig. 5 Graph representing the selected intersection at 7:30 AM (left side), and at 9:37 PM (right side). The current traffic volume is displayed on each of the connections

4 Conclusion

Using methods of statistical data analysis and graph theory, it was possible to build a fully functional mathematical model of the intersection of Powstańców Warszawy Avenue and Batalionów Chłopskich Avenue in Rzeszów. The model development process revealed significant advantages and disadvantages of various types of regression models, which ultimately allowed the most appropriate model variant and set of input parameters to be selected.

Using linear regression we can create a highly accurate model of a small road network. However, its functionality is limited due to the dataset used as it only contains measurements taken in a span of 1 day. A broader timeframe would allow to create a more statistically significant model which would be of much more use in real-life implementations. Research has shown that the polynomial model, although the simplest to prepare, has very significant drawbacks, which became apparent in generations 9 and 11. The most important of these is the huge susceptibility to overtraining, i.e. high fit in some areas and huge divergence in others. The trigonometric model has been shown to lack some of the weaknesses of the polynomial model, including being much less susceptible to overtraining, allowing it to fit the training points accurately without significant discrepancies. A second advantage is its cyclicality with the right choice of explanatory variables. This means that one can naturally 'extend' the model for subsequent days without losing continuity.

The results obtained for the polynomial-trigonometric model showed that it does not outperform the purely trigonometric model in terms of fit, and in doing so loses some of its desirable properties.

Acknowledgements This publication is based upon work from COST Action "Optimizing Design for Inspection ODIN, CA 18203", supported by COST (European Cooperation in Science and Technology).

References

1. Bondy, J.A., Murty, U.S.R.: Graduate Texts in Mathematics 244: Graph Theory. Springer (2008)
2. Jungnickel, D.: Algorithms and Computation in Mathematics, vol. 5: Graphs, Networks and Algorithms. Springer (2005).
3. Montgomery, D.C., Peck, E.A., Vining, G.G.: Introduction to Linear Regression Analysis, 5th edn. John Wiley & Sons, Inc. (2012)
4. Harrell, F.E.: Regression Modeling Strategies with Applications to Linear Models, Logistic and Ordinal Regression, and Survival Analysis, 2nd edn. Springer (2015)
5. Polynomial Regression. https://online.stat.psu.edu/stat462/node/158/. Accessed 20 Dec 2022
6. Fourier Series and Integrals. https://math.mit.edu/~gs/cse/websections/cse41.pdf. Accessed 22 Dec 2022
7. Genuer, R.: Jean-Michel Poggi. Springer, Random Forests with R (2020)
8. Linear regression analysis in Excel. https://www.ablebits.com/office-addins-blog/linear-regression-analysis-excel/. Accessed 15 Oct 2022
9. Carlberg, C.: Regression Analysis Microsoft Excel. Pearson Education, Inc. (2016)

10. Pearson Correlation Coefficient. https://online.stat.psu.edu/stat501/lesson/1/1.6. Accessed 7 Dec 2022
11. Adjusted R Squared Formula. https://www.educba.com/adjusted-r-squared-formula/. Accessed 8 Jan 2023
12. Dymora, P., Mazurek, M.: Influence of model and traffic pattern on determining the self-similarity in IP networks. Appl. Sci. **11**, 190 (2021). https://doi.org/10.3390/app11010190

Performance Analysis of a Real-Time Data Warehouse System Implementation Based on Open-Source Technologies

Paweł Dymora, Gabriel Lichacz, and Mirosław Mazurek

1 Introduction

The main goal of the paper is to present a data warehouse system implementation that would update and feed the data warehouse, which has historical data, continuously from the production database with minimal latency. The system would load the data from, for example, the databases of smaller branches of the company, to the data warehouse so there would be the most up-to-date view of the entire data set. This would make it possible to precisely analyze the data and implement accurate data mining algorithms, in particular those creating predictions. To this end, an attempt was made to build such a system based on available open-source technologies, and performance tests were carried out to determinc whether the created system meets the properties of a real-time system.

The article consists of six chapters. Each focuses on a different aspect of real-time data warehouse system design. Chapter number 1 focuses on the purpose of the paper. The second presents the problem of storing large data sets, an overview of the topic, and an explanation of the basic concepts and the essence of the task. The third focuses on the concept of data warehouses, their implementation, and the technologies used today. The fourth describes the structure of the system and the principle of its operation. In the fifth, an analysis of the performance test results is carried out. Finally, the sixth chapter presents the conclusions drawn from the execution of the tests.

P. Dymora (✉) · G. Lichacz · M. Mazurek
Rzeszów University of Technology, 35-959 Rzeszów, Al. Powstańców Warszawy 12, Warszawy, Poland
e-mail: pawel.dymora@prz.edu.pl

W. Zamojski et al. (eds.), *Dependable Computer Systems and Networks*, Lecture Notes in Networks and Systems 737, https://doi.org/10.1007/978-3-031-37720-4_6

2 Data Warehouses and the Problem of Storing Large Data Sets

As recently as 20 years ago, enterprise-supported databases were only able to accept, at most, hourly updates on the status of a cash register or, in the case of larger businesses, a single day's shop status [1]. Technology has evolved significantly since then and continues to do so. Since Bill Inmon introduced the concept of data warehousing in 1991, analytical systems of this type have developed and grown in popularity enough to become an integral part of all major businesses [2]. There is also a growing demand for more and more accurate data, as well as for predicting future trends to make more accurate business decisions. All this makes the processing of huge data sets extremely important. It has led to the dynamic development of the field known as Big Data in recent years [2]. Analysis of data from data warehouses is a large part of this area. It is an important topic mainly from the point of view of optimizing resources, minimizing losses, and making better decisions.

There are many commercial solutions on the market for building data warehouses. However, a significant number of them are too costly for a large number of companies, very labor-intensive to implement, or present a high level of complexity for the needs of a given enterprise. Open-source solutions, i.e. under an open license, are therefore a good alternative. It can be configured according to one's needs and the cost of usage is low, outside of the hardware (or cloud usage) [3].

Data warehouses are database management systems that are used to efficiently collect, store and process large amounts of data. They improve analysis, which in turn facilitates business decisions, optimizes the production process, and reduces unnecessary costs [2, 3]. There are many approaches to designing and building a data warehouse, depending on the project and specific needs. Models proposed by the pioneers of data warehousing are, for example:

- Business approach (top-down): involves designing and building a data warehouse with business needs in mind. It takes into account the various data sources of the business and builds the structure of the warehouse from only the data necessary and selected in the ETL process. Proposed by Ralph Kimball.
- Technical (bottom-up) approach: the opposite of the top-down approach, it focuses on the relevance of the main enterprise data warehouse. It is the main source of data for other, smaller, thematic databases. This solution was proposed by Bill Inmon.

The main features by which a data warehouse can be defined are:

- Storing large amounts of data, mainly historical, from multiple sources such as CRM, transactional, or production systems.
- They are designed to perform analysis and complex queries, making it easier to analyze the stored data and subsequently make better business decisions.
- They are generally updated frequently.
- Warehouses are standardized and allow the integration of data from multiple sources.

- They are protected from unauthorized access, as they usually store sensitive data.

There are differences between databases (including data warehouses) in a development environment and in a production environment. A development database is a database that is used to create, test, and develop new features or applications. It is a database that is used by developers or testers and is not shared with other external users. A production database, on the other hand, is a database that is used to store and share data in production, i.e. in a business environment. It is the main source of data for the business and is used by all users and applications [2–4].

In big data storage environments, there is a significant synchronization problem. The main data warehouse is often fragmented in a distributed environment and is connected to many external and internal sources from which data is read, sent, and updated. In an ideal system, synchronization would take place instantly and with zero delays. However, this is not possible, and over the years many ways have been developed to deal with this problem. Data can be shared and sent at specific times or dedicated tools can be used for such purposes—an approach mentioned in the article [4, 5].

Data synchronization is particularly important when the system is connected to analytical tools. Some fields are particularly sensitive to the timeliness of data—such as stock markets. There is a need to access new data as quickly as possible and create reports, charts, and other data presentation methods based on this data to make effective and accurate decisions. Also in the case of prediction, the timeliness of the data is of considerable importance—if the data were out of date, the prediction could be calculated for a time period that has already passed. For the system to run as fast as possible and offer minimal delays it is necessary to minimize human intervention. As much automation as possible is necessary—if every action and query had to be performed by a qualified engineer, it would be extremely time inefficient [6, 7].

Real-time data warehouses on the surface seem to be the ideal solution in every business case. However, such systems bring with them a certain amount of requirements that not every company can meet. Such solutions require qualified personnel to build a system and maintain it. In addition, they are expensive in terms of hardware—the servers on which such data warehouses run must be sufficiently powerful, otherwise, there may be excessive queuing of actions (e.g. receiving data from external sources) or the system may crash altogether. If this is not the case, it may be better to opt in the course of decision-making for a warehouse that isn't a real-time one, but updates data at certain times of the day—for example, twice a day or at the time of least load on the servers. In this way, unwanted system downtime can be avoided. In view of all the above issues, it was decided to undertake the construction of a data warehouse system operating with minimal latency, so-called real-time [7, 8]. This is a very future-oriented and extremely fast-growing area. If such a system would be run in a production environment, a server would be needed, preferably in a cloud environment, which would facilitate administration. Linux system of any distribution (e.g. Debian) installed in said environment, appropriate network security, and a Java JDK version 8, in addition to the main component platforms of the planned system, namely Apache Druid, Apache Kafka, Apache Hadoop, and Apache Hive [9–13, 15].

3 Implementation and Effectiveness Testing of the Real-Time Data Warehouse Environment

The system is based on the close cooperation of four systems: Apache Hadoop, Apache Hive, Apache Druid, and Apache Kafka. They all run on a single virtual machine set up as a server (an IP address is `192.168.1.112`). Hive runs on a Hadoop cluster, while Druid and Kafka run independently.

Data in the form of.csv files, which simulate the flow of data from external sources (e.g. from production centers to the main enterprise system) is absorbed by the Kafka `topic` called `druid_stream`. A stream of data is created between Druid and Kafka through which data from the Kafka `topic` is sent directly to the data warehouse in Druid. External data tables are created in Hive which is directly from the Druid broker. The SQL support in Apache Druid is an underdeveloped solution, which is why the system uses the extensive HiveQL in Hive to perform advanced queries and analyses. A client connecting with a Python-written connector can query the main data warehouse in Druid via Hive [8–13, 15].

A model of the constructed real-time data warehouse system is shown in Fig. 1. The arrows indicate the direction of data transfer. The IP addresses listed there are from the local environment where operations were performed.

The above platforms and applications were installed and the subsequent processes on them were conducted on a virtual machine with the following parameters: CPU: 4-core, RAM: 8192 MB, Virtual VDI Drive: 500 GB (real size: 37 GB), and operating system: Debian 11. The virtual machine was running on a computer with the

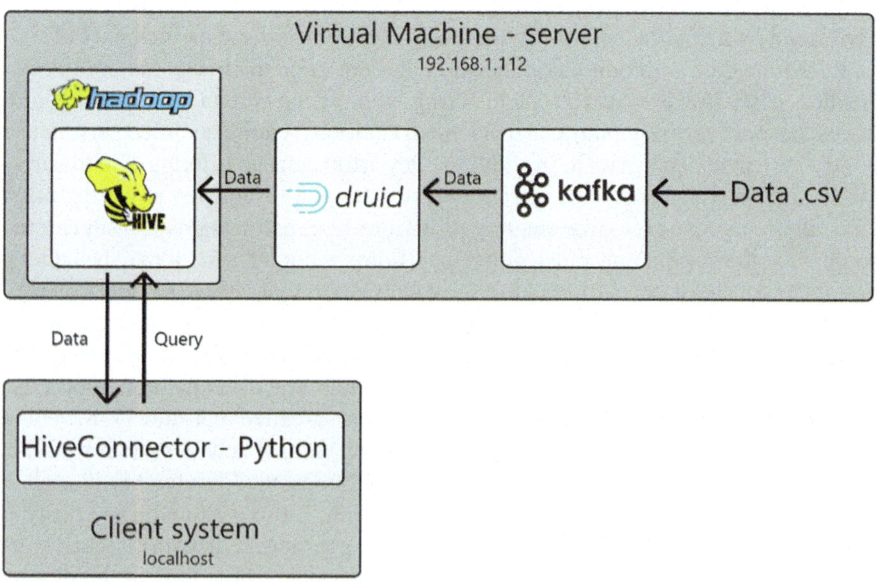

Fig. 1 System implementation architecture

parameters: CPU: i5-10400F 2.9 GHz, RAM: 32 GB 3200 MHz, drives 2 × 1 TB HDD and 120 GB SSD, GPU: GTX 1060 6 GB, and operating system: Windows 10.

3.1 Methodology and Dataset

For the tests, data slices from the main table of the airline flight operator of 5,000, 500,000, 1 million, 2 million, and 5 million records were used. The tests aim to verify that the system is a real-time system, i.e. one that operates with minimal delays.

Each file is subjected to the test of adding all its data rows to the data warehouse nine times in two modes. The first is to add them without disrupting the system - i.e. it is the only operation running on the virtual machine and its host computer. The second mode is the execution of database queries while inserting the data. In both cases, the usage of hardware resources is also measured. Modules are written in the Python language.

The dataset used in the tests is from Kaggle website: https://www.kaggle.com/datasets/robikscube/flight-delay-dataset-20182022, licensed under CC0: Public Domain. It presents detailed information about commercial flights in the United States of America. They have been split into several tables through a normalization process. In addition to increased readability, performance has also been improved—removed repetition of the same data columns. This has reduced the size of the data by more than half.

3.2 Performance Tests of the Implemented Real-Time Data Warehouse Environment

Time in conducted tests is measured using the `time()` function, belonging to the Python time library, from the moment of insertion of the data file into the Kafka stream, all the way until all of its data appears in the data warehouse. A function that inserts data records into the data warehouse has been written for this purpose. As arguments, it uses a link to the Druid broker, the IP address of the machine on which Kafka is running and the name of the `csv` file to be inserted. Initially, the function selects the appropriate mode, i.e. how many rows of data will be entered based on the given `csv` file name. It further determines via a function derived from Druid Connector how many rows are currently in the table in the data warehouse and how many should be present after the operation. At this point, the starting time is saved. The next line executes the command that adds the data file to the Kafka stream - this is a bash command via SSH from within Python. The next part of the function checks the current amount of data in the data warehouse table and waits until the predicted number of rows is reached. When this happens operation ending time is saved and the function returns how long the operation took.

For conducting tests under load, a script that simulates the system load was used (Fig. 2)—`TestHive.py`. It executes data warehouse queries in an infinite loop. The script connects to the Hive database and retrieves query results from external data tables (stored by the Druid broker). At the same time as the scripts mentioned above, a script measuring RAM consumption and CPU usage is also run. Both values are expressed as percentages. The script uses the `psutil` library to measure resource consumption, `pandas` to store the data in a data frame and `matplotlib` to draw graphs. Resource measurement is performed every second (function `time.sleep(1)`), and the script creates plots when it is interrupted, i.e. when the script that adds data to the data warehouse finishes its execution.

At the same time as the scripts mentioned above, a script measuring RAM consumption and CPU usage is also run. Both values are expressed as percentages. The script uses the `psutil` library to measure resource consumption, `pandas` to store the data in a data frame and `matplotlib` to draw graphs. Resource measurement is performed every second (function `time.sleep(1)`), and the script creates plots when it is interrupted, i.e. when the script that adds data to the data warehouse finishes its execution.

3.3 *Elaboration of the Test Results*

The final results of the carried-out test are presented in Table 1.

Figures 3, 4 and 5 presents the summary results for data loading time in the system load simulation test, CPU and RAM usage during the inserting of one million records of data into the data warehouse. When a file containing 5,000 rows of data was added to the Kafka stream, it appeared entirely in the data warehouse in an average of 1.92 s for the test without a simulated system load and 1.88 s for the test with a simulated load. The median is slightly lower because there was an anomaly in the first attempt to add the file—the data appeared in the data warehouse almost 2 times later than in the rest of the attempts. The same situation occurred in the interference test. The longest `INSERT`-type operation took 3.80 s to execute (no load) and the shortest only 1.44 s (with simulated load). The performance for small data packages can be considered satisfactory, as even when the system load was continuously simulated, the execution times for operations were negligible.

For a data file containing 500,000 records of data, the times oscillated between 17.40 s and 36.07 s, both extreme values coming from the test without simulated load. The maximum value can be considered an outlier, as it differs significantly from the other times obtained, the average of which is 21.99 s for the test without interference and 22.49 s with interference, respectively. Overall, the times without load simulation are marginally better, but the maximum value makes the average significantly higher.

The times for the test with one million rows of data, which is the most relevant one—the essence of data warehousing is storing and operating on large data sets, are correct, ranging from 34.31 s without disruption to 58.65 s with a simulated load. In

```python
def performanceTest(druidSqlUrl, kafkaHostIp, csvName) -> 'string':
    """
    Performance test function - data ingestion

    Parameters:
    druidSqlUrl (string): Druid database host URL
    kafkaHostIp (string): Server IP where the kafka is running
    csvName (string): csv file name to insert

    Returns:
    end-start (string): how long ingestion operation took
    """
    if csvName == 'data_5tys.csv':
        toAddRowsNumber = 5000
    elif csvName == 'data_500tys.csv':
        toAddRowsNumber = 500000
    elif csvName == 'data_1mln.csv':
        toAddRowsNumber = 1000000
    elif csvName == 'data_2mln.csv':
        toAddRowsNumber = 2000000
    elif csvName == 'data_4mln.csv':
        toAddRowsNumber = 4000000
    elif csvName == 'data_5mln.csv':
        toAddRowsNumber = 5000000
    else:
        return "Wrong file name provided"

    countBeforeOperation = int(druidCountTable(druidSqlUrl, 'old_data')[13:-3])
    countAfterOperation = countBeforeOperation + toAddRowsNumber

    cmd = '''cd /opt/kafka_2.13-2.7.0 && ./bin/kafka-console-producer.sh
        --broker-list localhost:9092
        --topic druid_stream < /home/stud/Downloads/import_data/''' + csvName
    client = SSHClient(kafkaHostIp, 'root', 'osource')

    # Operation start
    start = time.time()
    host_out = client.run_command(cmd) # Data ingestion into Kafka

    count = int(druidCountTable(druidSqlUrl, 'old_data')[13:-3])
    while count < countAfterOperation:
        count = int(druidCountTable(druidSqlUrl, 'old_data')[13:-3])
    end = time.time()
    # Operation end

    return end-start
```

Fig. 2 Python function to perform system performance tests

Table 1 Summary table of performance test results

Data loading time		Min (s)	Max (s)	Avg (s)	Std (s)
Without disruption	5,000	1.49	3.8	1.9278	0.7166
	500,000	17.4	36.07	21.9867	6.08
	1 million	34.31	37.26	35.8222	1.0099
	2 million	61.41	117.01	77.5244	21.2734
	4 million	139.88	282.27	182.4156	43.4328
	5 million	x	x	x	x
With disruption	5,000	1.44	3.77	1.8822	0.7169
	500,000	19.12	28.81	22.4933	3.2827
	1 million	36.83	58.65	45.0722	6.7269
	2 million	60.54	159.61	76.9433	31.2537
	4 million	x	x	x	x
	5 million	x	x	x	x

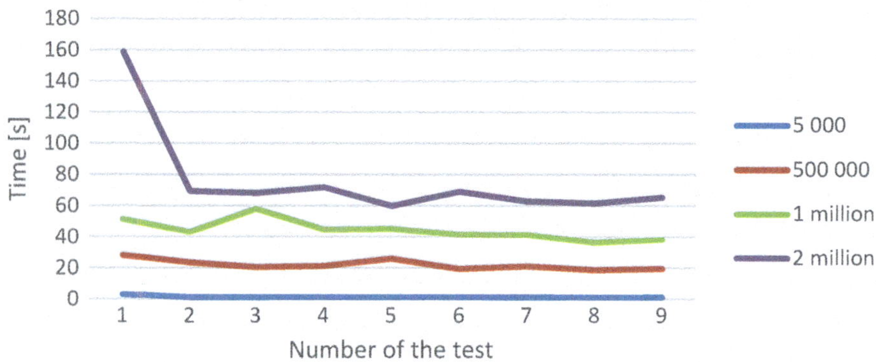

Fig. 3 Data loading time in system load simulation test

this case, the biggest difference between the tests without and with disruption was evident, with the average for the former being 35.82 s and with the simulated load 45.07 s.

For this test, a similar trend to that of the test with five hundred thousand records of data in the shaping of the resource use curve over time can be observed. There is a sudden increase in CPU usage at the very beginning of the test, and then an oscillation from around 10–35% by the end of the test. Here, however, RAM uses decreased well before the halfway point of the test—in the case of the test with simulated load, this occurred after just 1.5 min, and without simulation after 2.5 min.

The test of four million rows of data was only completed during the test without system load simulation. The average time to add a data file to the data warehouse was about 182 s. The maximum time was almost 5 min and the minimum time was

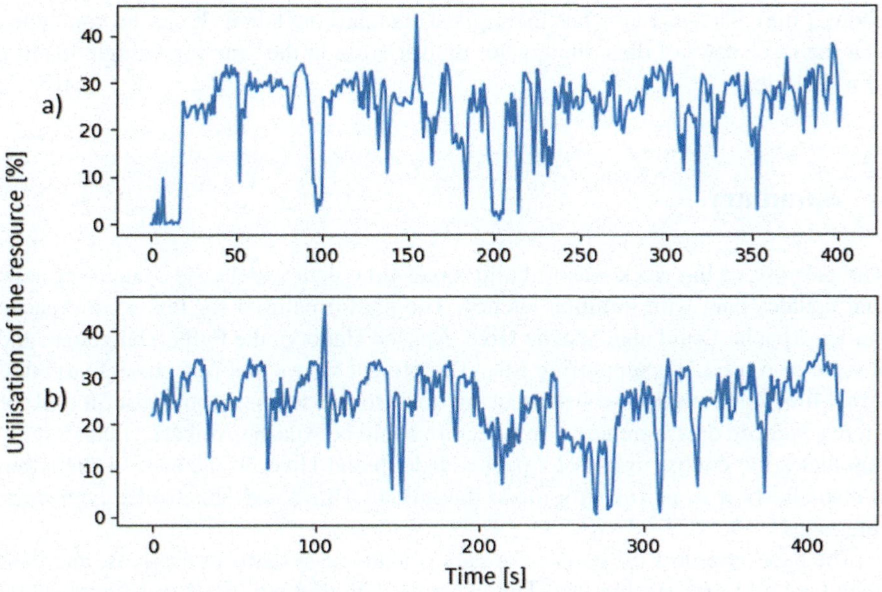

Fig. 4 CPU usage during inserting one million records of data to the data warehouse, **a** no simulated system load, **b** simulated system load

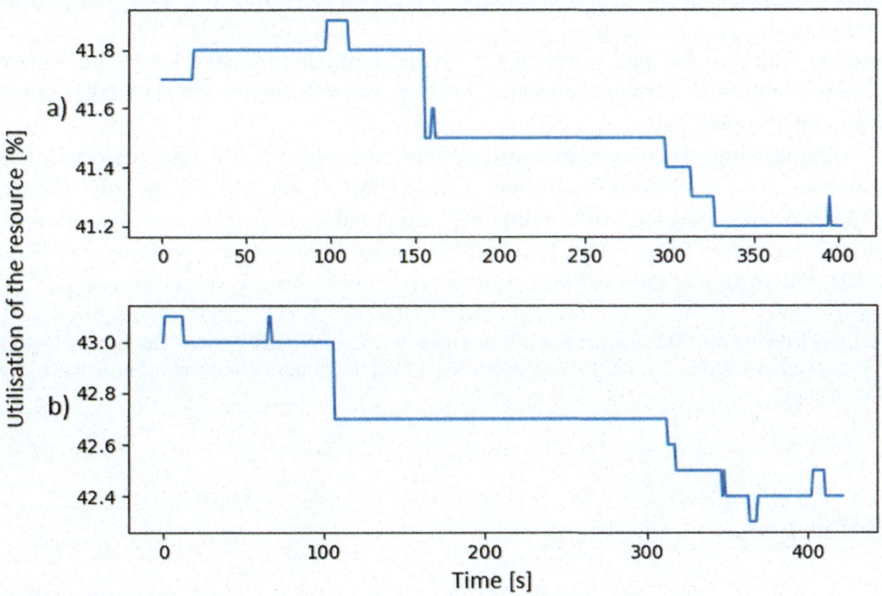

Fig. 5 RAM usage during inserting one million records of data to the data warehouse, **a** no simulated system load, **b** simulated system load

about 2 min. This test also has the highest standard deviation. It can be concluded that the more rows of data, the greater the variation in the time for a single loading of a data file.

4 Summary

The purpose of the work was to build a real-time data warehouse system, i.e. one that updates data with minimal latency. The platforms used for this were Apache Kafka, Apache Druid and Apache Hive. Apache Hadoop, the Python language, and R were also used in a supporting role. The project shows that it is possible to build a real-time data warehouse system using open-source tools. Another useful tool for more advanced data warehouse architecture could be Apache Ambari, which largely automates the configuration of Apache Hadoop and Hive. In the case of the latter, it also allows it to be run in a mode that allows Druid and Hive to be even more integrated.

The paper presents the essence of data warehouse systems in business, and their analytical and data storing use. The example data that was used and it's meaning are presented. Methods and functions in Python to facilitate interfacing with the warehouse, through the Apache Druid and Apache Hive interfaces, included in the corresponding repository, are created and presented. During the tests, the performance of the system was tried and tested in the form of multiple calls to functions loading data into the data warehouse with and without simulated load. The system runs with minimal latency even under loading one million rows of data takes about a minute at most.

Summarising, the data warehouse system created with simulated external data sources can be considered real-time. The loading of one million records of data, under the simulated load of continuously executed queries to the data warehouse, closed under one minute and four million under five (this one without simulated load). The results of the load tests thus proved correct for a system of this type.

Acknowledgements This publication is based upon work from COST Action "Optimizing Design for Inspection ODIN, CA 18203", supported by COST (European Cooperation in Science and Technology).

References

1. Kimball, R., Strethlo, K.: Why Decision Support Fails and How to Fix it. ACM SIGMOD (1995)
2. Dhaouadi, A., Bousselmi, K., Gammoudi, M.M., Monnet S. Hammoudi, S.: Data Warehousing Process Modeling from Classical Approaches to New Trends: Main Features and Comparisons, Data (2022)

3. Gupta, A., Sahayadhas, A.: A comprehensive survey to design efficient data warehouse for betterment of decision support systems for management and business corporates. Int. J. Manage. (2020)
4. Gryglewicz-Kacerka, W., Kacerka, J.: Model hurtowni danych zasilanej w czasie rzeczywistym, na podstawie technologii Oracle (2013)
5. Nadj, M., Schieder, C.: Quo Vadis real-time business intelligence? A Descriptive Literature Review and Future Directions, Research Papers (2016)
6. Golfarelli, M., Rizzi, S.: Data Warehouse Design: Modern Principles and Methodologies. Mcgraw-Hill Education-Europe (2009)
7. Nambiar, A., Mundra, D.: An overview of data warehouse and data lake in modern enterprise data management. Big Data Cogn. Comput. (2022)
8. Dymora, P., Mazurek, M.: Performance assessment of selected techniques and methods detecting duplicates in data warehouses. Theory and Applications of Dependable Computer Systems. DepCoS-RELCOMEX 2020. Advances in Intelligent Systems and Computing, vol. 1173. Springer, Cham (2020). https://doi.org/10.1007/978-3-030-48256-5_22
9. https://druid.apache.org/docs/latest/design/index.html. Accessed 20 Mar 2022
10. https://docs.microsoft.com/pl-pl/azure/hdinsight/hadoop/hdinsight-use-hive. Accessed 05 Mar 2022
11. https://blog.cloudera.com/apache-hive-druid-part-1-3. Accessed 11 Mar 2022
12. https://www.oracle.com/pl/database/what-is-oltp. Accessed 04 Nov 2022
13. https://www.kaggle.com/datasets/robikscube/flight-delay-dataset-20182022. Accessed 04 Nov 2022
14. https://www.djangoproject.com. Accessed 04 Nov 2022
15. https://www.oracle.com/pl/database/what-is-a-data-warehouse. Accessed 10 Dec 2022

Hammering Test on a Concrete Wall Using Neural Network

Atsushi Ito, Yuma Ito, Jingyuan Yang, Masafumi Koike, and Katsuhiko Hibino

1 Introduction

In recent years, social infrastructures such as buildings, bridges, and roads have been aging, and the demand for inspection is expected to increase [1]. Infrastructure inspection involves a wide variety of maintenance tasks and requires detailed diagnosis and inspection related to structural strength and seismic resistance. Robots and drones are being used to inspect parts that require scaffolding and are difficult for people to enter, such as large bridges and walls of high-rise floors [2]. In inspecting defected areas that cannot be visually observed, such as the inside of a building wall, a method that detects sound changes by hammering is often used. Although this is a low-cost testing method, it requires experience in distinguishing sounds caused by defected areas. However, the decision relies on personal senses and may lead to differences in judgment. Concrete walls are not the only target of sound testing; tile walls are also subject to sound testing. Tiles are mainly used for the walls of apartments, of which there are approximately 6.65 million in Japan [3]. In order to inspect this number of buildings, it is essential to make hammering testing more efficient and simplified. There are several types of buildings and structures as shown in Fig. 1. In this paper, our target is a structures with exposed concrete.

We have developed a technique that uses neural networks (NN) and also uses Transfer Learning (TL) [4] to distinguish the sound of hitting tile walls and have

This research is supported by JSPS Kakenhi (17H02249,18K111849, 20H01278, 20H05702, 22K12598).

A. Ito (✉) · Y. Ito · J. Yang
Chuo University, Tokyo, Hachioji 192-0351, Japan
e-mail: atc.00s@g.chuo-u.ac.jp

M. Koike
Utsunomiya University, Utsunomiya, Tochigi 321-8585, Japan

K. Hibino
PORT DENSHI Corporation 1-3-8 Shimizugaoka, Fuchu, Tokyo 183-0015, Japan

W. Zamojski et al. (eds.), *Dependable Computer Systems and Networks*, Lecture Notes in Networks and Systems 737, https://doi.org/10.1007/978-3-031-37720-4_7

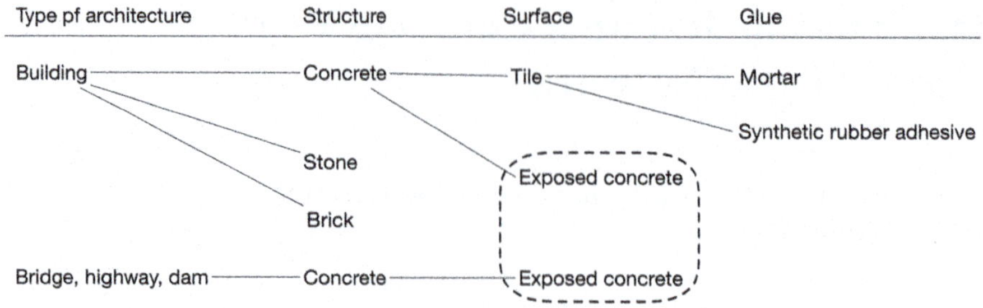

Fig. 1 Types of structure and surface

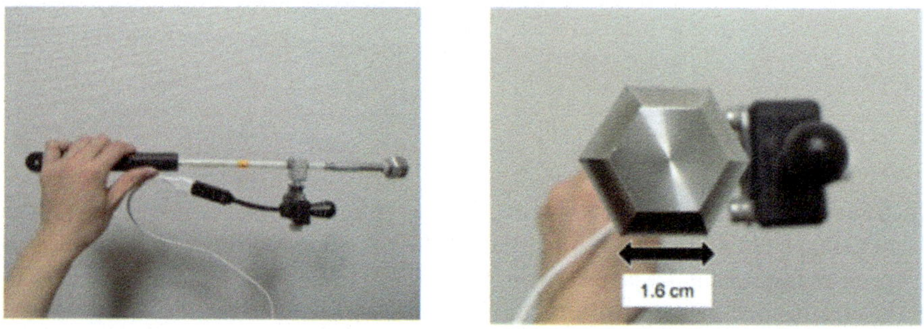

Fig. 2 Rolling hammer (KoroKoro)

demonstrated the effectiveness of this technique on tile walls, which are often used in apartments [5].

TL is a way to develop a learning model quickly by modifying the last layer of existing learning model and used frequently in many applications. One of the famous environment to use TL is a Teachable Machine [1]. The Teachable Machine uses MobileNet [12] as the basis of the learning.

The device we used for the research was a rolling hammer called KoroKoro (Fig. 2). KoroKoro is a Japanese onomatopoeia that describes the way something rolls. The device has a metal ball with a hexagonal edge. The distance between the two edges is 1.6 cm. This device produces a hammering sound by rolling on the wall. Hammering sounds are recorded with a microphone near the ball. Figure 3 shows an example of a hammering sound obtained using a KoroKoro. The merit of using KoroKor is that it is possible to hit a wall about 15 times in a second (approximately every 0.07 s). On the contrary, it is possible to get a hammering sound once a second if we use a standard hammer. The inspection flow is as follows.

(Step 1:) Start recording and roll the KoroKoro along the wall.
(Step 2:) When one strike is detected, the recording is stopped, and the GPU captures the sound.
(Step 3:) Compute FFT on GPU and use the neural network (NN) to make decisions (Then go to Step 1).

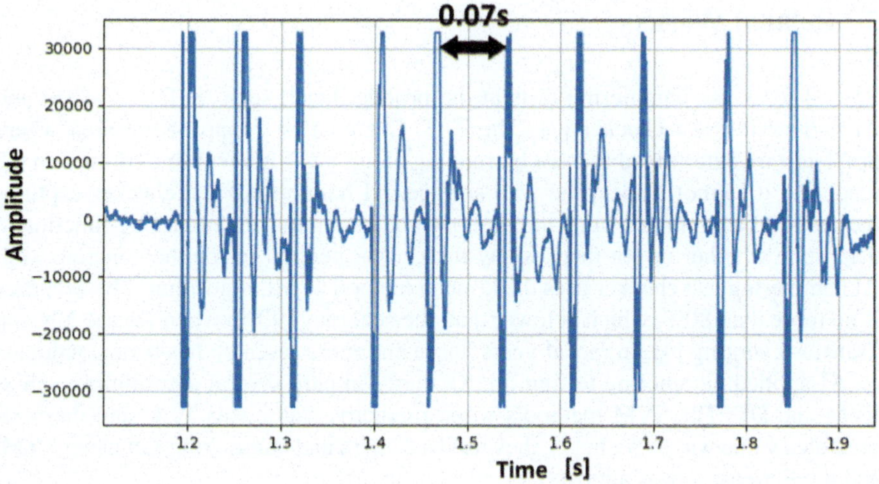

Fig. 3 Recorded sound by using KoroKoro

Table 1 Difference of tile wall and concrete wall

	Tile	Concrete
Surface	Different surface textures, material There is a joint between tiles. The size of each tile is almost the same	Not rough and smooth No joint and uniform
Location of flacking	At the boundary between a tile and wall	Inside concrete
Depth of flacking	Tile thickness is about the same	The depth of flaking varies a lot
Size of flacking	Less than a tile size	The size of flaking varies a lot
KoroKoro	Sometimes KoroKoro is trapped at a joint and slips on a slippery surface	KoroKoro roles uniformly

In this paper, we discuss the outline of the research on the hammering test of concrete walls. There are differences between tile and concrete walls, as shown in Table 1. A significant difference is that the tiles' flaking (cavity) location is almost the same depth, a few millimeters down. However, concrete walls vary from 20 mm to 100 mm or even more in some cases. For this reason, we made test blocks with flacking with different depths of flaking (25 and 55 mm). In contrast, concrete walls do not vary in material and texture from place to place, but the depth at which flacking varies. However, developing a learning model using training data of multiple depths of flacking takes much work. So, this paper describes the results of a study to build a learning model that is not affected by the depth of flacking.

2 Related Works

Several tools use hammering tests at the product level, such as T. T. Car [6] and AI hammering test checker [7]. The T. T. Car creates a map of the area where problems are determined to exist by running along measurement lines drawn on the road, and it cannot inspect the wall surface. AI hammering test checker captures the sound of a hammer from a microphone and uses a machine-learning function to determine whether or not there is any flacking or peeling inside the concrete wall. AI hammering test checker uses the k-mean method as ML algorithm. The accuracy is no more than 80%, which is lower than the accuracy (90%) when we use NN and Transfer Learning [4]. In recent years, more and more research has been conducted on AI-applied hammering testing. In [8], they compare several algorithms, such as SVM and DT. The SVM method's worst predictive value was 72%, and the best-predicted value was 99%. In [9], they used a camera and hammering test using SVM, and the F-measure was almost 0.73.

3 Trial on Concrete Wall

This section describes the results of the verification of the accuracy of the hammering test on the specimen shown in Fig. 4.

3.1 A Test Using the Specimen that has Flacking 55 mm Under the Surface

First, we randomly selected 56 training data and 8 test data from data collected from a specimen with a 55 mm depth flacking, created a learning model using Transfer Learning, and conducted an experiment with 150 epochs. Because of the efficient learning model development feature of Transfer Learning, 56 training data is enough. Then we tested 8 test data (normal 4, flacking 4). As a result, we obtained an accuracy of 91.2%, as shown in Fig. 5.

3.2 A Test Using the Specimen that has Flacking 25 mm Under the Surface

Next, we evaluated the generality of the model developed using data collected from a test concrete block with a 55 mm depth flacking was verified by inputting 10 test data obtained from a specimen with a flacking at 25 mm. As shown in Fig. 5, the accuracy was 97.3%. The result shows the versatility of the generality of a model

Fig. 4 Test sample of a concrete wall

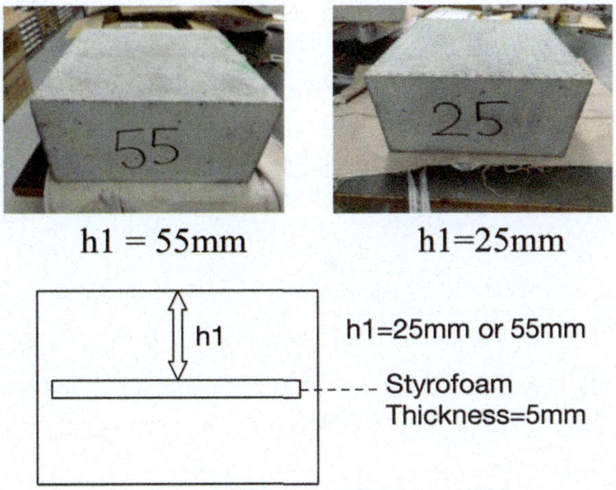

developed from the concrete block with a 55 mm depth flacking. We developed a learning model using Teachable Machine [10, 11]. The result in Sect. 3.1 (test data of flacking 55 mm depth with the learning model of flacking 55 mm depth) and the result of this section (test data of flacking 25 mm depth with the learning model of flacking 55 mm depth) differ about 6%. We assume that the test data of flacking 55 mm depth contain unclear sound. The detail is discussed in Sect. 4.

3.3 A Test of Generality Using the Different Specimen

Then, we evaluated the model using the test specimens in Fig. 6. This specimen flacks 40, 60, 80, and 100 mm deep. We tested data collected from the test concrete block with a 40 mm depth flacking. Unfortunately, the result could have been better (about 50%).

4 Analysis of Trial Using New Specimen in Fig. 6

We analyzed the hammering sound data in detail to find the reason for the unexpected result. We picked up typical waves from the sound data.

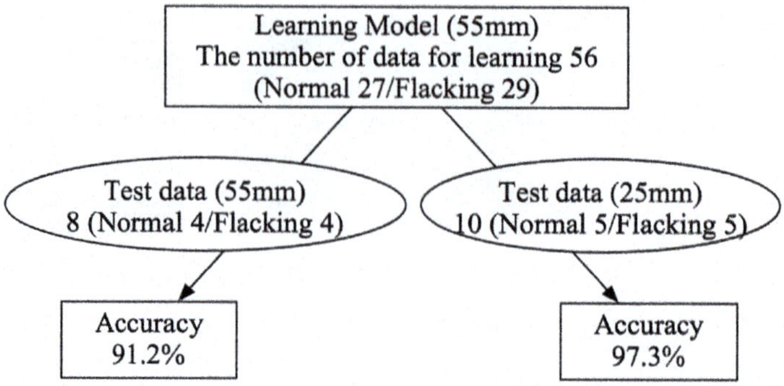

Fig. 5 Accuracy of test data (55 and 25 mm) using a learning model developed using learning data of depth of flacking of 55 mm

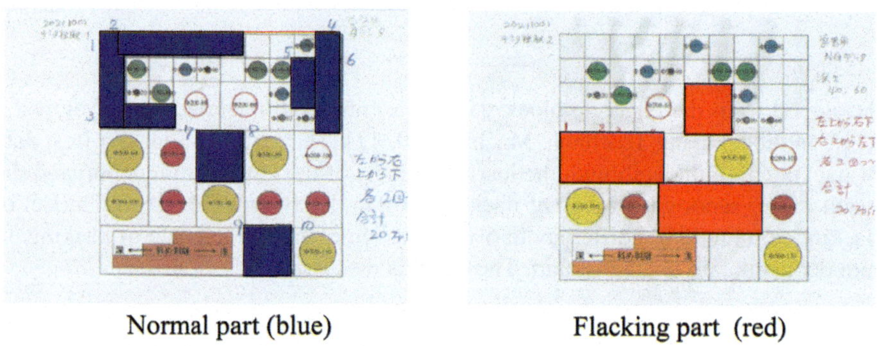

Fig. 6 Different sample

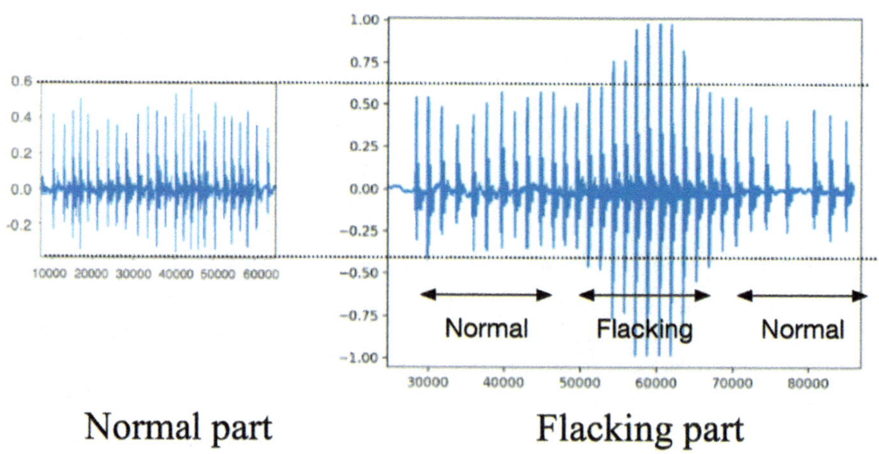

Fig. 7 Normal part/flacking part

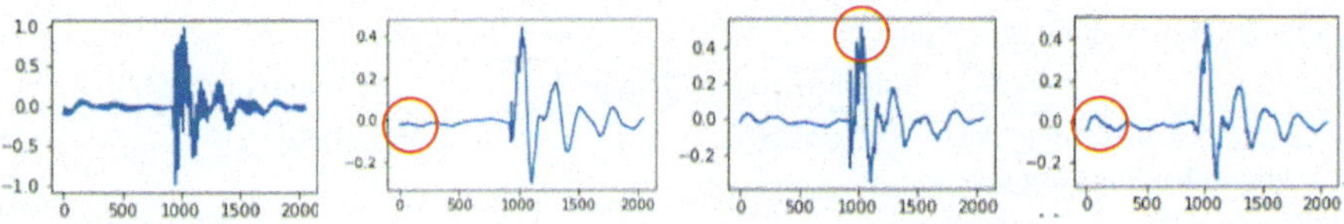

(a)type1(flacking) (b) type2 (normal) (c) type3 (edge) (d) type3 (edge)

Fig. 8 Comparison of sound (wav) of normal and flacking part

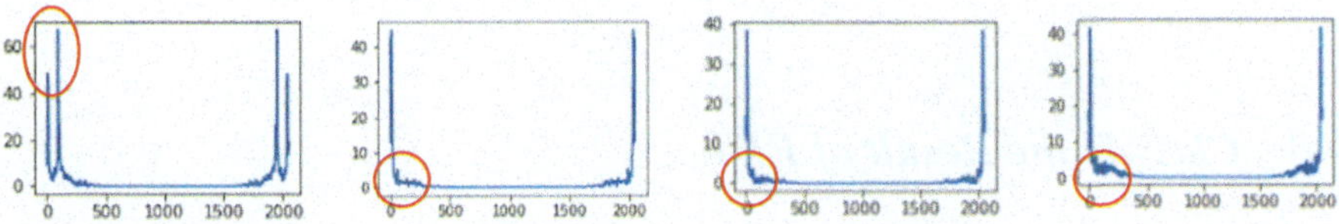

(a)type1(flacking) (b)type2(normal) (c)type3 (edge) (d)type3(edge)

Fig. 9 Comparison of FFT of normal and flacking part

4.1 Sound at the Impact

Figure 7 shows the output from KoroKoro. A regular part's sound in Fig. 7 was smaller than flacking part, and the loudness is almost the same. However, the sound of the flacking part (center of the figure) was larger at the peak are not flat. In this case, the real flacking part is the center (loud sound), and both sides (flat part) are normal areas. So, the loudness of the sound is one key to separating flacking.

4.2 Classify the Wave Shape (wav)

Figure 8 shows the typical patterns of sound. Based on the above assumption, we classified data into four patterns. The data of the flacking part (type 1), the normal part (type 2), and the boundary of the normal and flacking part (type 3) to understand the characteristics of the hammering sound in those parts. The sound of the flacking part (type 1) was larger since the peak was 1.0. Moreover, the waveform was thicker since the flaking part contained a reverberation sound. Then we compared the sound of the normal part and the edge part. The loudness of the peak is similar in type 2 and type 3. However, the type 3 contains some reverberation sound. The waveform of type 3 contains some noise compared to type 2 but needs to be clarified.

Table 2 Trial without type 3 data

	Before removing type 3	After removing type 3
The number of learning data (Normal/Flacking) (removed type 3)	40 (20/20)	40 (20/20)
The number of test data (Normal/Flacking)	20 (10/10)	20 (10/10)
Accuracy	90%	95%

4.3 *Classify the Result of FFT*

The waveforms after the FFT were compared are shown in Fig. 9. As shown in type 1, the flaking part usually contains two peaks because the cavity reflected a reverberation sound in the wall. On the other hand, the normal part contains no second peak (type 2). However, the edge part does not have clear characteristics. Such part contains some additional vibration but is smaller than type 1. So it is not easy to distinguish them.

4.4 *Removing Ambiguous Data*

It is challenging to distinguish type 3 data as flacking, so we tried to remove such data from data for learning. We tested the learning model using transfer learning. We used the Teachable Machine to create a model using Transfer Learning. The result is shown in Table 2. After removing type 3 data, the learning model showed good performance.

4.5 *Testing Using CNN and 1DCNN*

We tested the CNN and 1DCNN to develop a learning model for wav and FFT data. The result is shown in Table 3. We used two test patterns, one using raw data (wav), and another using FFT data.

The result showed different characteristics. When we used wav data, 1DCNN showed the best accuracy. 1DCNN is good at pattern matching for the wave shape as a picture. When we used FFT data, both wav and FFT data displayed similar accuracy. FFT data is more abstracted and does not require high-cost calculation. A hyperparameter for our learning model is described in Tables 4 and 5.

Table 3 Result of tests using CNN and 1DCNN

	CNN (wav)	CNN (FFT)	1DCNN (wav)	1DCNN (FFT)
The number of learning data = 1724 (Normal/Flacking: 1407/317) The number of test data = 1338 (Normal/Flacking: 927/411)	73%	81%	83%	79%

Table 4 Learning model using CNN

Dense (input)	256, relu
Dense (output)	2, sigmoid
Optimizer	Adam
epochs	30

Table 5 Learning model using 1DCNN

Conv1D (input)	2048/64, relu
max_pooling1D	1024/64, relu
Dense	1024/64, relu
Conv1D	512/64, relu
max_pooling1D	512/6, relu4
Conv1D	512/32, relu
Flatten	16384, relu
Dense (output)	2, sigmoid
Optimizer	Adam
epochs	30

5 Conclusion

This paper describes a method for efficiently detecting concrete wall flacking by combining NN with hammering testing. Since flacking occurs at different depths in a concrete wall, it is necessary to provide a generic learning model that can be used for flacking at different depths. This paper shows that the learning model for flacking at one depth is valid for flacking at another depth. Furthermore, the data at the boundary between areas of flacking and normality are shown to be a cause of misjudgment. It was shown that training by removing ambiguous data from the boundary area before learning improves accuracy. In a further study, we plan to verify the generality of the learning model using specimens of various depths flacking, as well as to investigate more general methods of learning model development.

Acknowledgements We thank Mr. Utagawa of Sato Kogyo Corporation for supporting us in collecting the sound sample.

References

1. http://www.mlit.go.jp/hakusyo/mlit/h25/hakusho/h26/html/n1131000.html. Last accessed 16 June 2022
2. TERRA DRONE https://www.terra-drone.net/industry/inspection_infrastructure/. Last accessed 16 June 2022
3. https://www.mlit.go.jp/common/001351557.pdf. Last accessed 16 June 2022
4. Pan, S.J., Yang, Q.: A survey on transfer learning. IEEE Trans. Knowl. Data Eng. **22**(10), 1345–1359 (2010). https://doi.org/10.1109/TKDE.2009.191
5. Fukumura, T., Aratame, H., Ito, A., Koike, M., Hibino, K., Kawamura, Y.: Improvement of sound classification method on smartphone for hammering test using 5G network. Int. J. Netw. Comput. **12**(2) (2022)
6. Car, T. T.: Hammering Inspection System. http://www.daon.jp/. Last accessed 16 June 2022
7. AI Hammering Checker. http://www.port-d.co.jp/p_pdc-100.htm#toi_jmp. Last accessed 16 June 2022
8. Güçlüer, K., Özbeyaz, A., Göymen, S., Günaydin, O.: A comparative investigation using machine learning methods for concrete compressive strength estimation. Mater. Today Commun. **27**, 102278 (2021). https://doi.org/10.1016/j.mtcomm.2021.102278
9. Ushiroda, K., Louhi Kasahara, J.Y., Yamashita, A., Asama, H.: Multi-modal classification using domain adaptation for automated defect detection based on the hammering test. In: 2022 IEEE/SICE International Symposium on System Integration (SII), pp. 991–996 (2022). https://doi.org/10.1109/SII52469.2022.9708607
10. Teachable Machine. https://teachablemachine.withgoogle.com/train/audio
11. An introduction to Teachable Machine—AI for dummies. https://blog.etereo.io/an-introduction-to-teachable-machine-ai-for-dummies-61d1f97f5cf
12. Howard, A., et al.: Searching for Mobilenetv3 (2019). arXiv:1905.02244

Artificial Intelligence Methods in Email Marketing—A Survey

Anna Jach

1 Introduction

Marketing activities based on sending email messages are everlastingly popular around the world, although the technology itself is several dozen years old. Recently, the email industry has been successfully supporting solutions based on artificial intelligence methods, which is a kind of combination of a well-known tool with something relatively new. According to Jain and Aggarwal [1], in 2020, marketing was in fourth place among the economic sectors devoting the most resources to the use of AI technology in the industry. The constantly evolving tools create a promising field for the development of digital marketing, especially in terms of user behavior research [2]. Predictive analysis for forecasting future recipients' actions together with AI-based time sending optimization is the desired key for marketers to maintaining the highest email deliverability and reply rate. This paper provides an overview of the methods based mainly on machine learning and artificial intelligence used in modern marketing over the last few years.

There were six primary research fields taken under consideration. The most commonly used one is text classification—mainly sentiment analysis and spam detection. From the point of view of email marketing, it allows for more effective work and easier reach to the target customer. In this field, Bayesian methods are frequently used.

Prediction of the number of replies and openings of messages is achieved by using such tools as neural networks or decision trees. This method allows for defining communication patterns for a given group of receivers.

Finding optimal sending time is clearly self-explanatory—this method help to decide when to send a message to obtain the best results, i.e. the highest possibility

A. Jach (✉)
Wroclaw University of Science and Technology, Wroclaw, Poland
e-mail: anna.jach@pwr.edu.pl

W. Zamojski et al. (eds.), *Dependable Computer Systems and Networks*, Lecture Notes in Networks and Systems 737, https://doi.org/10.1007/978-3-031-37720-4_8

of receiving a reply. In this field, there is a variety of tools used by researchers, from decision trees and random forests to recurrent neural networks.

Email tracking—the aim here is counting open and click rates, which is a crucial factor for forecasting email campaign performance.

Threats identification makes avoiding dangerous scams or phishing possible. It is a complex field, where deep learning methods are successfully used.

The last revised method is email sending volume forecasting. Using recurrent neural networks, in this case, allows estimating server load to maintain email providers servers in a good state.

2 Review of Available Solutions

The last few years have been a period of dynamic development of methods based on machine learning. This chapter summarizes the literature review from 2012 to 2021. It was decided to choose this timeframe so that the paper does not contain outdated information or describe methods that are no longer valid.

2.1 Text Classification

Recognizing unsolicited messages (spam) is a big problem due to the dynamically evolving strategies of the units that are the source of this mailing type. With this in mind, Choudhary and Dhaka proposed a genetic algorithm for efficient message classification [3]. In the first stage, a database of 500 email messages was created, 300 of which were spam messages. Only the content of the message was taken into account, without the headers. Successively, a dictionary of words belonging to the spam category was created and on its basis, each individual (message) was equipped with appropriate genes. During the algorithm's operation, 12% of individuals were recombinated, and only 3% were mutated.

In the second stage, a message database consisting of 2448 messages was used, and the word dictionary was additionally divided into 7 categories. The percentage of crossing and mutation among individuals was analogous to that in the first stage.

The authors managed to achieve a classification accuracy of 81.7%. The results presented in the work clearly show the proportional relationship between the size of the created dictionary and the algorithm's effectiveness.

Nayak et al. present a more comprehensive approach to the problem of spam detection [4]. They propose a combination of two methods—Bayesian classification and the J48 decision tree. In addition, unlike Choudhary and Dhaka, the set of features also included, inter alia, information extracted from the message header, parameters taken from BCC (blind carbon copy) and CC (carbon copy), and data on the sending domain. The training of both classifiers was performed with the use of separate training data sets. The accuracy achieved by the synthesis of the two classification methods is clearly higher compared to the predecessors and amounts to 88.1%.

Gibson et al. also raised the problem of distinguishing spam in email correspondence [5]. The team of researchers presents five methods: naive Bayesian classifier, SVM, decision tree, random forest, and multilayer perceptron, which are supported by evolutionary algorithms (including particle swarm optimization). 7 sets of training data used in the training of classifiers were prepared. The authors provide an exhaustive description of the learning process that allows for its easy reconstruction.

Particularly noteworthy is the detailed description of the experiments carried out and the methods used. The best classification efficiency, in this case, was achieved by the naive Bayes classifier cooperating with the genetic algorithm.

Zhao and Cai propose the classification of texts in Chinese [6] using a hybrid approach using deep neural networks. Several sets of deep networks (including convolutional, recurrent, and LSTM) were created to perform the task. The authors presented a rich analysis of existing solutions, and the categorization of statements in Chinese, pointing to an underappreciated research area in comparison with the analogous problem in English.

The contextual method of classifying email messages was in turn implemented by Wasi et al. [7]. A high categorization accuracy of 94% was achieved using a graphical representation of emails and exploratory algorithms. The authors clearly and reliably present the process of creating a classifier. The practical application of the created model is the detection of events included in the content of business messages and assigning them to appropriate classes.

Deshmukh and Dhavale propose a real-time email classification system [8]. The methods tested were decision tree, logistic regression, k-nearest neighbors algorithm, random forest, naive Bayes classifier, and Support Vector Machine. The authors consider the impact of individual methods of preprocessing text data on the effectiveness of the tested solutions. As in the examples cited above, also in this case the naive Bayesian classifier turned out to be the best method.

Tourani et al. tried to identify the sentiment of messages in two mailing lists [9]. They aimed to detect sadness or happiness among developers. 3 research questions were asked, regarding the accuracy of existing sentiment analysis methods, types of sentiment, and differences in expression between developers and users. It found out, that tools available on the market cannot tackle classifying conversations rich in technical terms, achieving low accuracy. The authors distinguished 6 categories of positive and 4 categories of negative sentiment during the process.

Kumar et al. have done an exploratory analysis of spam email classifiers [10]. The authors presented the proposed framework in detail and used the large dataset of emails obtained from the UCI Machine Learning Repository. The TANAGRA data mining tool was used to analyze the data. After examining 19 classifiers, they found that the random tree achieves the highest classification accuracy and can be considered for spam detection in the first place.

Sahmoud and Mikki propose building a spam detector using BERT [11], which is a pre-trained language model based on transformer architecture. One of the key features of this solution is its ability to handle multiple NLP tasks with only minor modifications to the model. The paper explains how the BERT model is fine-tuned to build an effective spam detector, where the email is tokenized using a BERT tokenizer and input tokens are converted to integer IDs based on the tokenizer vocabulary file. The authors used various datasets in the training process and managed to achieve impressive performance—from over 97% up to over 99% for each of them. Due to the nature of those sets, they proved that the BERT model allows to build a very precise spam detector for any type of business.

2.2 Prediction of the Number of Replies and Openings of Messages

Predictive methods bring with them the possibility of forecasting the results of email campaigns. Luo et al. in their work [12] developed models to predict the number of opened messages for two companies from different industries. In both cases, the surveyed campaigns were sent to recipients from the sender's target market (i.e. the group of buyers that a given company wants to acquire). The C4.5 decision tree and Support Vector Machine were used for classification. The training sets contained a set of features extracted from the recipient profile, including information about location, type of operating system, software used, and the type of device.

The achieved F1-score (i.e. harmonic mean precision and recall) was equal to 0.78. The C4.5 decision tree outperformed the SVM. The results achieved by the authors indicate that it is reasonable to train a separate model for each business client.

Biloš et al. present action research—a controlled experiment with the use of AB tests [13]. The subject of the research was email campaigns sent on the Croatian B2C market for a company in the retail industry. Tests were carried out to determine the subscriber's behavior depending on changes made to the message or to the shipping method. They consisted in sending 2 versions of the message to two sets of recipients. The advantage of the analyzed work is measuring the behavior of recipients in real conditions. This approach can help you spot small changes in campaign performance. In this case, too few tests were carried out to be able to draw conclusions about subscriber behavior. Out of ten tests, two showed a significant difference in the results (the subject of the message was modified), but it is not a sufficient number to confirm the thesis made regarding the influence of the structure of the subject of the message on the number of its openings.

The Monte Carlo random walk algorithms and hierarchical Bayesian estimation were used by Kumar to analyze the behavior of message recipients [14]. The publication puts forward numerous hypotheses regarding the impact of the newsletter structure on the user's decision to purchase and the probability of subsequent interaction with the message after its first opening. An example is a hypothesis regarding the number of links in the content—the authors assumed that more URLs favored reopenings and replies to a message that has been fully validated through research.

Yang et al. proposed a method of predicting whether a specific message will receive a reply and the time frame in which the reply will be sent [15]. The work uses binary and multiclass classification. The parameters of the email message, making an impact on the response, such as the length of the subject and content, the presence of attachments, the date of sending, and many others, were characterized. The recipient's past behavior was also taken into account. Based on the collected data, 10 sets of features were created that were used to design five predictive models with an accuracy not exceeding 46%.

2.3 *Finding Optimal Sending Time*

The problem of predicting the optimal message transmission time is undertaken by [16–18]. Abakouy et al. [18] propose the following algorithms: KNN, SVM, neural networks, and decision trees. The work is of an overview character, and its purpose is to choose the best method used for the segmentation of recipients. Dividing recipients into appropriate groups is a common method of optimizing email deliverability levels.

Singh et al. propose the use of survival analysis based on recurrent neural networks to determine the best point in time to send a message [16]. The calculated time intervals allowed to achieve the opening rate at the level of 16.48–19.24%. The use of specific intervals when sending emails as part of the studied sample resulted in obtaining over twenty-seven thousand additional message openings.

Paralič et al. present the use of CRISP-DM methods, a naive Bayesian algorithm, a random forest, and a decision tree for the prediction of sending time [17]. The work is a proposal of behavior models for any group of recipients based on historical data. The research used subsets of data describing events when the recipient's reaction was positive (i.e. a response was received). The results of the research conducted by the authors indicate that the decision tree allows for the highest accuracy among all the analyzed methods.

2.4 *Email Tracking*

Email tracking is a popular method of controlling user interactions with the message, such as opening it or clicking on a link in the content. The problem of recognizing messages with tracking included was taken up by Haupt et al. [19]. The work contains a very detailed description and analysis of the essence of tracking. The authors studied machine learning methods, including multilayer perceptron, random forest, and stochastic gradient boosting.

The structure of training data used in the work was described very meticulously, taking into account the division into the industries from which they came.

The aim of the authors was to select and then get rid of the part responsible for tracking from the HTML code of the email message—unwanted from the recipient's perspective, but most desired by marketing teams. In all the studied cases, the random forest showed the best parameters.

A broad investigation of the usage of email tracking on the German market has been conducted by Fabian et al. [20]. Remotely observation of users' behavior is prevalent there because it allows gaining information about when and where the recipient opened the message. They considered two types of tracking—based on hidden pixels and links placed in the email content. The authors also created a prototype of software for email tracking detection and managed to achieve relatively good accuracy—over 80% in recognizing both types. The software they propose uses a variety of thresholds and weighted criteria, such as switching between upper and lower case, switching from characters and numbers, or using keywords.

Another example of using this technique describes a comprehensive paper written by Kalantari [21]. The author was focused mainly on the privacy aspects of tracking in email communication. It has been stated that some part of senders follows' open rate of messages to adjust their sending rate and the number of emails that come from marketers to subscribers can vary depending on recipients' reading behavior. External images (pixels) and personalized links are the most common in the email content. A doubt regarding security risk may arise here because to check if a particular recipient has downloaded the image or clicked on the link, the sender does not need any consent. However, according to the results obtained by the author, most email marketers strictly follow the privacy rules, and using the opt-in lists and adding an unsubscribe link are common practices.

2.5 Threats Identification

Threat detection is an important research aspect in the area related to email marketing. Methods, which include phishing or so-called business compromises, aimed at exposing the recipient to financial or face losses, are constantly developing, so their detection is not a trivial problem.

Deep neural networks are widely used in this sector [22–24]. Vinayakumar et al. propose the use of various network architectures, including the proprietary deep structure of DeepSpamPhishEmailNet. In this case, when identifying fraudulent messages, the best results were achieved with the use of a convolutional neural network. In turn, Bagui et al. present the use of deep learning methods, as well as the naive Bayesian classifier or Support Vector Machines [25] for the implementation of an analogous classification task. Based on the collection of over a dozen thousand email messages for each of the analyzed approaches, the accuracy exceeded 90% each time.

Kaddoura et al. used a deep neural network to recognize spam messages [26]. Only the text of the message was classified. The public, balanced Enron dataset of over 32,000 messages divided into two categories was used. The network architecture designed by the authors with 7 hidden layers allowed to achieve F1 values at the level exceeding 0.98. In their short paper, Mohey and Mohsen present an analysis comparing the results obtained by methods such as SVM, the naive Bayesian classifier, or the decision tree in spam detection [23].

Alhogail and Alsabih propose the use of graph convolutional neural networks to detect phishing messages [27]. The authors used a public email collection on which they performed procedures in the field of natural language processing, which increased the classification accuracy. The test results of the created model show an accuracy of 98.2% and a very low false-positive correlation coefficient.

Another method of detecting phishing based on machine learning methods is presented by Shirazi et al. [2]. The work uses classifiers available within the Scikit-learn library: random forest, gradient boosting and a Support Vector Machine with a radial and linear basis function kernel. The authors' approach is to look for similarities between the false and the original website, and the role of the trained classifier is to decide whether a given website is impersonating a reliable source. The data sample contains features pertaining to both the visual and textual parts of the website in question. The committee's models of classifiers achieved 98% accuracy in recognizing fraudulent sites, which is an impressively high result.

Another dangerous phenomenon that appears in the email environment is the so-called business email compromise—a cyberthreat based on social engineering. Kurematsu et al. [28] dealt with the issue of identifying this type of messages, using decision trees, naive Bayes classifier, SVM, and the KNN method. The best results were obtained by SVM-based models, and a slightly lower accuracy was achieved for the Bayesian classifier. The authors also propose identifying the sender but argue that it is an exceptionally complex task.

A similar problem has been raised by Yaseen et al. A proposed approach to identify fraudulent emails was a committee of eleven classifiers [29]. Each of them was trained and assessed by means of tenfold cross-validation. It is worth noting that the feature vectors in the training set contain not only data related to the content of the message, but also related to the senders' domains. The created model is distinguished by excellent classification accuracy, the F1 parameter reaches 99.30%, and the AUC (area under curve) 99.9%.

Jáñez-Martino et al. attempted to test the reliability of email addresses using machine learning methods [30]. The quality of the sender's address, along with the content of the message, is a key factor in building trust among the recipients of scams. In essence, the problem comes down to a two-class classification. The performance of four popular classifiers was tested: naive Bayesian classifier, Support Vector Machine, logistic regression, and random forest. The best results were obtained by the Bayesian classifier with over 88% accuracy and the F1-score of 0.808.

2.6 Email Sending Volume Forecasting

Researchers are also interested in modeling the load caused by traffic generated in servers that support the sending of email messages [26, 31]. Om et al. propose recurrent neural networks and LSTM networks to predict the size of the sending volume. The authors attempted to model email traffic for four universities, achieving significantly better results with LSTM than with recurrent networks. The work published in 2020 presents results that surpass many models described in the literature on the subject [31].

Boukoros et al. also tried to forecast the volume of email traffic within university servers. In this case, a recurrent neural network containing two hidden layers was used [32], and the problem was treated as a prediction of time series. Several categories were considered: volume of outgoing emails, the volume of incoming emails, the volume of system messages, and spam. The prediction accuracy obtained for each of them is within the range of 80–90%, which can be considered a satisfactory result.

3 Summary

During the preparation of the literature review, 98 research works were considered, of which only 32 were selected. The works published between 2005 and 2022 were analyzed. Early articles, published in 2005–2010, i.e. at the time when email marketing began to gain the interest of researchers, aimed at solving very basic problems, and due to the dynamic development of the field, often also out of date. Therefore, from the current perspective, the presented results were not helpful. Publications from the last few years, on the other hand, present methods of solving contemporary problems arising in the email industry. Artificial intelligence and machine learning methods are widely used in email marketing. There is a great number of applications: from naive Bayesian classification and BERT model for sentiments analysis, SVM and decision trees for reply rate prediction, to recurrent and deep neural networks for time sending optimization and threats identification. However, despite the huge diversity of methods working in the industry, we will surely have the opportunity to see the development in the email marketing field in the coming years. The email world is constantly evolving and without any doubt will demand new approaches to old problems. In recent years privacy considerations gain importance, so it will be reasonable to apply AI-based methods to ensure compliance with such legal acts as GDPR in Europe or CCPA in the USA.

The main challenge for the whole industry, besides such basic issues as improving time sending frequency or reply rate prediction, is personalization. There is an untold potential for using AI tools to create email marketing campaigns that fit perfectly for each recipient. As the email world evolves, new problems and challenges will undoubtedly arise, and email marketers will need to adopt new approaches to old problems.

References

1. Jain, P., Aggarwal, K.: Transforming marketing with artificial intelligence. Int. Res. J. Eng. Technol. **7**, 3964–3976 (2020)
2. Shirazi, H., Zweigle, L., Ray, I.: A machine-learning based unbiased phishing detection approach. In: Proceedings of the 17th international joint conference on e-business and telecommunications (2020). https://doi.org/10.5220/0009834204230430
3. Choudhary, M., Dhaka, V.S.: Automatic e-mails classification using genetic algorithm. Int. J. Comput. Sci. Inf. Technol. **6**, 5097–5103 (2015)
4. Nayak, R., Jiwani, S.A., Rajitha, B.: Spam email detection using machine learning algorithm. Mater. Today: Proc. (2021)
5. Gibson, S., Issac, B., Zhang, L., Jacob, S.M.: Detecting spam email with machine learning optimized with bio-inspired metaheuristic algorithms. IEEE Access **8**, 187914–187932 (2020). https://doi.org/10.1109/ACCESS.2020.3030751
6. Zhao, R., Cai, Y.: Research on online marketing effects based on multi-model fusion and artificial intelligence algorithms. J. Ambient. Intell. Humaniz. Comput. (2021). https://doi.org/10.1007/s12652-021-03216-7
7. Wasi, S., Jami, S.I., Shaikh, Z.A.: Context-based email classification model. Expert. Syst.: J. Knowl. Eng. **33**, 129–144 (2016)
8. Deshmukh, S., Dhavale, S.: Automated real-time email classification system based on machine learning. In: Bhalla, S., Kwan, P., Bedekar, M., Phalnikar, R., Sirsikar, S. (eds.) Proceeding of International Conference on Computational Science and Applications, pp. 369–379. Springer, Singapore (2020)
9. Tourani, P., Jiang, Y., Adams, B.: Monitoring sentiment in open source mailing lists: exploratory study on the apache ecosystem. In: Proceedings of 24th annual international conference on computer science and software engineering, pp. 34–44 (2014)
10. Kumar, R., Poonkuzhali, G., Pandiarajan, S.: Comparative study on email spam classifier using data mining techniques. Lect. Notes Eng. Comput. Sci. **2195**, 539–544 (2012)
11. Sahmoud, T., Mikki, M.: Spam detection using BERT (2022). https://doi.org/10.48550/arXiv.2206.02443
12. Luo, X., Nadanasabapathy, R., Nur Zincir-Heywood, A., Gallant, K., Peduruge, J.: Predictive analysis on tracking emails for targeted marketing. In: Japkowicz, N., Matwin, S. (eds.) Discovery Science. Lecture Notes in Computer Science, vol. 9356, pp. 116–130. Springer, Cham (2015)
13. Biloš, A., Turkalj, D., Kelić, I.: Open-rate controlled experiment in email marketing campaigns. Market – Tržište **28**, 93–109 (2016)
14. Kumar, A.: An empirical examination of the effects of design elements of email newsletters on consumers' email responses and their purchase. J. Retail. Consum. Serv. **58**, 102349 (2021). https://doi.org/10.1016/j.jretconser.2020.102349
15. Yang, L., Dumais, S.T., Bennett, P.N., Awadallah, A.H.: Characterizing and predicting enterprise email reply behavior. In: Proceedings of the 40th International ACM SIGIR Conference on Research and Development in Information Retrieval, Aug. 2017. https://doi.org/10.1145/3077136.3080782
16. Singh, H., Sinha, M., Sinha, A.R., Garg, S., Banerjee, N.: An RNN-survival model to decide email send times (2020). arXiv:2004.09900
17. Paralič, J., Kaszoni, T., Mačina, J.: Predicting suitable time for sending marketing emails. In: Świątek, J., Borzemski, L., Wilimowska, Z. (eds.) Information Systems Architecture and Technology: Proceedings of 40th Anniversary International Conference on Information Systems Architecture and Technology—ISAT 2019, vol. 1051, pp. 189–196. Springer, Cham (2020)
18. Abakouy, R., En-Naimi, E.M., El Haddadi, A.: Classification and prediction based data mining algorithms to predict email marketing campaigns. In: Proceedings of the 2nd International Conference on Computing and Wireless Communication Systems (ICCWCS'17), pp. 1–5. Association for Computing Machinery, New York (2017)

19. Haupt, J., Bender, B., Fabian, B., Lessmann, S.: Robust identification of email tracking: a machine learning approach. Eur. J. Oper. Res. **271**, 341–356 (2018)
20. Fabian, B., Bender, B., Weimann, L.: E-mail tracking in online marketing—methods, detection, and usage. In: Wirtschaftsinformatik Proceedings 2015, Mar. 2015. https://aisel.aisnet.org/wi2015/74/. Accessed 04 Jan 2022
21. Kalantari, S.: Open about the open-rate?. In: IFIP Advances in Information and Communication Technology, pp. 187–205 (2021). https://doi.org/10.1007/978-3-030-72465-8_11
22. Ravi, V., Kp, S., Poornachandran, P., Soman, A., Elhoseny, M.: Deep learning framework for cyber threat situational awareness based on email and URL data analysis. In: Hassanien, A., Elhoseny, M. (eds.) Cybersecurity and Secure Information Systems. Advanced Sciences and Technologies for Security Applications, pp. 87–124. Springer, Cham (2019)
23. Mohey, H., Mohsen, S.: Using machine learning techniques for predicting email spam. Int. J. Instr. Technol. Educ. Stud. **2**(4), 19–23 (2021). https://doi.org/10.21608/ihites.2021.204000
24. Rutkowski, A., Czoków, M., Piersa, J.: Wstęp do sieci neuronowych. Wykład 14, support vector machine. https://www-users.mat.umk.pl//~rudy/wsn/wyk/wsn-wyklad-16-SVM.pdf. Accessed 04 June 2022
25. Bagui, S., Nandi, D., Bagui, S., White, R.J.:Classifying phishing email using machine learning and deep learning. In: 2019 International Conference on Cyber Security and Protection of Digital Services (Cyber Security), pp. 1–2 (2019). https://doi.org/10.1109/CyberSecPODS.2019.8885143
26. Kaddoura, S., Alfandi, O., Dahmani, N.: A spam email detection mechanism for English language text emails using deep learning approach. In: 2020 IEEE 29th international conference on enabling technologies: infrastructure for collaborative enterprises (WETICE), pp. 193–198 (2020). https://doi.org/10.1109/WETICE49692.2020.00045
27. Alhogail, A., Alsabih, A.: Applying machine learning and natural language processing to detect phishing email. Comput. Secur. **110**, 102414 (2021)
28. Kurematsu, M., Yamazaki, R., Ogasawara, R., Hakura, J., Fujita, H.: A study of email author identification using machine learning for business email compromise. In: Fujita, H., Selamat, A. (eds.) Advancing Technology Industrialization Through intelligent Software Metodologies, Tools and Techniques, vol. 318, pp. 205–216. IOS Press (2019)
29. Yaseen, Y.A., Qasaimeh, M., Al-Qassas, R., Al-Fayoumi, M.A.: Email fraud attack detection using hybrid machine learning approach. Recent Adv. Comput. Sci. Commun. **14**, 1370–1380 (2019)
30. Jáñez-Martino, F., Alaiz-Rodríguez, R., González-Castro, V., Fidalgo, E.: Trustworthiness of spam email addresses using machine learning. In: Proceedings of the 21st ACM Symposium on Document Engineering, Aug. 2021. https://doi.org/10.1145/3469096.3475060
31. Om, K., Boukoros, S., Nugaliyadde, A., McGill, T.J., Dixon, M., Koutsakis, P., Wong, K.W.: Modelling email traffic workloads with RNN and LSTM models. Hum.-Centric Comput. Inf. Sci. **10**, 39 (2020)
32. Boukoros, S., Nugaliyadde, A., Marnerides, A., Vassilakis, C., Koutsakis, P., Wong, K.W.: Modeling server workloads for campus email traffic using recurrent neural networks. Neural Inf. Process. 57–66 (2017). https://doi.org/10.1007/978-3-319-70139-4_6

Detection of Oversized Objects in a Video Stream Using an Image Classification with Deep Neural Networks

Przemysław Jamontt, Juliusz Sarna, Jakub Wnuk, Marek Bazan, Krzysztof Halawa, and Tomasz Janiczek

1 Introduction

Detecting big rocks that are extracted by coal mining machinery and falling onto its conveyor belt is a crucial task in the process of the operating of such machines. This is due to the fact that when such situations occur the operation of the conveyor belt has to be stopped, otherwise allowing an extracted oversized piece of rock to go downstream will cause serious damage to the excavation machine. Therefore such unwanted rocks must be detected while still travelling on the conveyor belt in a way that there is sufficient time to stop its operation and enable the removal of the unwanted rocky object. Bucket-wheel excavators used in opencast mines are huge in size and their damage is costly. Watching the conveyor belt is monotonous and tedious work for the operator, who may overlook a dangerous rock [2]. Therefore, it is advisable to use an automatic detection system.

Four different approaches to tackle the problem of detection of an oversized rock on a conveyor belt are present in the literature. These approaches cover

- digital image processing methods [11, 14, 15]
- lidar methods [19],
- vibration monitoring [3],
- machine learning [21] , shallow neural networks [4, 6] and deep learning [18, 20].

P. Jamontt · J. Sarna · J. Wnuk · M. Bazan (✉) · K. Halawa · T. Janiczek
Wrocław University of Science and Technology, 27 Wybrzeże Wyspiańskiego st., 50-370 Wrocław, Poland
e-mail: marek.bazan@pwr.edu.pl

P. Jamontt
e-mail: 252868@student.pwr.edu.pl

K. Halawa
e-mail: krzysztof.halawa@pwr.edu.pl

T. Janiczek
e-mail: tomasz.janiczek@pwr.edu.pl

W. Zamojski et al. (eds.), *Dependable Computer Systems and Networks*, Lecture Notes in Networks and Systems 737, https://doi.org/10.1007/978-3-031-37720-4_9

Due to the many advantages and good results, deep neural networks are very often used for many image classification tasks [17]. Papers [7, 10, 13] describe tasks that bear some resemblance to the rock detection problem.

In this paper, we propose a method based on deep neural networks exploiting transfer learning to classify the whole frame as containing a dangerous rock or not. Such an approach is applied since there is no need to localise the rock in the frame and we can resign from object localisation. Figure 1 shows an example of a rock on the conveyor belt of a bucket-wheel excavator.

The deep learning network is then used as a crucial tool of tracking a moving conveyor belt to stop it if the rock has been detected in a certain number of successive frames.

The detection of dangerous rocks has to be performed in a variety of lighting conditions, so the networks were trained to detect boulders both in daylight and under artificial lighting. The great difficulty stems from the fact that the rocks could be partly covered with excavated material and could be difficult to see even by a focused person. In addition, the rocks had different shapes, sizes and textures.

The main contribution of this paper is showing the performance of two different deep learning models for detection of frames with rocks falling on a conveyor belt. Moreover, the method based on a moving average to trigger alarms when consecutive frames contain a stone was developed.

All research was conducted on recordings from a real mine. The effectiveness of the proposed solution was tested on recordings made in various lighting conditions.

The rest of the article is organised as follows. Section 2 describes the preparation of the training set. Section 3 presents the examined models with deep neural networks. Sections 4 and 5 are devoted to the process of training the network and the applied inference method based on the results obtained from several frames. Section 6 discusses the limitations of the proposed method. Section 7 explains what the authors intend to improve in further research. A summary is included at the end of the work.

2 Preparing a Dataset

The dataset contained 67 videos showing a conveyor belt on a mining excavator in different weather conditions captured by a `cctv` camera. All the videos capture a short interval of normal conveyor operation, the occurrence of a stone extracted by the mining excavator and passed to the conveyor belt and stopping of the conveyor operation. The videos cover the period from 60 to 45 s before the event, i.e. without a rock, until the moment when the conveyor belt is stopped with a visible rock, i.e. about 30 s from the moment of falling onto the conveyor belt. All frames from videos were manually annotated into 2 classes: `rock present` and `no rock`. In practice, only frames with a visible rock were annotated with a label `rock present` and the rest of the frames were given a `no rock` label. The view in the image was cropped to contain only a conveyor belt that resulted in a 580×400 image resolution per frame.

Fig. 1 Camera view with a visible rock on the conveyor belt

There were 3847 samples with a rock present and about 11 times more without a rock. Building classification models directly on such an imbalanced dataset was not successful, and therefore some method of handling imbalanced had to be used. To tackle the problem of an imbalanced dataset down-sampling was applied. The final solution was trained on a dataset obtained from a custom generator (see Sect. 4 for how the custom generator delivers a dataset) that in each training epoch a size of positive and negative classes were equal. Additionally, during the training of models, augmentation of data in training sets was done.

3 Models

A transfer learning approach was chosen to resolve the given task. A few models pre-trained on `ImageNet` [5] were tested. Two of them gave notable results. These are VGG16 [12] and ResNet50 [9].

3.1 The VGG16 Model

The VGG16 is a deep convolutional network model [12] developed by Karen Simonyan and Andrew Zisserman for the ImageNet challenge. The standard input image size for this network is 224×224 pixels. The collected images, therefore, were rescaled to this resolution before being fed into the network. The top classification layers were removed and two dense layers were added with dropout equal

Table 1 The VGG16-based model

Layer (type)	Output shape	Param #
Vgg 16 (Functional)	(None 7, 7, 512)	14714688
Flatten (Flatten)	(None, 25008)	0
Dense (Denst)	(None, 256)	6422784
Dropout (Dropout)	(None, 256)	0
Dense_1 (Dense)	(None, 1)	257
Total params: 21, 127, 729 Trainable params: 6, 423, 041 Non-trainable params: 14,714,688		

Table 2 The ResNet50-based model

Layer (type)	Output shape	Param #
Resnet 50 (Functional)	(None , 2048)	23587712
Dropout (Dropout)	(None, 2048)	0
Dense (Denst)	(None, 256)	524544
Dense_1 (Dense)	(None, 1)	257
Total params: 21, 112, 513 Trainable params: 24, 059, 393 Non-trainable params: 53,120		

to 40%, c.f. Table 1. The model was trained and tested with the use of 5-fold cross-validation. The results with the highest accuracy were obtained using RMSprop as an optimiser. Since we provided that the final sets used for training were balanced (see Sect. 4) as a measure of the quality of the solution the accuracy measure was chosen.

3.2 The ResNet50 Model

The ResNet50 is a residual network model developed by He et al. [9]. Similarly to the VGG16 model, images have to have a resolution of 244×224 pixels. Therefore the rescaling to such a size was performed.

The top classification layers were removed and replaced with two dense layers with a dropout equal to 20% preceded by global pooling (see Table 2). The model was trained and tested with the use of 5-fold cross-validation. Similarly to the VGG16, the RMSprop optimiser turned out to give the best-performing solutions.

Table 3 Learning rate values chosen in a two-phase training process for two models

Training phase	Model	
	VGG16	ResNet50
Only head training	10^{-3}	$5 \cdot 10^{-5}$
All weights training	$2 \cdot 10^{-5}$	10^{-5}

4 Training Process

A custom function that provided cross-validation on 5 folds was created to test models. Each fold was made up of a test, validation and training set in the proportion 1:1:3 with respect to the number of frames with a rock present. Folds were given videos exclusively—frames from the same video are not spread across different folds to prevent information leaks.

An unequal number of samples in classes were obtained when all the available data was taken. It resulted in problems with the training of models. A custom function contained generators which provided that the number of positive samples i.e. with a rock present and without a rock, is equal in each training epoch—see Table 4. All samples with a rock present were used, and the same number of frames without a rock were at random. During each of the epochs, all positive samples and the other subset of available negative samples were provided. It enabled us to exploit as effectively as possible negative samples, since they were distributed over the whole training process.

For both models, RMSprop optimiser [16] gave the best results. For both models, training was performed in two phases. The first phase covered training of only the final added last layers with the frozen convolution base with a higher learning rate for 5 epochs. The second phase relied on unfreezing all the weights and performing optimisation for the next 5 epochs. The learning rate values used in training are presented in Table 3. As the solution we chose the model from the second phase with the best performance across the last 5 epochs on the validation set.

The counts of frames in different folds were close to each other , but since they were taken from videos with a different length, it turned out that the test sets of whole videos constructed with such an assumption contained a different number of videos. The number of videos in a test set for a given fold is shown in Table 5.

In the training process, data augmentation of random up-down and right-left shifts was used with such parameters as the rotation range from –9 to 9°, width and a height shift range of 10% with a fill mode of the added space set to the nearest true pixel.

The final quality of the obtained models measured with the accuracy for all folds is given in Table 6.

Table 4 Number of positive and negative samples in the whole dataset. For the training process the same number of negative examples selected at random as a number of positive ones was chosen

Fold no.	Images with rocks	Images without rocks
1	807	5782
2	729	6346
3	776	6954
4	743	7544
5	792	6825

Table 5 Number of videos in a test set for a given fold. Then the number of videos differs since the videos are of different lengths. The number of positive samples is maintained on approximately the same level for all folds

Fold no.	Count of videos
1	14
2	13
3	13
4	10
5	11

Table 6 Accuracy of models in [%] on folds that were used for cross-validation. The total is the percentage accuracy obtained from cross-validation

Fold no.	VGG16	ResNet50
1	72.35	82.72
2	87.88	86.80
3	78.81	85.82
4	83.93	90.08
5	76.10	79.57
Total	**79.81**	**85.00**

5 Usage of Detection Models for Videos

The moving average was chosen as the final decision-making mechanism.

Two parameters were specified in order to implement a moving average approach, i.e. a `window length` which defines the `number of preceding frames` to be taken for calculating a moving average and the `detection threshold`. The moving average was calculated from the prediction results obtained by the neural network on the `window length` preceding frames.

The output from VGG16 and ResNet50 in our model is between 0 and 1, where 0 corresponds to a class with a rock present, whereas 1 refers to a class of frames without a rock. For most of the frames from the test set, the results were scattered around the

Table 7 Comparison of performance of the VGG16 and ResNet50 models applied to a rock detection algorithm based on a moving average thresholding of successive frames (In the first column are the numbers of fields used for cross-validation)

Fold no	VGG16	ResNet50
1	8/14	11/14
2	10/13	11/13
3	10/13	12/13
4	7/10	8/10
5	10/11	9/11
Total	**73.77%**	**83.61%**

limits of the range, i.e. slightly above 0 and slightly below 1. The detection threshold was set to 0.1 and the moving average window length was set to 30. When the value of the moving average over the preceding frames dropped below the detection threshold the alarm was triggered. The use of a moving average over the preceding frames allows the elimination of isolated false negative decisions of the model and to trigger the alarm when the model sees a considerable number of true positive samples (with the output close to 0 in our configuration) in a window.

The VGG16 and ResNet50 models that achieved 79.81 and 85.00% on test dataset evaluation, respectively, were tested on videos which were the source of frames used for the test dataset. This means that a given model did not see them previously. Next, for both models two parameters were chosen with a manual grid search to be the best fit for given model. The results are presented in Table 7.

6 Limitations

The most frequent false positive detection of the rock occurred when single piles of sand were coming from a mining bucket onto the clean conveyor belt. This issue can be present due to the lack of data that shows the starting point of work and clean conveyor belt state to the model (Fig. 2).

The colour diversity of the rocks which are the object of the search is not clearly different from the sand where it needs to be detected. This can be due to settling dust or being covered by sand, resulting from the continuous discharge of dirt from the coal mining machine.

The dataset is not differentiated in terms of atmospheric conditions, time of day, and year. Data used for training the models is limited only to favourable weather conditions without any rain, snow, or fog, which makes the model vulnerable in working environment.

One possible way of mitigating this could be a custom-crafted augmentation. Currently it uses only traditional rotation and shift of images. Much better results

Fig. 2 Examples of a false positive predictions

should be achieved by colours augmentation to simulate different day, hour and weather conditions. The simulation of dust clouds and fog can be done by drawing some kind of noise over the images. Perlin noise could be good for this purpose, but this needs further experiments.

The problem with false detections can be mitigated by isolating what is causing them, and retag the data to mark these elements. This would allow to building of multiclassification models. They can be forced in such a way as to recognise certain events, and prevent the model from misrecognizing them as rocks.

7 Future Work

Our plan for future work is twofold. First, denoising methods to improve quality of the detection on a single frame need improvement. In addition, the method may make effective use of omitted frames while balancing positive and negative samples for better robustness for different weather conditions. Second, the method may be compared with methods for object detection in images such as YOLO [1] or Mask R-CNN [8]. This, however, requires some work on annotating the rocks with boxes in all frames where rocks are present. Additionally, it seems interesting to study in detail the approaches consisting in feeding several successive frames to the network inputs at the same time. Doing a lot of additional experiments with different network architects and hyperparameter values can help improve the results.

8 Conclusions

The examined problem is of great practical importance and may allow for the reduction of the operating costs of an opencast mine. All of the videos come from a real mine.

In this paper we presented the solution of a mechanism based on deep learning to detect dangerous oversized rocks on coal mining machine conveyor belt. The method uses transfer learning models that are fine-tuned on a limited number of videos that cover the events of alarms after the human guided stopping of the conveyor belt. The research showed the applicability for this task of two deep neural network models, i.e. VGG16 and ResNet50, respectively, that are used in the framework of the moving average to filter a false positive and false negative. The overall approach enables us to correctly detect real rocks in more than 83% of cases.

It is worth emphasising that the detection of rocks was performed both in daylight and in artificial night lighting. More videos recorded in the mine would probably allow for a better selection of hyperparameters and further improvement of the results.

Acknowledgements The authors would like to thank Produs S.A. and the Bełchatów coal mine for their cooperation and research data.

References

1. Bochkovskiy, A., Wang, C.Y., Liao, H.Y.M.: Yolov4: Optimal Speed and Accuracy of Object Detection (2020). arXiv:2004.10934
2. Branton, P.: Process control operators as responsible persons. In: Person-Centred Ergonomics: A Brantonian View Of Human Factors, p. 171 (2003)
3. Buckley, J.: Monitoring the vibration response of a tunnel boring machine: application to real time boulder detection. Colorado School of Mines (2015)
4. Cabello, E., Sánchez, M.A., Delgado, J.: A new approach to identify big rocks with applications to the mining industry. Real-Time Imaging **8**(1), 1–9 (2002)
5. Deng, J., Dong, W., Socher, R., Li, L.J., Li, K., Fei-Fei, L.: ImageNet: A large-scale hierarchical image database. In: CVPR09 (2009)
6. Fahle, W., KÖRber, S.: Exploration and selective extraction of stones and rocks in overburdens of lusatian open-pit lignite mines. Universitas Publishing House Petroşani–RomÂnia **10**, 27 (2009)
7. Fu, Y., Aldrich, C.: Online particle size analysis on conveyor belts with dense convolutional neural networks. Miner. Eng. **193**, 108019 (2023)
8. Fujita, H., Itagaki, M., Ichikawa, K., Hooi, Y.K., Kawano, K., Yamamoto, R.: Fine-tuned pre-trained mask r-cnn models for surface object detection (2020). arXiv:2010.11464
9. He, K., Zhang, X., Ren, S., Sun, J.: Deep residual learning for image recognition. In: 2016 IEEE Conference on Computer Vision and Pattern Recognition (CVPR), pp. 770–778 (2016). https://doi.org/10.1109/CVPR.2016.90

10. Ma, X., Zhang, P., Man, X., Ou, L.: A new belt ore image segmentation method based on the convolutional neural network and the image-processing technology. Minerals **10**(12), 1115 (2020)
11. Saran, G., Ganguly, A., Tripathi, V., Kumar, A.A., Gigie, A., Bhaumik, C., Chakravarty, T.: Multi-modal imaging-based foreign particle detection system on coal conveyor belt. Trans. Indian Inst. Metals **75**(9), 2231–2240 (2022)
12. Simonyan, K., Zisserman, A.: Very deep convolutional networks for large-scale image recognition (2014). arXiv:1409.1556
13. Singh, T., Jhariya, D., Sahu, M., Dewangan, P., Dhekne, P.: Classifying minerals using deep learning algorithms. In: IOP Conference Series: Earth and Environmental Science, vol. 1032, p. 012046. IOP Publishing (2022)
14. Sun, J., Su, B.: Coal-rock interface detection on the basis of image texture features. Int. J. Min. Sci. Technol. **23**(5), 681–687 (2013)
15. Suresh, M., Abhishek, M.: Kidney stone detection using digital image processing techniques. In: 2021 Third International Conference on Inventive Research in Computing Applications (ICIRCA), pp. 556–561. IEEE (2021)
16. Tieleman, T., Hinton, G., et al.: Lecture 6.5-rmsprop: Divide the gradient by a running average of its recent magnitude. COURSERA: Neural Netw. Mach. Learn. **4**(2), 26–31 (2012)
17. Wang, W., Yang, Y., Wang, X., Wang, W., Li, J.: Development of convolutional neural network and its application in image classification: a survey. Opt. Eng. **58**(4), 040901–040901 (2019)
18. Wang, Y., Guo, W., Zhao, S., Xue, B., Zhang, W., Xing, Z.: A big coal block alarm detection method for scraper conveyor based on yolo-bs. Sensors **22**, 1592–1596 (2022). https://doi.org/10.3390/s22239052
19. Xing, Z., Zhao, S., Guo, W., Guo, X., Wang, Y., Bai, Y., Zhu, S., He, H.: Identifying balls feature in a large-scale laser point cloud of a coal mining environment by a multiscale dynamic graph convolution neural network. ACS Omega **7**(6), 4892–4907 (2022)
20. Zhang, K., Wang, W., Lv, Z., Fan, Y., Song, Y.: Computer vision detection of foreign objects in coal processing using attention cnn. Eng. Appl. Artif. Intell. **102**, 104242 (2021)
21. Zhang, Z., Yang, J., Wang, Y., Dou, D., Xia, W.: Ash content prediction of coarse coal by image analysis and ga-svm. Powder Technol. **268**, 429–435 (2014)

Reliability Model of Bioregenerative Reactor of Life Support System for Deep Space Habitation

Igor Kabashkin and Sergey Glukhikh

1 Introduction

The development of deep space habitation has created the need for autonomous transportation systems that can operate for extended periods in isolated environments. These systems include space stations for long-term habitation and manned spacecraft for deep space exploration [1]. A key factor in the success of these missions is minimizing the risks that could disrupt the crew's well-being. To prolong the duration of autonomous operation, efforts are underway to develop closed life support systems (LSS) that regenerate waste products and sustain the crew [2]. The focus is on completely closed life support systems that operate independently from the outside world [3]. However, the dependability of such systems is critical for crew survival during extended missions. NASA research has shown that reliable models of life support systems should be developed early in the design process. This paper proposes one of the approaches to solving this issue.

2 Related Works

Today the challenges of providing long-term life support for crew members are being actively investigated, particularly for Mars exploration programs [4]. Life support systems are crucial for these missions and must be reliable during both the journey and the stay on the Martian surface. A proposed life support system for Mars missions is generally described in [5], which outlines the primary requirements from the perspective of human needs. Paper [6] provides a description of a LSS based on

I. Kabashkin (✉) · S. Glukhikh
Transport and Telecommunication Institute, Lomonosov 1, Riga 1019, Latvia
e-mail: kiv@tsi.lv

W. Zamojski et al. (eds.), *Dependable Computer Systems and Networks*, Lecture Notes in Networks and Systems 737, https://doi.org/10.1007/978-3-031-37720-4_10

molecular sieve that removes carbon dioxide, regulates air purity, produces oxygen, and purifies water suitable for human consumption.

A space station for long mission operation is designed with the assumption that failed components of any system can be restored by the crew, and crew members are trained in eliminating individual failures [7]. NASA accumulates data on failures and conducts dependability analysis based on the experience gained from various missions. Reliability calculations are typically adjusted based on experimental data, and a database of reliability is formed to manage risks during mission implementation [8]. This is done using a special application tailored to the specific needs of space flights [9], which can define the dependability parameters and their consequences based on available data and focused on the particular properties of transport means, such as a probabilistic risk assessment considering the failure modes of the Space Shuttle [10]. This methodology can be used for single reliability model for all spacecraft systems. A similar approach is used in other areas, including aviation [11], and has been applied to important projects such as the design of Airbus [12] and the lunar module [13]. In addition to this approach, other methods, such as fault tree analysis, are also used [14].

The current LSS used in relatively short-term space missions, such as those for the International Space Station (SS) [15], unfortunately lack super-dependability in the conditions of long-term and remote missions where resupplying necessary life-support supplies becomes impossible. This has been confirmed by experience gained from working on space stations where deviations were required to supply crew members with necessary resources due to failures in regular LSS to prevent disasters during actual flights [16]. Today architecture of LSS consist from independent functional components and can have various alternative designs and operations [2, 17]. Therefore, dependability models must be developed early in the design stage to assess the risks of the LSS during the mission life cycle for various possible LSS architectures [18, 19].

In the paper, an approach is outlined for creating an integrated LSS ecosystem for long-term space station operation with LSS included five main components of biotechnological regeneration: oxygen, methane, carbon dioxide, food, and water. The article proposes developing dependability models for components of such biotechnological reactor using Petri net modeling.

3 Bioregenerative Reactor of the Life Support Systems

Reducing expendable stocks is a practical concern of the of biotechnological life support system (BLSS) development for space stations, as it minimizes the need for expensive, complex, and unsafe shipments by space stations. For autonomous space stations, the architecture of bioregenerative reactor consists of five main components: oxygen, methane, carbon dioxide, food, and water [3]. Figure 1 illustrates the integration of the main closed biotechnological components of BLSS, the details and functionality of which is described in [3].

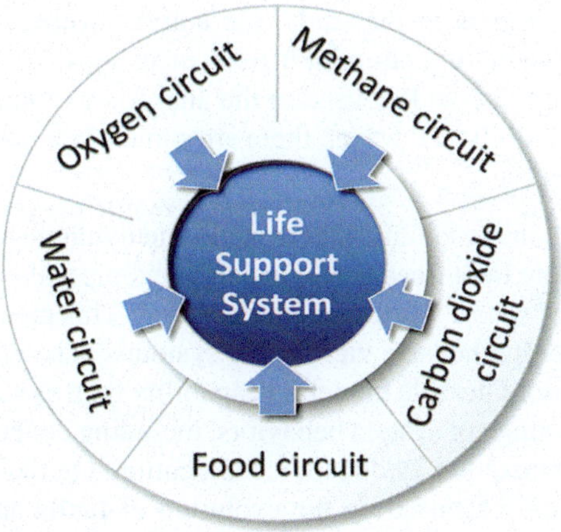

Fig. 1 Architecture of biotechnological life support system

4 Model Formulation and Solution

For long-term space missions in conditions of limited or impossible replenishment of life-support products, the reliability of LSS operation becomes critical for the mission [20]. This is due to two factors. On the one hand, a highly reliable local supply of everything necessary for crew life support can reduce the necessary reserve stocks of biomaterials. On the other hand, reliability is required for the life support process itself.

When building reliable SS technical systems, the method of fault tree analysis (FTA) has become widespread [21]. This method is also widely used in the practice of analyzing the dependability of aviation and space systems [16].

When developing a reliability-based technical design of the LSS bioregenerative reactor in accordance with the process described above, it is necessary to create a model that satisfies several requirements:

- allowed to determine reliability indicators depending on the operating time,
- allows to consider the real distribution of failures of individual architectural elements of the structure,
- identified structural elements that were weak in terms of reliability depending on the operating time,
- made it possible to easily and at minimal cost transform the reliability model of the system when changing the design of its individual subsystems and components.

To solve these problems in the study, the authors were faced with the choice of applying analysis using dynamic fault trees or using Petri nets. Both Petri nets and dynamic fault trees are widely used in the analysis of complex systems and the evaluation of their reliability. However, there are some advantages of using Petri nets over dynamic fault trees:

- Petri nets allow for the modeling of both static and dynamic behaviors of a system. In contrast, dynamic fault trees only consider the temporal evolution of faults and do not capture the system's structural information. This means that Petri nets can provide a more comprehensive view of the system's behavior.
- Petri nets have a more flexible modeling capability than dynamic fault trees. They can model a wide range of system behaviors, including concurrent and distributed processes, synchronization, and resource allocation. On the other hand, dynamic fault trees are mainly focused on the evolution of faults and the impact on the system's behavior.
- Petri nets provide a natural way to perform quantitative analysis, such as determining the probability of failure, time-to-failure, and system availability. Dynamic fault trees can also perform these analyses, but they are often more complicated and require additional techniques such as Monte Carlo simulation.
- Petri nets are graphical models that are easy to understand and interpret, even by non-experts. This makes them an effective tool for communicating the results of reliability analysis to stakeholders and decision-makers. Dynamic fault trees, on the other hand, may be less accessible to non-experts due to their formal notation and complex analysis techniques.

Overall, Petri nets provide a more comprehensive, flexible, and easy-to-use approach for modeling and analyzing complex systems' reliability compared to dynamic fault trees. In connection with the above, further reliability analysis was performed using dynamic models based on Petri nets.

In [22], an approach was proposed that allows one to transform the FTA based model into one of the classes of Petri nets—Evaluation Petri Net (E-net).

A Petri nets can be represented by four-tuple: $N = (P, T, A, M)$, where: P is a set of places, T is a set of transitions, $A \subseteq (P \times T) \bigcup (T \times P)$ is a set of arcs, M is an initial marking.

Process of E-net model development consists of three steps [22]:

1. Design of the fault tree with general approach [23].
2. Transformation of fault tree into the E-Net model with formal procedures.
3. Simulation experiment.

***Step 1**. Fault tree construction*

In accordance with the general methodology of the FTA [23], in [3], fault trees were built for each of the BLSS subsystems. The fault trees for each of the components of the biotechnological life support system shown in Figs. 2, 3, 4, 5, and 6 [3]. The events that characterize the failures of the corresponding elements are indicated in the figures by the symbols $EI.J$, where E is abbreviation of word "event", I is the serial number of the analyzed subsystem, J is the number of the event of the subsystem under consideration.

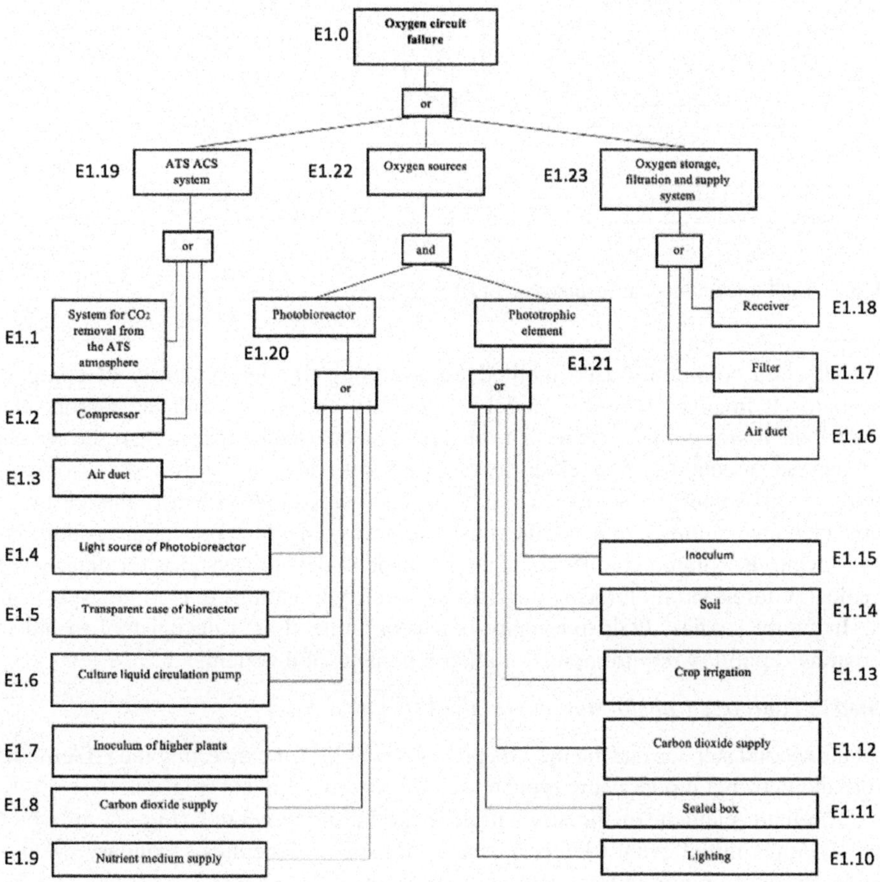

Fig. 2 Fault tree for oxygen component of BLSS [3]

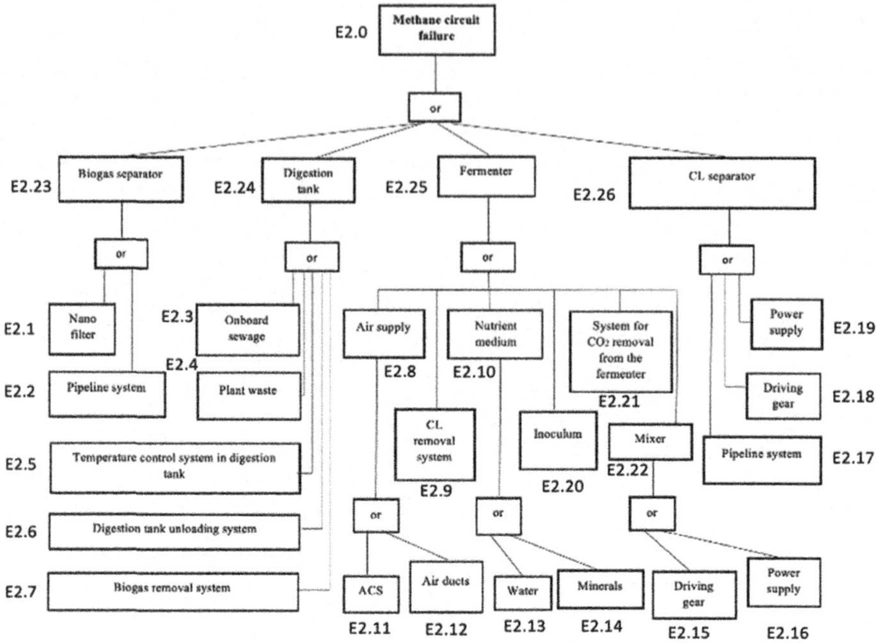

Fig. 3 Fault tree for methane component of BLSS [3]

FTA is a commonly used method for assessing the dependability of complex systems. It involves building models to identify and understand the factors that contribute to undesirable events. The analysis can be used to determine if the system under study meets specified reliability requirements and to assess the risks associated with potential undesirable events. Fault tree analysis can also identify the least reliable elements of the system and the most likely ways in which pre-failure conditions may occur. Recommendations can then be made to develop systems for diagnosing critical failures and monitoring the risks of security violations in complex systems.

In future studies, fault tree analysis can be utilized as a design tool to aid in creating reliability requirements for all components of a system.

Step 2*. Fault tree transformation into the E-net model*

At the second step we use the method proposed in [22] for converting the FTA model into the Petri net model using typical transformation elements (Fig. 7).

The main element of the simulation is the failure generator (Fig. 7), in which t1 generates initial event (defect appearance), and t2 generates secondary event—appearance of failure with the law of its distribution.

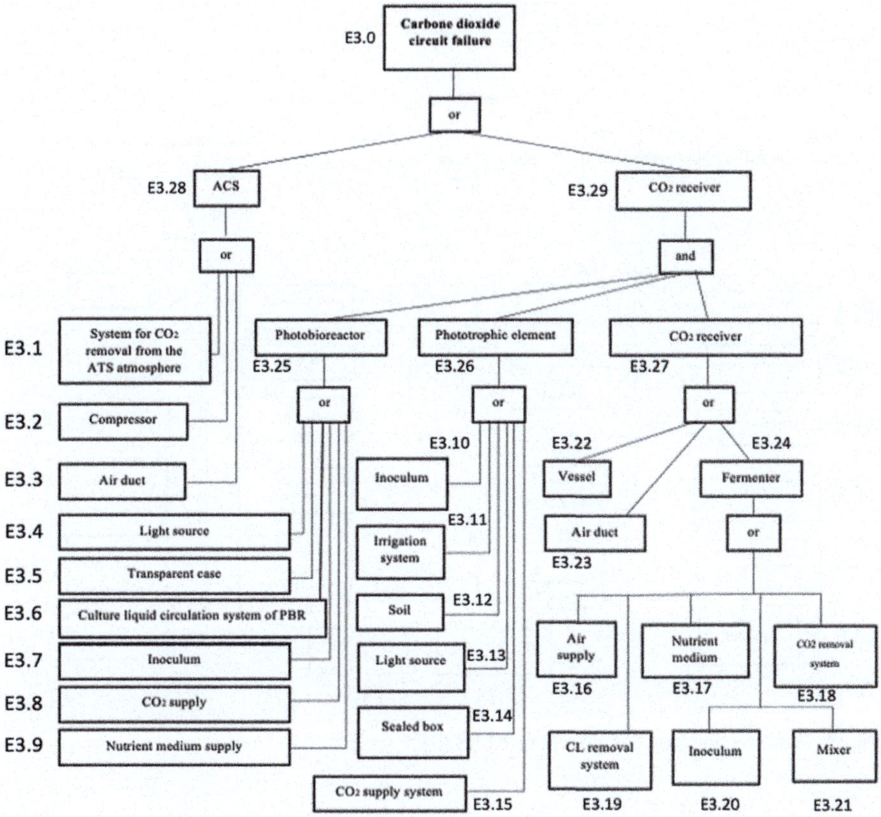

Fig. 4 Fault tree for carbon dioxide component of BLSS [3]

The transformation method proposed in [22] and a set of basic elements of transformation (Fig. 7) make it possible to convert the FTA models of LSS subsystems (Figs. 2, 3, 4, 5, and 6) into Petri nets, which are shown in the Figs. 8, 9, 10, 11, and 12.

The resulting models of Petri nets of individual subsystems of the CSC are easily combined using the model element OR from the table in Fig. 7. As a result, we obtain a common Petri net for studying the dependability of the LSS as a whole.

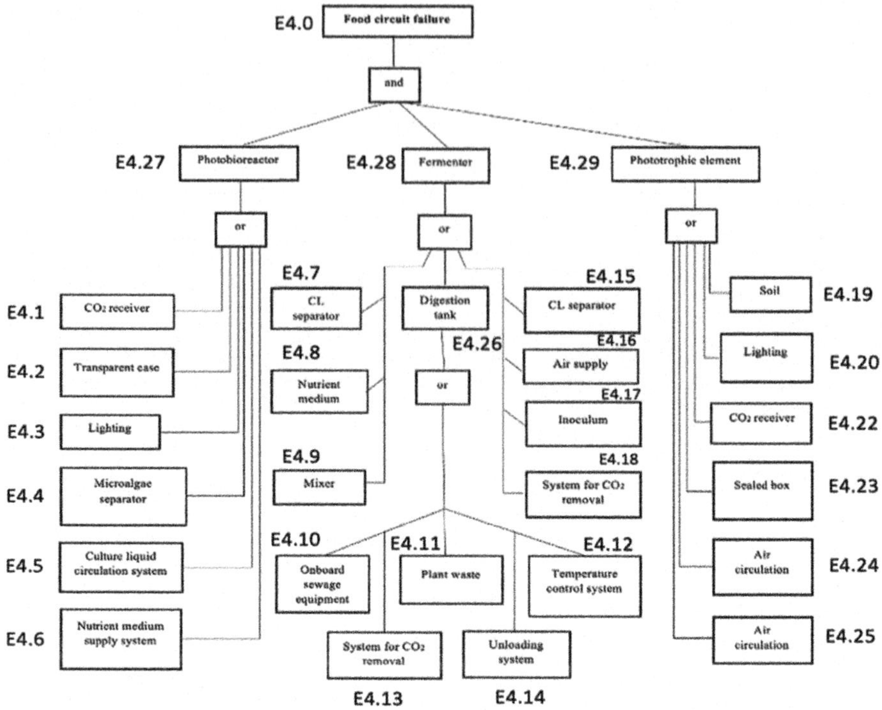

Fig. 5 Fault tree for food component of BLSS [3]

Step 3. *Simulation experiment*

It easy to transform obtained reliability models based on Petri nets if it is possible to refine individual subsystems with any level of construction details. In this case, a similar Petri net is built for the corresponding level of the hierarchy, and in the original high-level model, the state S2 (primary failure of element with two states at Fig. 7) is replaced by the existing Petri net with greater architectural details.

Applied software tools [24] make it possible to carry out experiments for Petri nets of various configurations.

5 Conclusions

For deep space habitation transportation systems must possess high reliability, stability, and ease of maintenance, particularly in relation to LSS. The bioregenerative reactor of LSS is one of the important subsystems for such autonomous means. Dependability of BLSS and its closed cycle components is critical for mission of

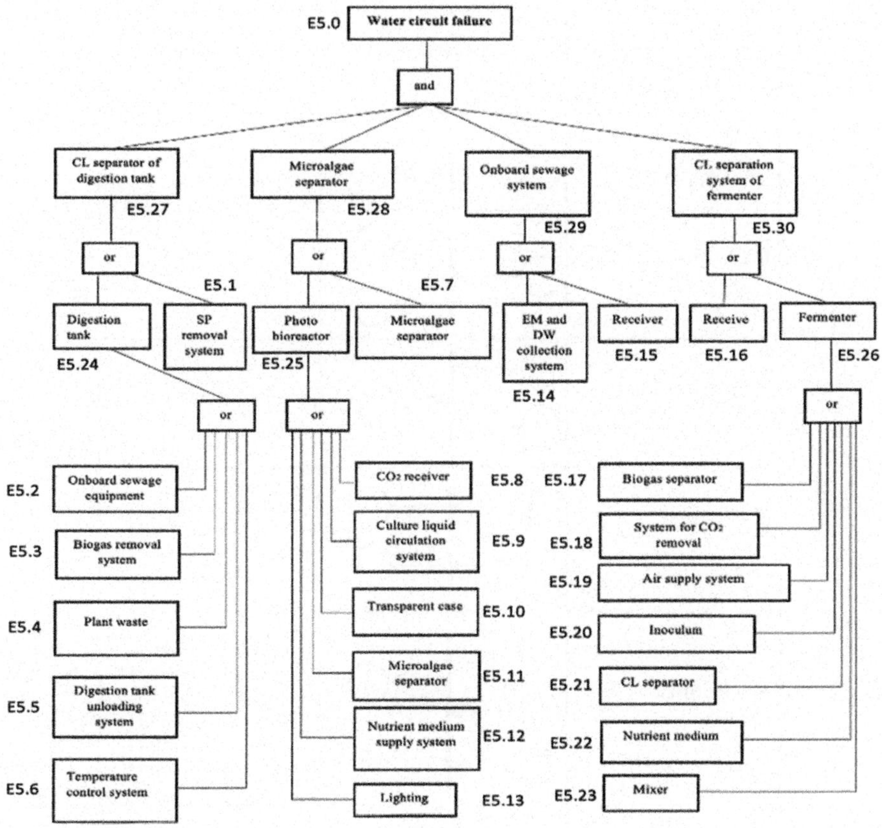

Fig. 6 Fault tree for water component of BLSS [3]

such autonomous transport systems. In the paper the Petri net model is proposed for analysis of dependability on the base of fault trees of studied subsystems of bioregenerative reactor.

The procedure for transforming a fault tree into a Petri net makes it possible to move from static reliability models to their dynamic counterparts. The reliability model of the BLSS based on Petri net makes it possible to analyze the dependability of the bioregenerative reactor of LSS and to study the influence of the design features of the LSS, taken at the early stages of design, on the dependability of the LSS, depending on the time of its operation, with different real distribution of failures of individual architectural elements of the LSS structure and with different possible detail levels of its architecture.

Class of Elementary Networks	*Graph*	*Marking of the Graph*
T - Simple Transition		$(1,0) \rightarrow (0,1)$
F - Multiplication	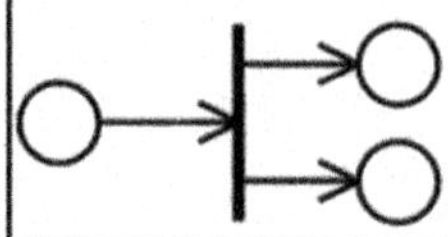	$(1,0,0) \rightarrow (0,1,1)$
J - Integration	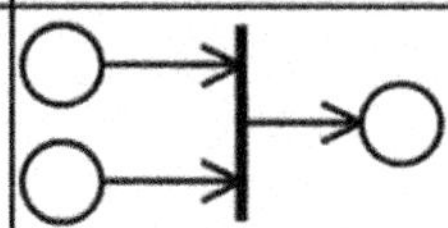	$(1,0,0) \rightarrow (0,1,1)$
X - Switch	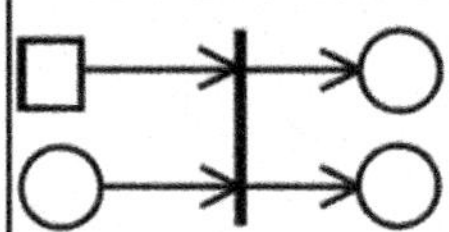	$(0,1,0,0) \rightarrow (e,0,1,0)$ $(0,1,0,1) \rightarrow (e,0,1,1)$ $(1,1,0,0) \rightarrow (e,0,0,1)$ $(1,1,1,0) \rightarrow (e,0,1,1)$
Y - Priority Sample	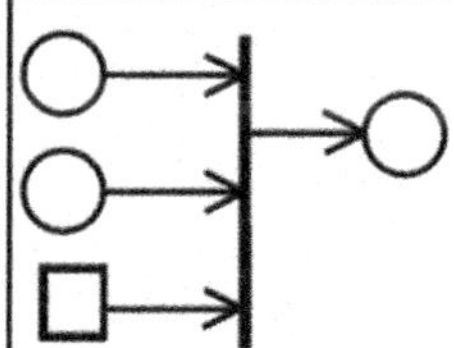	$(0,0,1,0) \rightarrow (e,0,0,1)$ $(0,1,0,0) \rightarrow (e,0,0,1)$ $(0,1,1,0) \rightarrow (e,0,1,1)$ $(1,0,1,0) \rightarrow (e,0,0,1)$ $(1,1,0,0) \rightarrow (e,0,0,1)$ $(1,1,1,0) \rightarrow (e,1,0,1)$

Fig. 7 Set of elementary networks for Petri net construction on the base of fault tree [22]

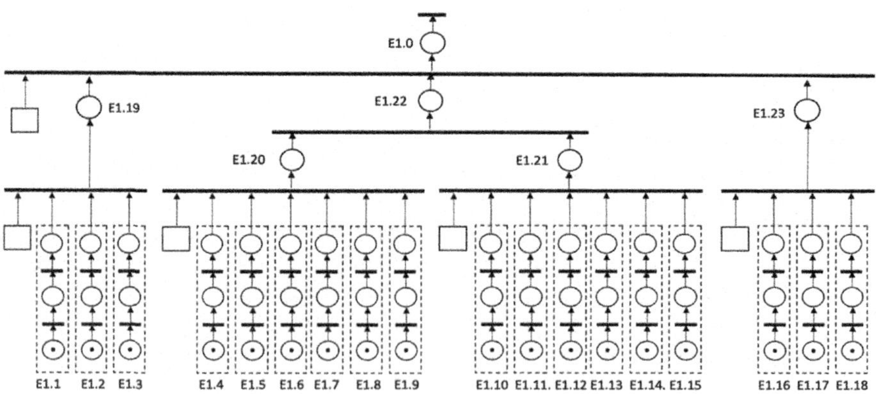

Fig. 8 E-net model of oxygen circuit

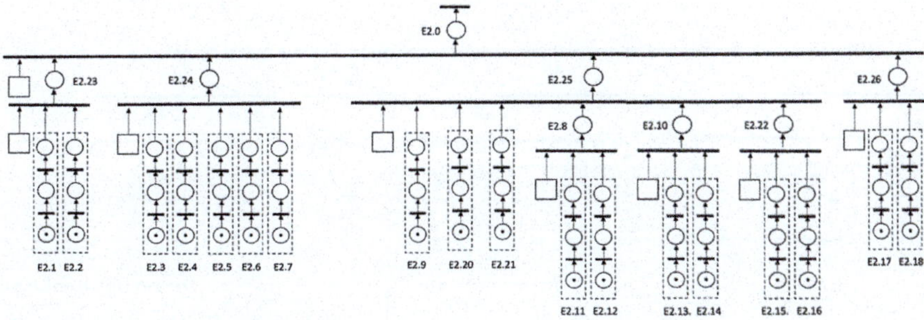

Fig. 9 E-net model of methane circuit

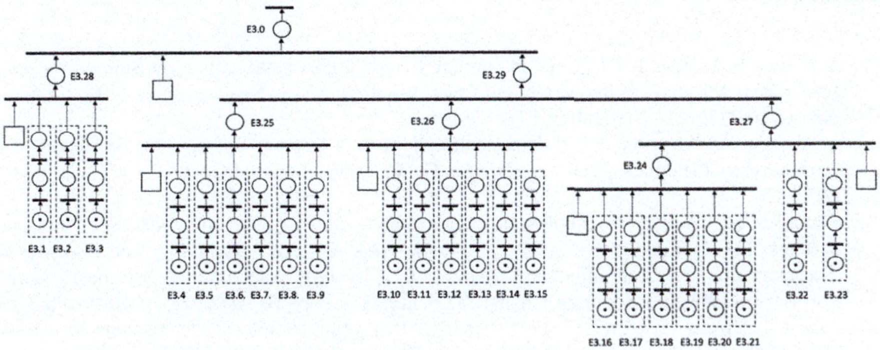

Fig. 10 E-net model of carbon dioxide circuit

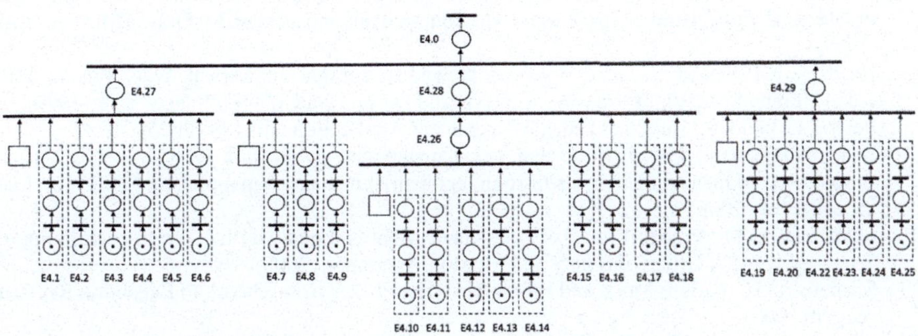

Fig. 11 E-net model of food circuit

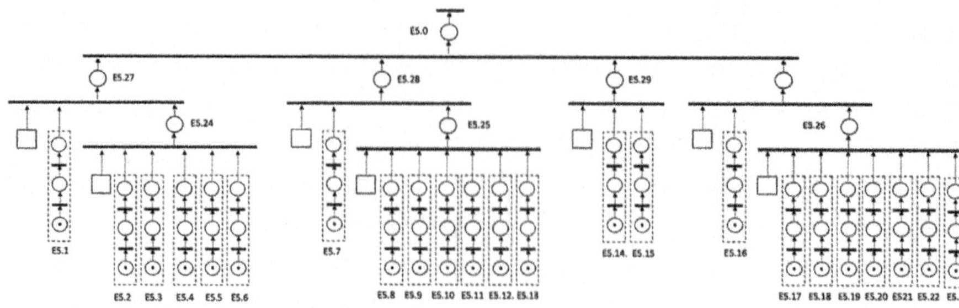

Fig. 12 E-net model of water circuit

References

1. Verseux, C., de Vera, J.-P.P., Leys, N., Poulet, L. (eds.): Bioregenerative Life-Support Systems for Crewed Missions to the Moon and Mars. Frontiers Media SA, Lausanne (2022). https://doi.org/10.3389/978-2-83250-301-0
2. Seedhouse, E., Shayler, D.J. (eds.): Handbook of Life Support Systems for Spacecraft and Extraterrestrial Habitats, 1200 pp. Springer, Cham (2020). https://doi.org/10.1007/978-3-319-09575-2
3. Glukhikh, S.: Closed biotechnological cycles in life support systems of autonomous transport systems. In: Kabashkin, I., Yatskiv, I., Prentkovskis, O. (eds.) Reliability and Statistics in Transportation and Communication. RelStat 2021. Lecture Notes in Networks and Systems, vol. 410, pp. 389–398. Springer, Cham (2022). https://doi.org/10.1007/978-3-030-96196-1_36
4. Jones, H., Ewert, M.: Ultra reliable closed loop life support for long space missions. In: AIAA-2010-6286, 40th International Conference on Environmental Systems (2010)
5. Connolly, J.F.: Mars design example. In: Larson, W.K., Pranke, L.K. (eds.) Human Spaceflight: Mission Analysis and Design. McGraw-Hill, New York (1999)
6. Doll, S., Eckart, P.: Environmental control and life support systems (ECLSS). In: Larson, W.K., Pranke, L.K. (eds.) Human Spaceflight: Mission Analysis and Design. McGraw-Hill, New York (1999)
7. Heydorn, R.P., Railsback, J.W.: Safety of crewed spaceflight. In: Larson, W.K., Pranke, L.K. (eds.) Human Spaceflight: Mission Analysis and Design. McGraw-Hill, New York (1999)
8. Perera, J., Field, S.: Integrated Risk Management Application (IRMA) (2005)
9. Risk management: Futron integrated risk management application (FIRMA). http://www.futron.com/riskmanagement/tools/futronintegratedriskmanagementapplication.htm. Last accessed July 2008
10. Stamatelatos, M.: Probabilistic risk assessment: what is it and why is it worth performing it? Tech. rep., NASA Office of Safety and Mission Assurance (2000)
11. Stamatis, D.H.: Failure Mode and Effect Analysis: FMEA from Theory to Execution Revised. Asq Pr (2003)
12. Lievens, C.: System Security. Caepadues Editions, Toulouse (1976)
13. Bussolini, J.J.: High Reliability Design Techniques Applied to the Lunar Module, Lecture Series no. 47 on Reliability on Avionics Systems, September 1971 (1971)
14. Blokdyk, G.: Fault Tree Analysis a Complete Guide. 5STARCooks (2021)
15. Likens, W.C.: A preliminary investigation of life support processor reliabilities. In: International Conference on Life Support and Biospherics, Huntsville, AL, 18–20 Feb 1992 (1992)
16. Russell, J.F., Klaus, D.M.: Maintenance, reliability and policies for orbital space station life support systems. Reliab. Eng. Syst. Saf. **92**(6), 808–820 (2007)
17. Lobascio, C., Lamantea, V., Vittorio Cotronei, V., et al.: Plant bioregenerative life supports: the Italian CAB Project. J. Plant Interact. **2**(2), 125–134 (2007). https://doi.org/10.1080/17429140701549793

18. Hurlbert, K., Bagdigian, B., Carroll, C., et al.: Human health, life support and habitation systems. Technology Area 06, National Aeronautics and Space Administration (2010)
19. NASA Technology Roadmaps. TA 6: Human Health, Life Support, and Habitation Systems. NASA (2015)
20. Lange K., Anderson, M.: Reliability impacts in life support architecture and technology selection. In: 42nd International Conference on Environmental Systems, San-Diego, California, 15–19 July 2012, AIAA 2012-3491 (2012)
21. Li, X., Li, F.: Reliability assessment of space station based on multi-layer and multi-type risks. Appl. Sci. **11**, 10258 (2021). https://doi.org/10.3390/app112110258
22. Kabashkin, I.: Reliability model of intelligent transport systems. In: IEEE 7th International Conference on ITS Telecommunications, Sophia Antipolis, pp. 1–4 (2007). https://doi.org/10.1109/ITST.2007.4295911
23. Guide to Reusable Launch and Reentry Vehicle Reliability Analysis. Federal Aviation Administration, ver. 1.0 (2005)
24. Petri nets tools database. https://www.informatik.uni-hamburg.de/TGI/PetriNets/tools/quick.html. Last accessed 16 Jan 2023

Safety Assessment of Maintained Control Systems with Cascade Two-Version 2oo3/1oo2 Structures Considering Version Faults

Vyacheslav Kharchenko, Yuriy Ponochovnyi, Ievgen Babeshko, Eugene Ruchkov, and Artem Panarin

1 Introduction

At present a whole class of devices, hardware and software components important from the functional safety point of view involves the adoption of dedicated information and control systems. As a basis of such systems FPGAs and microprocessor platforms are utilized for critical domains [1]. An example is the RadICS modular platform [2] which is used to build fault-tolerant FPGA architectures meeting strict dependability, reliability, and functional safety requirements nuclear international standards and normative documents [3, 4]. This necessitates the development of adequate and complete models for evaluating the parameters of reactor trip systems in nuclear field both at the design stage and during their operation.

Currently, various classes of such models are being developed and used: set-theoretic [5], Bayesian [6], FTA [7], Markov [8], semi-Markov [9], FMECA [10] and its modifications [11] and combinations [12]. The use of such models is advised by industry standards for electronic and programmable components and platforms from IEC 61508 series [13]. However, the scenarios of failures of hardware (HW) and software (SW) channels of redundant and diverse systems are constantly expanding, as new threats associated with malicious interference in the operation of these systems,

V. Kharchenko · I. Babeshko (✉)
National Aerospace University KhAI, Kharkiv, Ukraine
e-mail: e.babeshko@csn.khai.edu

V. Kharchenko
e-mail: v.kharchenko@csn.khai.edu

Y. Ponochovnyi
Poltava State Agrarian University, Poltava, Ukraine

E. Ruchkov · A. Panarin
Research and Production Company Radiy, Kropyvnytskyi, Ukraine
e-mail: rev@radiy.com

W. Zamojski et al. (eds.), *Dependable Computer Systems and Networks*, Lecture Notes in Networks and Systems 737, https://doi.org/10.1007/978-3-031-37720-4_11

the threat of hostilities, sabotage and cyberattacks appear [14]. In addition, when evaluating a traditional two-version system [5, 7], it is often assumed that the diagnostic and supervision means for both subsystems have perfect reliability.

The development of the Industry 4.0 concept and the application of Industrial IoT technologies lead to an increase in the requirements for the level of safety integrity SIL2-SIL3 [15]. All these factors determine the need to develop an adequate, complete model that will allow determining the resulting indicator with a high level of accuracy.

This work presents a macro model of operation for information and control system such as NPP RTS accommodating various failure types, in particular, faults of diagnostic and supervision means in both (main and diverse) subsystems. The paper continues investigating the microprocessor and FPGA-based ICSs outlined in [16–18].

The object of the study is the process of ICS operation under conditions of failures caused not only by hardware and software version anomalies, faults of the diagnostic and supervision means and separated recovery of the ICS components.

The structure of the work is as follows: Sect. 2 provides a reliability block diagram of RTS ICS, as well as discusses the main directions of expanding its state space. In Sect. 3, a macro model of the ICS operation and its separate part (sub-model) is built, which describes the manifestation of absolute faults of software versions. Analysis of modeling results has been the objective of Sect. 4. Finally, in Sect. 5 we summarize findings, provide guidance on the usage of developed models and discuss the future work.

2 Block Diagram and Failure Model of Two-Version ICS

The ICS reliability block diagram presented in Fig. 1 is a modified 1oo2 redundant architecture, in which each channel has an additional 2oo3 redundancy and built-in diagnostic and supervision means. Subsystems (main and diverse RTS subsystems) generate one-bit signals with priority for the shutdown signal. If the output channels retain identical states due to version errors (main and diverse software subsystems), their detection relies on built-in diagnostic means. Such redundant architecture contains a certain margin of reliability, which allows it to adequately withstand failures.

Using the RadICS platform provides a modular implementation of RTS [2]. This allows using the central diagnostic module (D) and means built into other modules to check the functionality of the signals from these modules. The output signal comparison module (module «≠») additionally increases reliability and safety, but at the same time it expands the space of diagnostic states, as well as general ICS state space.

Figure 2 illustrates the directions of system state space.

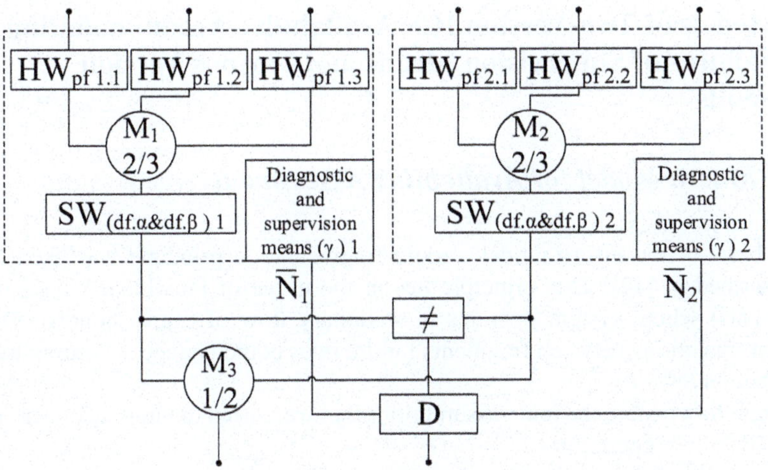

Fig. 1 ICS block diagram for modeling manifestation of physical faults (pf), design faults (df; absolute β and relative α) and faults of diagnostic and supervision means (γ)

Fig. 2 Graph of RTS ICS states

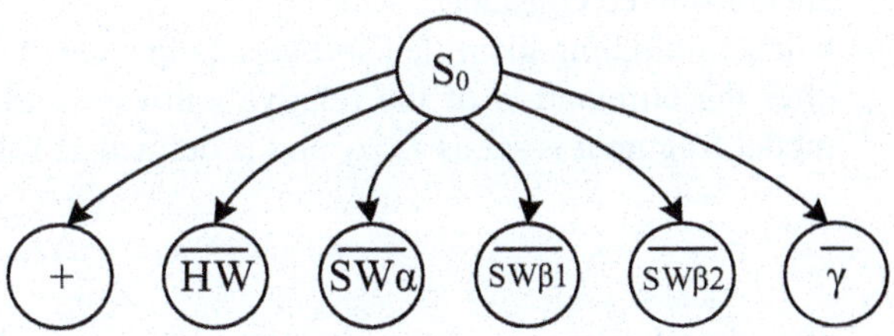

The first transition illustrates operable space. The remaining 5 transitions describe the inoperable states caused by the following failures: HW channels, SW versions (α—relative and β—absolute faults); γ for diagnostic and supervision means.

The generalized model of faults and failures was described in detail in the [16]. Recall that the related software design faults (α) are manifested separately in each version of the software, the model of their manifestation is described in the previous studies [16, 17] as well. Absolute faults appear simultaneously in both versions of software, the following details are introduced in the study: β1—absolute faults that cause different values of the output signal, β2—the most dangerous faults that cause the same values of the output signal and are not detected by the comparison system. The model of the manifestation of absolute faults will be investigated in this work.

3 Models of Two-Version ICS Availability Accommodating Failures of Supervision Means and Various Version Faults

3.1 Macro Model for Availability Assessment

As a basis of the macro model a multi-fragmentation principle was used, which is described in [12]. This principle lies in the usage of individual blocks (model fragments) where system operates in stationary flow of events context. Only in case of transitions between fragments failure rates could change according to given distribution laws.

The following are the main assumptions that were used to build the macro model of the RTS ICS operation:

- the flow of events, which causes the transition of the system from one operable state to another within one fragment, has the properties of stationarity, ordinariness and the absence of an aftereffect, the parameters of the model within one fragment are considered constant;
- when eliminating identified software faults, new faults are not introduced;
- after the elimination of the relative software fault, the software failure rate in model fragment is set as $\lambda_{\alpha\, i+1}$ and is defined as follows:

$$\lambda_{\alpha\, i+1} = \lambda_{\alpha\, i} - \Delta\lambda_{\alpha} \quad (1)$$

- after the elimination of the absolute software fault, the software failure rate in model fragment is set as $\lambda_{\beta\, i+1}$ and is defined as follows:

$$\lambda_{\beta\, i+1} = \lambda_{\beta\, i} - \Delta\lambda_{\beta} \quad (2)$$

The macro model is presented in Fig. 3 as a collection of fragments. When building the model, the design features of the ICS were taken into account, namely the use of the same hardware in both diverse channels (the hardware in each channel is characterized by the reliability parameters λp and μp).

Macro graph shown in Fig. 3 includes:

- 24 basic fragments (total 24*11 = 264 states);
- 24 transitions that model manifestation and elimination of relative DD (24*10 = 240 states);
- 23 transitions that model manifestation and elimination of absolute DD (23*5 = 115 states);

i.e., in total, a complete directed graph should include 619 states.

The built macro model simulates the operation of the system with two relative design faults that appear separately in each software version (their manifestation is

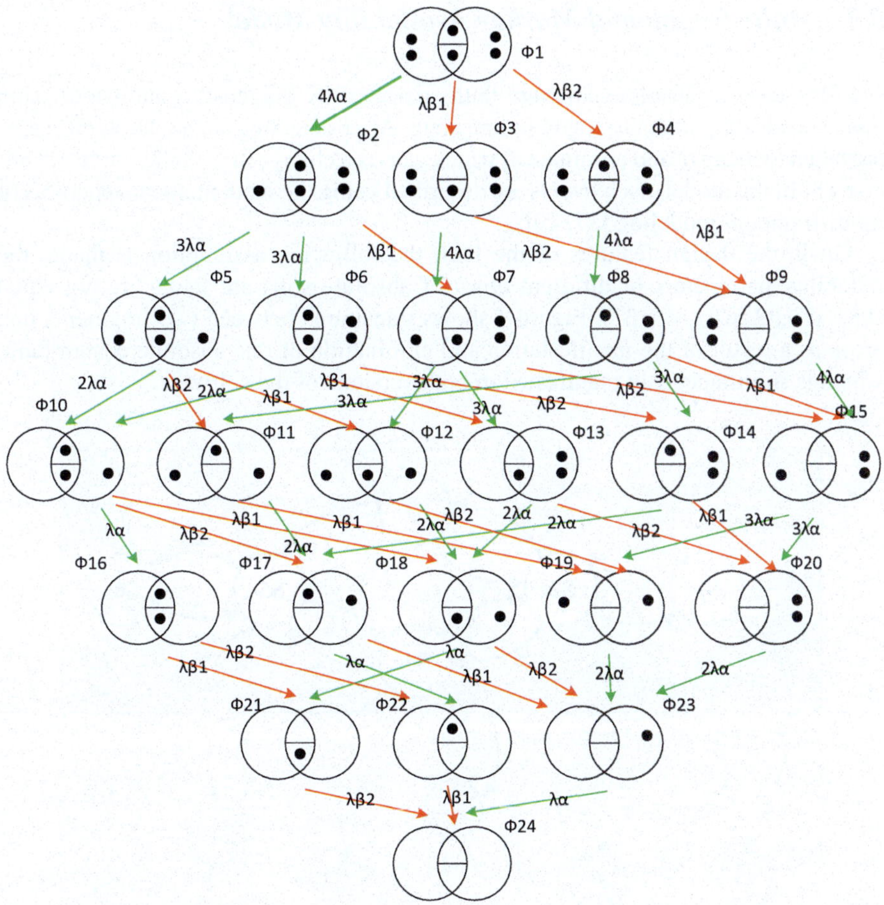

Fig. 3 Macro graph describing ICS operation with two version software configuration $\{n\alpha 1 = 2, n\alpha 2 = 2, n\beta 1 = 1, n\beta 2 = 1\}$

illustrated by the transitions of green color); and one design fault appearing in both versions of the software at the same time (illustrated by red transitions in Fig. 3).

To build a complete directed graph of the model, automation tools of the Matlab environment and the grPlot library [19] were used.

We emphasize that working with models of this size is significantly complicated and requires consolidation of individual blocks and detailing of the system operation within individual fragments. A detailed description of software version relative faults manifestation processes of supervision means was carried out in [12]. Therefore, only the model of manifestation of absolute software versions is considered further.

3.2 Multi-fragmental Markov Availability Model

Manifestation of absolute software faults detected by diagnostics and supervision means leads the system to a safe failure state. After detecting such a fault, measures are taken to localize and eliminate it, which causes a change in the failure rate parameter $\lambda\beta$. In the model, such events are described using the mathematical apparatus of multi-fragment modeling [17, 19].

Given the design features of the ICS, the following assumption is made: the reliability parameters of different types of absolute software faults are not equal ($\lambda\beta1 \neq \lambda\beta2$, $\mu\beta1 \neq \mu\beta2$). Figure 4 shows transitions between two fragments that arose as a result of the manifestation and elimination of one absolute design fault, which is simultaneously manifested in both versions of the software.

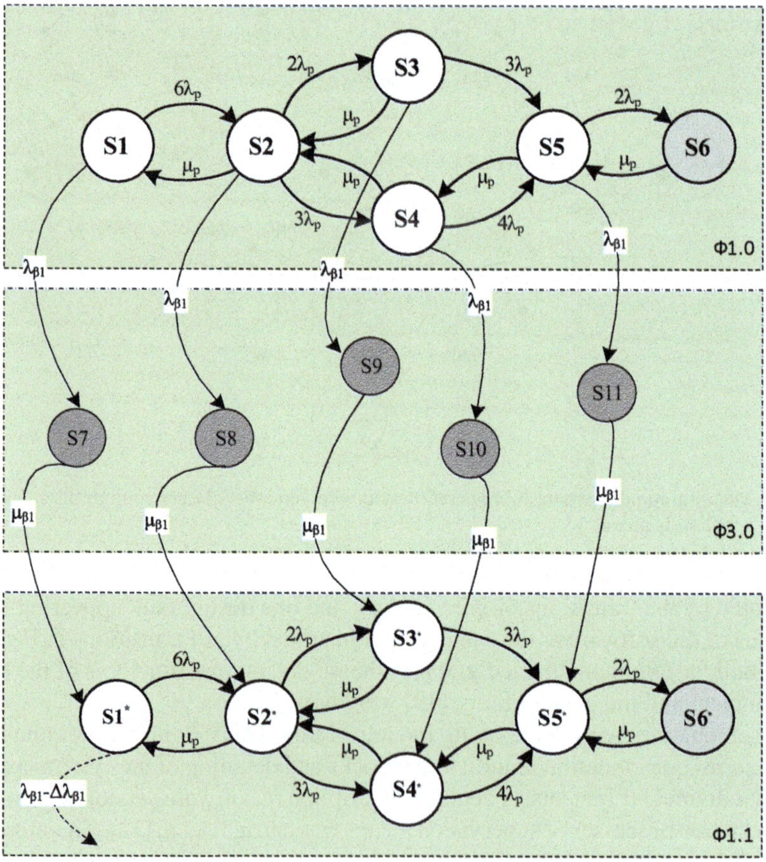

Fig. 4 Two-fragment directed graph that models manifestation of absolute design fault by transitions between fragments

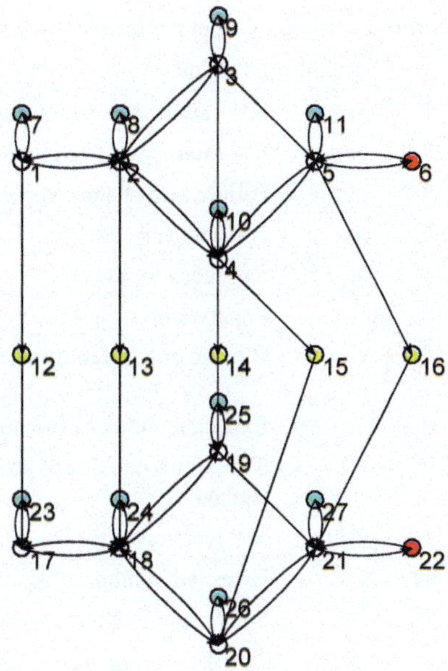

Fig. 5 Directed graph of model with fragments resulting from relative design fault in one of the software versions

When constructing the directed graph in Fig. 5, the following color marking is used: white color is used to represent operational states, red color is used for safe failure states caused by hardware channel failures, the blue color is used for safe failure states caused by control system failures. The yellow color is used to indicate safe failure states caused by the manifestation of absolute faults in the design of software versions.

The availability function for the directed graph in Fig. 5 is defined as (3):

$$A(t) = \sum_{i=1}^{5} P_i(t) + \sum_{i=17}^{21} P_i(t) \tag{3}$$

Basic conditions are the following: t = 0, $P_1(0) = 1$, $P_2(0) \ldots P_{27}(0) = 0$.

4 Modeling and Analysis of Results

The primary input parameters of the Markov models were determined based on the data of certification tests an operation experience for similar systems [12, 16, 18] for the previous samples of the versions of the RTS ICS. Their values are presented in Table 1.

To construct the matrix of the Kolmogorov-Chapman differential equation system (DES) in Matlab, the matrixA function [15] was used. DES can be solved using analytical methods (substitution, Laplace transform, etc.). This approach can be

Table 1 Values of input parameters used for modeling

#	Symbol	Description	Value
1	λp	HW failure rate due to physical faults (PD)	1e−4 (1/h)
2	λα	SW failure rate due to relative design faults (DD)	5e−4 (1/h)
3	λγ	Failure rate of supervision means	1e−6 (1/h)
4	μp	HW recovery rate	1 (1/h)
5	μα	SW recovery rate	2 (1/h)
6	μγ	Supervision means recovery rate	0.25 (1/h)
7	Δλα	Change of SW failure rate after elimination of relative fault	1.25e−4 (1/h)
8	Nα	Expected number of relative design faults	4
9	λβ1	SW failure rate due to absolute design faults that cause different signals	1e−6 (1/h)
10	μβ1	SW recovery rate	0.0667 (1/h)
11	Nβ1	Expected number of absolute design faults of the first type	1
12	λβ2	SW failure rate due to absolute design faults that cause different signals	2e−6 (1/h)
13	μβ2	SW recovery rate	0.0714 (1/h)
14	Nβ2	Expected number of absolute design faults of the second type	1

applied to ICSs of small dimensions. In this work, a large-dimensional macro model (619 states) is considered, therefore, a universal approach to the numerical solution of the DES was chosen using the ode15s function [12, 20]. The simulation results are shown in Fig. 6.

To assess the impact of different types of (absolute and relative) design faults on the availability function, separate models were built for each of the indicated fault type. The input parameters of the simplified models are identical to the parameters of the macro model in Fig. 4. The results are presented as graphs of different colors in Fig. 7.

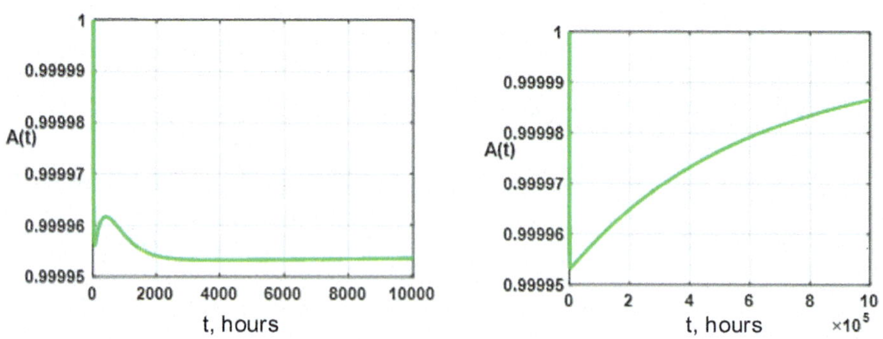

Fig. 6 ICS availability modeling results with the extension of time base to 10,000 h (**a**) and to 100,000 h (**b**)

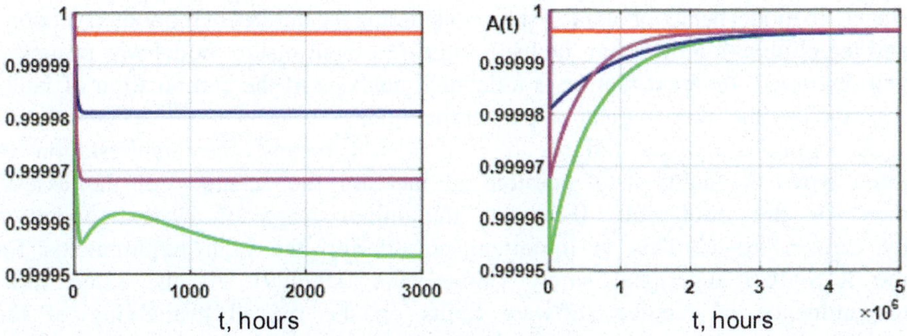

Fig. 7 ICS availability modeling results with the extension of time base to 3000 h (**a**) and to 500,000 h (**b**)

The graphs in Fig. 7 are marked as follows:

- for the model that accommodates the manifestation of HW physical faults, the failure of supervision means and the manifestation of relative SW design faults (red),
- for the model that accommodates the manifestation of SW design faults according to the beta1 parameter (blue),
- for the model that accommodates the manifestation of SW design faults according to the beta2 parameter (purple) and the general macro model (green).

The graphs of the ICS model (Figs. 6 and 7) illustrate the typical nature of the change in the availability function when it is reduced to the stationary coefficient A = 0.9999955 during the first 30 h of operation. During the next 2000 h, an improvement in system availability is observed due to the elimination of relative (α) software faults. Further, the availability of the system continues to increase, this is due to the elimination of absolute (β) software faults.

5 Conclusion

The paper provides results of the development and research of two-version safety system models with cascade 2oo3-1oo2 redundancy, which take into account the reliability of the operation of supervision means. The structure of the system with self-diagnosis functions is presented, which consists of a subsystem of supervision means, means of cross-channel comparison and analysis. Such a structure makes it possible not only to detect failure states of individual channels, but also to ensure the verification of supervision means and reduce the risk of dangerous and undetected failures.

The construction of Markov models describing the manifestation of physical and design faults and failures of diagnostic and supervision means was carried out in stages. To model behavior of the system considering the failures of individual versions and the elimination of design faults, a multi-fragment macro model was proposed and discussed. Its main feature is a detailed analysis of the combination of faults described by the inter-fragmental part of the model.

Accommodating the failures of supervision means, the manifestation of three types of faults, it is possible to increase the accuracy of the assessment of the availability factor at the initial stage of system operation by $\Delta A = 4e{-}6$. This is important considering the high requirements for the RTS ICS functional safety (more than 0.99999). At the same time, the influence of relative software faults on the overall availability of the entire system remains within $\Delta A = 2e{-}8$ in the interval [0...2000] h due to the two-version structure. However, the overlap of absolute version faults and diagnostic failures is more critical and is eliminated only after 30 000 h of operation time.

Further research should be devoted to the development of analytical models of availability and functional safety designed to assess the impact of possible attacks on the system [14, 21]. In case of cyberattacks or any intrusions, the results of the vulnerability analysis should be taken into account in such a way to parameterize and replan the Markov model or to implement a combined assessment methodology [12].

References

1. Yastrebenetsky, M., Kharchenko, V. (eds.): Cyber Security and Safety of Nuclear Power Plant Instrumentation and Control, p. 501. IGI-Global, PA, USA (2020). https://doi.org/10.4018/978-1-7998-3277-5
2. FPGA-Based Safety Controller (FSC) RadICS. Results of the IEC 61508 Functional Safety Assessment. V4R3, p. 26 (2020). https://www.exida.com/SAEL-Safety/rpc-radiy-fpga-based-safety-controller-fsc-radics
3. Guidance on using IEC 61508 SIL certification to support the acceptance of commercial grade digital equipment for nuclear safety related applications. Revision 1 (2011). https://www.nrc.gov/docs/ML2133/ML21337A380.pdf
4. IEC 61511-1:2016. Functional safety—safety instrumented systems for the process industry sector—part 1: framework, definitions, system, hardware and application programming requirements (2016). https://webstore.iec.ch/publication/24241
5. IAEA Safety Standards Series No. SSG-2 (Rev. 1). Deterministic safety analysis for NPPs (2019). https://www-pub.iaea.org/MTCD/publications/PDF/PUB1851_web.pdf
6. Zhao, X., Wang, X., Golay, M.W.: Bayesian network–based fault diagnostic system for nuclear power plant assets. Nucl. Technol. **209**(3), 401–418 (2023). https://doi.org/10.1080/00295450.2022.2142445
7. Kim, J.S., Han, S.H., Kim, M.C.: Direct fault-tree modeling of human failure event dependency in probabilistic safety assessment. Nucl. Eng. Technol. **55**(1), 119–130 (2023). https://doi.org/10.1016/j.net.2022.08.029
8. Liang, Q., Yang, Y., Zhang, H., Peng, C., Lu, J.: Analysis of simplification in Markov state-based models for reliability assessment of complex safety systems. Reliab. Eng. Syst. Saf. **221**, 108373 (2022). https://doi.org/10.1016/j.ress.2022.108373
9. Liang, Q., Peng, C., Li, X.: A multi-state semi-Markov model for nuclear power plants piping systems subject to fatigue damage and random shocks under dynamic environments. Int. J. Fatigue **168**, 107448 (2023). https://doi.org/10.1016/j.ijfatigue.2022.107448

10. Lo, H.-W., Liou, J.J., Yang, J.-J., Huang, C.-N., Lu, Y.-H.: An extended FMEA model for exploring the potential failure modes: a case study of a steam turbine for a nuclear power plant. Complexity **2021**, 1–13 (2021). https://doi.org/10.1155/2021/5766855
11. Babeshko, I., Illiashenko, O., Kharchenko, V., Leontiev, K.: Towards trustworthy safety assessment by providing expert and tool-based XMECA techniques. Mathematics **10**, 2297 (2022). https://doi.org/10.3390/math10132297
12. Kharchenko, V., Ponochovnyi, Y., Ivanchenko, O., Fesenko, H., Illiashenko, O.: Combining Markov and semi-Markov modelling for assessing availability and cybersecurity of cloud and IoT. Cryptography **6** (2022). https://doi.org/10.3390/cryptography6030044
13. IEC 61508. Functional safety of electrical/electronic/programmable electronic safety-related systems (2010). https://www.iec.ch/functional-safety
14. Pickering, S.Y., Davies, P.B.: Cyber security of nuclear power plants: US and global perspectives (2021). https://gjia.georgetown.edu/2021/01/22/cyber-security-of-nuclear-power-plants-us-and-global-perspectives
15. Gomes, F.C., de Andrade, A.A., Gasi, F.: Instrumentation and control systems applied to high-risk operating technologies: paving the way to the industry 4.0 at nuclear power plants. In: 2021 14th IEEE International Conference on Industry Applications (INDUSCON) (2021). https://doi.org/10.1109/induscon51756.2021.9529836
16. Kharchenko, V., Ponochovnyi, Y., Ruchkov, E., Babeshko, E.: Safety assessment of the two-cascade redundant information and control systems considering faults of versions and supervision means. In: New Advances in Dependability of Networks and Systems, pp. 88–98 (2022). https://doi.org/10.1007/978-3-031-06746-4_9
17. Kharchenko, V., Butenko, V., Odarushchenko, O., Sklyar, V.: Multifragmentation Markov modeling of a reactor trip system. J. Nucl. Eng. Radiat. Sci. **1**, (2015). https://doi.org/10.1115/1.4029342
18. Babeshko, E., Kharchenko, V., Leoniev, K., Ruchkov, E.: Practical aspects of operating and analytical reliability assessment of FPGA-based I&C systems. Radioelectron. Comput. Syst. **3**(95), (2020). https://doi.org/10.32620/reks.2020.3.08
19. Iglin, S.: grTheory—graph theory toolbox (2023). https://www.mathworks.com/matlabcentral/fileexchange/4266-grtheory-graph-theory-toolbox
20. Solve stiff differential equations and DAEs—variable order method—MATLAB ode15s (2023). https://www.mathworks.com/help/matlab/ref/ode15s.html
21. Lysenko, S., Kharchenko, V., Bobrovnikova, K., Shchuka, R.: Computer systems resilience in the presence of cyber threats: taxonomy and ontology. Radioelectron. Comput. Syst. **1**, 17–28 (2020). https://doi.org/10.32620/reks.2020.1.02

CPU Signal Rank-Based Disaggregation in Cloud Computing Environments

Jakub Kosterna, Krzysztof Pałczyński, and Tomasz Andrysiak

1 Introduction

The field of cloud computing is widely adopted by many companies. However, the problem of fluctuating workloads of Virtual Machines (VMs) has become increasingly difficult for cloud service providers (CSPs) to achieve the quality of service (QoS) for clients. Accurate workload prediction is crucial to provide high elasticity and cost-effectiveness for cloud resources. A clustering-based workload prediction method is proposed [4], which clusters tasks based on their workload patterns and trains a prediction model for each cluster to improve prediction accuracy. Another paper [2] focuses on the problem of missing data in remote sensing analysis, proposing a deep-learning-based framework for reconstructing missing data using available data from both earlier and subsequent timestamps while maintaining the causality constraint in spatiotemporal analysis. A case study validates the proposed forecasting ensemble. Finally, a recurrent neural network (RNN) is used to predict CPU utilization [3], proving its ability to accurately predict CPU utilization for short periods within 10,000 evaluations of the training data.

The [4] study focuses on predicting task workload in cloud computing environments. It is crucial to accurately predict the workload of virtual machines to efficiently manage cloud resources and achieve a high quality of service for clients. The research compares several state-of-the-art workload prediction methods and introduces a clustering-based method to improve prediction accuracy. The proposed method clusters tasks with similar workload patterns, build a prediction model for each cluster and achieves a prediction accuracy of around 90% for both CPU and

J. Kosterna (✉)
Warsaw University of Technology, Plac Politechniki 1, Warsaw 00661, Poland
e-mail: kosternaj@gmail.com

K. Pałczyński · T. Andrysiak
Faculty of Telecommunications, Computer Science and Electrical Engineering, Bydgoszcz University of Science and Technology, Bydgoszcz, Poland
e-mail: krzysztof@palczynski.dev

W. Zamojski et al. (eds.), *Dependable Computer Systems and Networks*, Lecture Notes in Networks and Systems 737, https://doi.org/10.1007/978-3-031-37720-4_12

memory based on trace-driven experiments using the Google cluster trace. Alternatively, [2] proposes Deep-STEP_FE, an ensemble forecasting model for missing data prediction in remote sensing analysis. The framework utilizes available data from earlier and subsequent timestamps while maintaining causality. The proposed model is validated by predicting missing NDVI imagery for two spatial zones in India from 2004 to 2011. It outperforms state-of-the-art deep-learning-based ST prediction models. Finally, [4] investigates recurrent neural networks' use to accurately predict CPU utilization for short periods in cloud computing environments. The paper shows that it is possible to predict CPU utilization with high accuracy for data sets with sudden extreme changes. The recurrent neural network trained with backpropagation through time was able to accurately predict CPU utilization within 10,000 evaluations of the training data. The research concludes that recurrent neural networks are a promising candidate for predicting CPU utilization with greater accuracy when compared to traditional approaches.

Although CPU consumption signals were used in many different data-science-related tasks, the problem of signal disaggregation was not studied for cloud computing purposes. Signal disaggregation in this work is defined as splitting the aggregated signal into additive components gathered from sources subjected to aggregation. Such operation simplifies Artificial Intelligence systems built for cloud environment maintenance due to the possibility of performing operations on aggregated signals (like forecasting or anomaly detection) and performing disaggregation in the final step for individual machine behavior analysis. The novelty of this research involves designing artificial intelligence techniques for CPU signal disaggregation. This operation is conducted using two different Deep Neural Networks; Multi-layer Perceptrons (MLP) and Bidirectional Dilated Residual Networks (BRDN) [5].

In Sect. 2, the research datasets are described, Deep Neural Networks are presented, and measurement metrics are defined. Section 3 presents results obtained by Neural Networks. In Sect. 4, the results presented in Sect. 3 are analyzed, and conclusions are summarized in Sect. 5.

2 Materials and Methods

2.1 Datasets

For this research purpose, the datasets for signal disaggregation were gathered. These datasets contained signals of individual machines and signals formed by adding signals from individual machines. The latter is considered an aggregated signal, and the goal of the designed system is to disaggregate it into individual components.

The task accomplishment acquired obtaining datasets from 10 customers from 5 sectors of historical data (logs) using more than 250 virtual machines over a 12-month period. Based on this, 10 training and 10 testing datasets were prepared within the following industries:

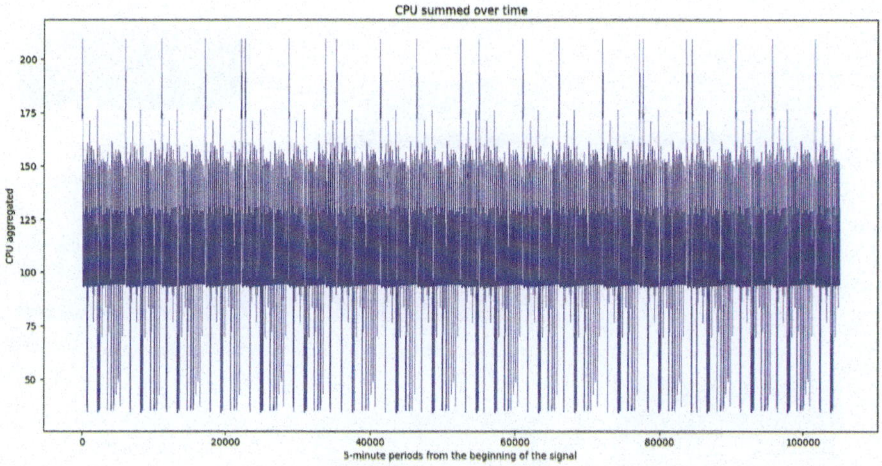

Fig. 1 CPU aggregated

- (B1)—web store hosting,
- (B2)—cloud resources optimization,
- (B3)—retail,
- (B4)—computing platforms,
- (B5)—cloud service sharing.

Analysis of the nature of the acquired data, in particular its informativeness and dynamics, as well as the potential impact of its resolution on possible system operations (i.e., effective disaggregation and prediction) indicated that the most appropriate resolution of the data (time series) would be a 5 min period. This allowed to isolate different time periods of machine operation, forming the basis for analyzing the use of cloud resources with different signal characteristics presenting a different profile of services provided by the customer.

The Fig. 1 shows example images of CPU utilization (industry B3—customer 1) in disaggregated form, Fig. 2—for VMs individually.

2.2 Data Pre-processing

Each dataset represented a multidimensional signal with one machine's CPU load in each channel and was sorted in descending order. As a result of the sorting, the first channel always contained the largest CPU load value of one machine among all the machines operating in a given time unit; the second channel contained the second largest value, and so on. This procedure was carried out to transfer the task from identifying a specific machine to determine the CPU load status of machines at a given moment. Identifying a machine is definitely more difficult than disaggregating the

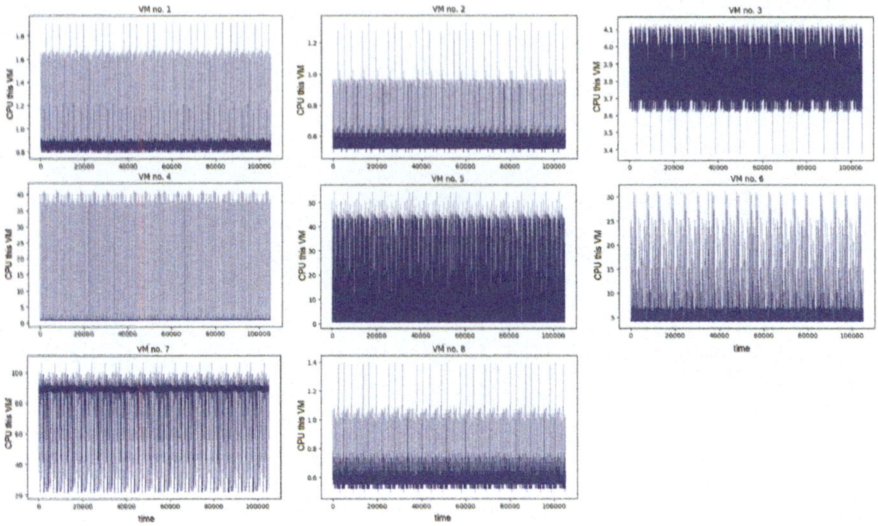

Fig. 2 CPU in VMs separately

signal into the degrees of machine usage at a given moment in time while providing no business gain. Such sorting of disaggregation values (from the highest CPU load to the lowest) resulted in the possibility of constructing as many neural networks as the number of machines the considered set had. Two neural solutions were considered for the disaggregation problem, i.e., Multi-layer Perceptron (MLP) and Bidirectional Dilated Residual Network (BDRN) [5].

For each acquired dataset, change points were found and detected with the *PELT* algorithm using the *ruptures* package [6] at a permanently selected *penalty* hyperparameter. The results of *PELT* algorithm was post-processed by human evaluation. In addition, sets consisting of periods of full days were divided into smaller periods with lengths that were multiples of 24 h. Finally, change points were determined for all datasets. Then, in the process of creating training and testing sets, the drawing with return implemented for the created divisions using change points was used. Such an operation was carried out until the lengths of the learning, and test sets exceeded a period of 12 months, to be precise, when the subsequent halves (the first for the training sequence, the second for the test sequence) did not reach this.

2.3 Machine-Learning Models

This section presents the machine-learning models used in this research in detail. The models used in this work are Multi-layer Perceptron (MLP) and Bidirectional Dilated Residual Network (BDRN).

2.3.1 Multi-layer Perceptron (MLP)

This network's architecture is typical for time series processing purposes. The usage of autocorrelation analysis allowed to determine the length of the signals' season as 24 hours. As a result, the network outputs a singular value representing a forecast. In this case, 24 * 12 = 288 values are provided on the input, corresponding to the aggregated signal values in the past 24 h. The *torch* [1] library was used to implement the solution. For each set, as many networks were trained as many virtual machines composed of the signal. As a result of the implementation of several experiments, the optimal hyperparameters of the MLP solution were determined as:

1. Hidden layers number—3
2. Neurons per hidden layer—48
3. Batch size—32
4. Epochs number—512
5. Dropout—0.1
6. Learning rate—0.002.

2.3.2 Bidirectional Dilated Residual Network (BDRN)

Another approach considered and studied was the more complex one described in [5]. The implementation of the entire solution was analogous to that of the multi-layer perceptron—the *torch* library was also used. However, the final results turned out to be comparable or slightly inferior to the multi-layer perceptron, while the training time was far superior to the simple MLP model. In the course of the study, it turned out that as the size of the learning set increases.

2.4 Metrics

For the purpose of evaluating the models of the algorithms: disaggregation of server load components and simulation (prediction) of synthetic logs of resource consumption for clients with no prior history, the MAPE metric that does not take into account observations $a \geq p$ was defined as:

$$A' = \{a | a \in A \cap a \geq p\} \tag{1}$$

$$MAPE_p = \frac{1}{|A'|} \sum_{t=1}^{|A'|} \left| \frac{A'_t - F_t}{\max(A', \epsilon)} \right| \cdot 100\% \tag{2}$$

where:

- n—observations number

Table 1 Learning times for both methods

Industry	MLP learning time (h)	BDRN learning time (h) (%)	BDRN / MLP
B1	0.36	14.21	39.56
B2	10.0	717.05	71.65
B3	5.14	678.07	131.86
B4	25.93	187.73	7.23
B5	26.21	176.60	6.73

- A—actual observations
- A'—subset of actual observations with values greater than or equal to p%.
- F—predicted observations
- ϵ—sufficiently small number to prevent division by 0.

$MAPE_{1\%}$ means calculating the $MAPE$ metric without including observations where the values were below 1%. This modification of a well-known metric was employed with the purpose of reduced penalization of close-to-zero signal forecasting.

3 Results

Calculation times for the implementation of the MLP and BDRN were compared (Table 1). It turned out that the BDRN is implemented in a much longer time. The research was conducted for a single dataset representing each industry.

The effectiveness of disaggregation using neural network and BDRN models was investigated, and it was found that there is only a marginal difference in effectiveness between the two (Table 2). However, the computation time required for the BDRN network was substantially higher than that of the neural network. As a result, the research focus was set on the neural network for further investigation into its disaggregation effectiveness.

In the preliminary phase of the research, a series of auxiliary experiments were performed to better understand the data and the quality of the models' performance. The prepared algorithm model for disaggregating server load components was tested on 10 datasets in 12-month ranges.

4 Discussion

The experimental results demonstrate that disaggregation of CPU workload in cloud computing environments is possible, provided the signal is pre-processed by con-

Table 2 MAPE measures for two datasets. The metrics used are standard MAPE metric, MAPE with restriction to 0.1%, and MAPE with restriction to 1%, respectively. Each industry (B1 ⋯ B5) is represented by two datasets acquired for it (Z1 and Z2)

Industry/dataset	MAPE (%)	$MAPE_{0.1\%}$ (%)	$MAPE_{1\%}$ (%)
B1/Z1	2.137	2.085	2.085
B2/Z1	9.604	9.693	9.874
B3/Z1	9.765	9.765	4.877
B4/Z1	5.809	5.808	3.675
B5/Z1	24.260	6.396	3.864
B1/Z2	13.136	13.115	13.115
B2/Z2	17.587	17.588	17.712
B3/Z2	24.117	24.117	23.461
B4/Z2	13.747	13.203	7.726
B5/Z2	26.670	6.396	4.264

verting it from individual machine signals to temporal ranks of descending order. However, keeping the information of the signal's machine origin complicates this task to the point of being unenforceable. The authors plan to investigate this phenomenon further to expand our understanding of disaggregation in cloud computing environments.

The present study investigated the effectiveness of two machine learning models for extracting information from time series data. Specifically, a simple multi-layer perceptron neural network was compared to a much more complicated BDRN (Bidirectional Recurrent Neural Network) model. Surprisingly, the results obtained from the multi-layer perceptron neural network were similar to those of the BDRN model, despite drastically reducing inference time. These results suggest that the current state of deep learning is unable to fully leverage its processing power to improve the extraction of information from time series data. However, this also opens up the possibility for new research to further investigate and improve the performance of deep learning models for time series analysis. Future research could focus on developing more efficient and effective neural network architectures or alternative machine learning approaches for time series analysis.

This study investigated the effectiveness of dedicated artificial intelligence models for the disaggregation of each channel of a time series signal. Specifically, each disaggregation channel had its dedicated trained model. However, creating one multi-channel model could result in faster training and inference times with comparable disaggregation effectiveness. The potential benefits of using a multi-channel model to disaggregate time series data warrant further investigation. In future work, the authors plan to investigate the effectiveness of a multi-channel model for disaggregation and compare its performance to the use of dedicated models for each disaggregation channel. Such research could provide insights into optimizing the design of artificial intelligence models for the disaggregation of time series signals.

5 Conclusions

This paper investigates the effectiveness of Multi-layer Perceptron and Bidirectional Dilated Residual Network (BDRN) models for time series disaggregation. Results show that both models have comparable effectiveness, but the computation time required for BDRN is substantially higher than that of neural networks. As a result, further investigation into the neural network's disaggregation effectiveness was conducted. In addition, dedicated artificial intelligence models were developed for each channel of the time series signal. A multi-channel model could potentially result in faster training and inference times without sacrificing disaggregation effectiveness. The authors plan to investigate the effectiveness of a multi-channel model in future work to optimize the design of artificial intelligence models for the disaggregation of time series signals.

References

1. Collobert, R., Kavukcuoglu, K., Farabet, C.: Torch7: A matlab-like environment for machine learning. In: Big Learn, NIPS Workshop (2011)
2. Das, M., Ghosh, S.K.: A deep-learning-based forecasting ensemble to predict missing data for remote sensing analysis. IEEE J. Sel. Top. Appl. Earth Obs. Remote. Sens. **10**(12), 5228–5236 (2017)
3. Duggan, M.: Predicting host CPU utilization in cloud computing using recurrent neural networks. In: 12th International Conference for Internet Technology and Secured Transactions (ICITST), pp. 67–72. IEEE (2017)
4. Gao, J., Wang, H., Shen, H.: Machine learning based workload prediction in cloud computing. In: 29th International Conference on Computer Communications and Networks (ICCCN), pp. 1–9. IEEE (2020)
5. Jia, Z., et al.: Sequence to point learning based on bidirectional dilated residual network for non-intrusive load monitoring. Int. J. Electr. Power Energy Syst. **129**, 106837 (2021)
6. Truong, C., Oudre, L., Vayatis, N.: Selective review of offline change point detection methods. Signal Process. **167**, 107299 (2020)

New Approach to Constructive Induction—Towards Deep Discrete Learning

Cezary Maszczyk, Dawid Macha, and Marek Sikora

1 Introduction

Deep neural networks gained incredible popularity in recent times due to their excellent performance in areas such as computer vision, and language modeling. Such models are however mostly considered as black-box meaning they lack interpretability [22]. Therefore their usage is limited to applications where we don't need to understand the internal model decision process. Interpretability of the machine learning model is an important feature in cases considering ethics such as law, medicine, or finance where there must exist a way to manually verify the models correctness [22]. That is a reason why explainable artificial intelligence (XAI) has become an important theme in ML research community [18]. Yet building deep explainable models for tabular datasets is still a major problem especially when it comes to knowledge discovery [12]. In such applications, white-box methods such as decision trees and decision rulesets are still being widely used and researched.

In this paper, attention was focused on Decision Rulesets models which are considered one of the most human-interpretable machine learning models which still achieve high predictive power. Most of the existing rule-induction algorithms and their implementations utilize only simple types of conditions in rule premises. Such conditions usually include attributes that equals certain value—for nominal attributes, sometimes including negation. For numerical attributes conditions usually take the

C. Maszczyk (✉)
Doctoral School, Silesian Univeristy of Technology, ul. Akademicka 2A, 44-100 Gliwice, Poland
e-mail: cezary.maszczyk@polsl.pl

C. Maszczyk · D. Macha · M. Sikora
Łukasiewicz Research Network - Institute of Innovative Technologies EMAG, ul. Leopolda 31, 40-189 Katowice, Poland

M. Sikora
Department of Computer Networks and Systems, Silesian University of Technology, ul. Akademicka 16, 44-100 Gliwice, Poland

W. Zamojski et al. (eds.), *Dependable Computer Systems and Networks*, Lecture Notes in Networks and Systems 737, https://doi.org/10.1007/978-3-031-37720-4_13

forms of attributes being lesser or greater than specified value or attributes belonging to values interval. Such conditions, referred to as simple conditions, may not be optimal for describing certain specific data relations, which was proven in later examples, leading to over-complex and less interpretable sets of rules. Usage of different types of conditions hereinafter referred to as complex conditions could therefore lead to potentially simpler, shorter, and more human-friendly rulesets.

A similar idea was utilized in the well-known concept of constructive induction in the 1990s [2, 20]. However, these methods did not gain much popularity due to limitations in computing power even though they allowed for more synthetic description of the data and sometimes better predictive results. Nowadays, however, this topic has been abandoned and receives little attention in research.

The contribution of this paper includes extending the sequential covering algorithm with the ability to generate complex conditions directly from a dataset. Multiple synthetic datasets are presented to prove how complex conditions could lead to shortened and more descriptive rulesets. Additionally, experiments were conducted on a set of 32 public datasets in order to examine the influence on rulesets complexity described by a number of rules and the average number of conditions in the rule. Additionally, the impact of complex conditions' presence on model accuracy has been investigated. We also share a repository with the full code of the algorithm available to be used as a convenient Python package. Current work was a preparation for developing a more advanced Deep Discrete Learning approach for rule induction, being able to generate M of N conditions and intermediate concepts.

2 Related Work

Rule induction is a well-established technique in the field of machine learning and data mining [7]. One of the main approaches to rule induction is sequential covering [4, 8, 16]. In comparison with other induction methods, rule sets obtained by the covering algorithms are characterized by good descriptive abilities and classification accuracy [13]. The main advantage of rule representation is its interpretability, although it should be remembered that this depends largely on its complexity [9].

The high complexity of the rule representation is often caused by the fact that most rule induction algorithms operate in a defined representation space and simple hypotheses, such as $attribute = value$ [2, 20]. Often there are more complex relationships in the data that cannot be described directly. This problem was attempted to be solved using data-driven [2] and hypotheses-driven [20] constructive induction. Constructive induction involves iteratively building models by changing the representation space in successive steps. However, this led to the creation of large representation spaces and often hardly interpretable rules [2, 20]. The constructive induction approach differs from the one proposed in this article, where the model is built once and more complex relationships (such as attribute comparisons, internal alternatives) are checked during induction.

Work on searching for complex dependencies is also being carried out towards Deep Discrete Learning [1, 15]. This approach allows to find complex conditions and so-called intermediate concepts, significantly reducing the rule representations. However, these are only early works and there is still much to be done to reach state-of-the-art performance [1].

One of the main goals of the work carried out was to make the developed algorithm available in the form of a ready-to-use library. Only a few of the rule induction algorithms are available as ready-to-use software. The examples are RuleKit [10], RIPPER [5] and M5Rules [11] contained in Weka [19], CN2 [4] included in the Orange suite [6] and AQ [14] implemented in Rseslib 3 [21]. However, none of them provides such an extensive space of complex conditions as the algorithm presented in this article, which is an extension of RuleKit [10].

3 Methods

3.1 Complex Conditions

Discrete Set Conditions

This type of conditions applies only to nominal attributes. Let a_i be a nominal attribute with a set of all possible values D_i. Then Discrete Set Conditions can be defined as follows:

$$a_i \in d \quad where\ d \subset D_i \tag{1}$$

Interval Conditions

For numerical attributes, the original RuleKit package only provides us with conditions of type:

$$\begin{gathered} a_i \in [l, r) \ \ if\ l \neq -\infty \ \ else\ a_i \in (l, r) \\ \text{where}\ l = -\infty \ \ or \ \ r = \infty \end{gathered} \tag{2}$$

Such literals allow you to specify only intervals whose one limit is infinite. A new type of condition has been added to the package that generalizes form from Eq. 2 so that both the left and right boundaries of the interval can be real numbers.

All possible interval candidates for an attribute are generated as follows:

$$\left\{ a_i \in \left[\left(a_{i,j-1} + \frac{a_{i,j} - a_{i,j-1}}{2} \right), a_{i,k} \right) \forall j \in [1, N-2) \right\} \forall k \in [j+2, N-1) \tag{3}$$

Attributes Relations Conditions

This type of conditions specifies relations between given attributes and covers all examples for which those relations are satisfied. They are generated only for attributes of the same type, either all nominal or all numerical. For nominal attributes only two relations are taken into account: $=$ and $\neq$. For numerical ones, possible relations are: $=$, $>$ and $<$. Previously mentioned relations may concern more than two attributes (up to three) if they are nominal ones.

Alternatives Conditions

This type of conditions is generated only for numerical attributes and takes the following form:

$$c_1 \vee c_2 \vee \ldots c_K \tag{4}$$

where c_i are either simple numerical conditions or interval conditions.

Negated Conditions

Additionally, negations of conditions of the following types: simple, Interval Conditions, and Attributes Relations Conditions were implemented. For the last one, they are only induced for nominal attributes.

3.2 *Algorithm*

The algorithm for rule induction with complex conditions is based on the sequential covering algorithm implemented in [10]. Its detailed description can be found in [16]. As part of the ongoing work, this algorithm was extended to generate also complex conditions described in Sect. 3.1. Thus, at the stage of induction of new conditions, complex conditions are additionally checked.

An important change introduced in the implementation of the algorithm is the generation of internal alternatives. It takes place when a numerical condition of type $a_i \in (l, r)$ is selected as the best condition at the induction stage. In such a case, the algorithm tries to extend this condition with alternatives with other generated intervals for the attribute to which the condition applies. Only non-overlapping intervals are checked. The search for alternatives is performed using a beam search algorithm.

4 Results and Discussion

4.1 Synthetic Examples

The introduction of complex conditions into the rule induction algorithm makes it possible to search for new relationships in the data. This makes it possible to obtain a more concise rule representation of the data set. This is extremely important because, as the research in [9] shows, the interpretability of a set of rules is largely affected by the number of rules and their length. However, it should be remembered that the introduction of complex conditions allows to reduce the size of the representation only if the relationships they describe are in the dataset. In order to demonstrate the advantages of introducing complex conditions into the algorithm, it was decided to carry out rule induction on synthetic sets containing dependencies that complex conditions can help describe. These sets are shown in Fig. 1.

The synthetic datasets were prepared as follows:

(a) class 1 (red) if $attr_1$ is in the range <3,7>, for other examples class 0 (blue)
(b) class 1 (red) if $attr_1$ is in the range <3,5> or <7,8>, for other examples class 0 (blue)
(c) class 1 (red) if $attr_1$ is smaller than $attr_2$, for other examples class 0 (blue)
(d) class 1 (red), if $attr_1$ has a value of "medium" or "low", for other examples class 0 (blue)
(e) class 1 (red) if $attr_1$ and $atrr_2$ have the same values, for other examples class 0 (blue)

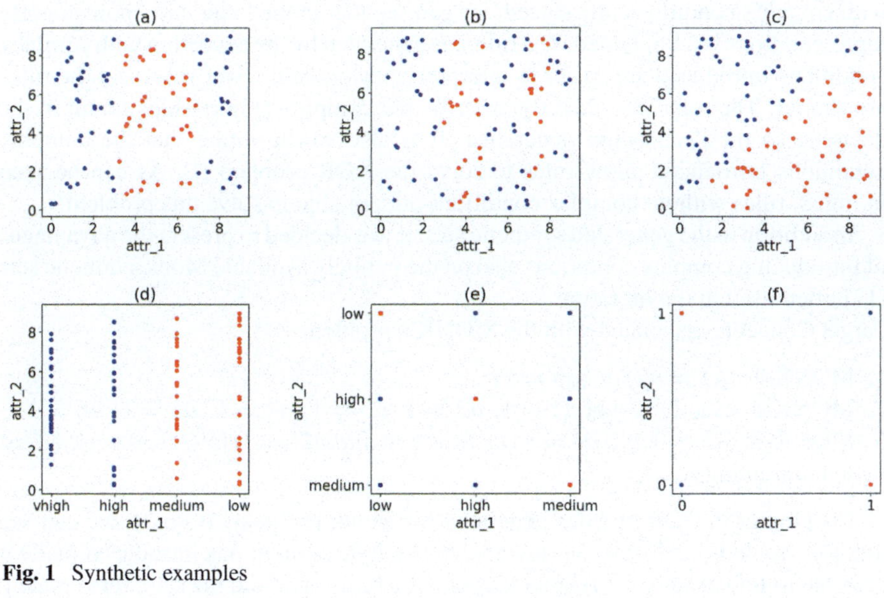

Fig. 1 Synthetic examples

	Rules with simple conditions	Rules with complex conditions
(a)	IF attr_1 = (-inf, 3.12) THEN class = {0} IF attr_1 = <6.96, inf) THEN class = {0} IF attr_1 = <3.12, 6.96) THEN class = {1}	IF attr_1 != <3.12, 7.04) THEN class = {0} IF attr_1 = <3.12, 7.04) THEN class = {1}
(b)	IF attr_1 = (-inf, 2.97) THEN class = {0} IF attr_1 = <7.99, inf) THEN class = {0} IF attr_1 = <5.15, 7.03) THEN class = {0} IF attr_1 = <6.98, 7.99) THEN class = {1} IF attr_1 = <2.97, 4.92) THEN class = {1}	IF (attr_1 != <2.97, 8.05) OR attr_1 = <4.92, 7.02)) THEN class = {0} IF (attr_1 = <2.97, 5.03) OR attr_1 = <6.98, 8.05)) THEN class = {1}
(c)	IF attr_2 = <7.02, inf) THEN class = {0} IF attr_1 = (-inf, 1.52) THEN class = {0} IF attr_2 = <1.89, inf) AND attr_1 = (-inf, 2.48) THEN class = {0} IF attr_2 = <5.36, inf) AND attr_1 = (-inf, 5.71) THEN class = {0} IF attr_2 = (-inf, 7.59) AND attr_1 = <5.71, inf) THEN class = {1} IF attr_2 = (-inf, 2.26) AND attr_1 = <1.17, inf) THEN class = {1} IF attr_2 = (-inf, 5.43) AND attr_1 = <4.08, inf) THEN class = {1}	IF attr_1 < attr_2 THEN class = {0} IF attr_1 > attr_2 THEN class = {1}
(d)	IF attr_1 = {vhigh} THEN class = {0} IF attr_1 = {high} THEN class = {0} IF attr_1 = {low} THEN class = {1} IF attr_1 = {medium} THEN class = {1}	IF attr_1 = {vhigh, high} THEN class = {0} IF attr_1 = {medium, low} THEN class = {1}
(e)	IF attr_2 = {high} AND attr_1 = {medium} THEN class = {0} IF attr_2 = {high} AND attr_1 = {low} THEN class = {0} IF attr_2 = {low} AND attr_1 = {medium} THEN class = {0} IF attr_2 = {medium} AND attr_1 = {medium} THEN class = {1} IF attr_2 = {low} AND attr_1 = {low} THEN class = {1} IF attr_2 = {high} AND attr_1 = {high} THEN class = {1}	IF attr_2 != attr_1 THEN class = {0} IF attr_2 = attr_1 THEN class = {1}
(f)	Not generated	IF attr_2 = attr_1 THEN class = {0} IF attr_2 != attr_1 THEN class = {1}

Fig. 2 Rules generated for synthetic examples problem

(f) XOR problem

Comparing the rule sets (Fig. 2) obtained using the rule induction algorithm with simple conditions only with the rule sets obtained using the rule induction algorithm with complex conditions introduced, we can see that in the latter case more concise representations were obtained. The rule sets generated by the algorithm with complex conditions introduced are easier to interpret: they contain fewer rules and the rules are shorter. The rules also directly describe the complex relationships found in the datasets, so they can better generalize. It is also worth noting that the complex conditions introduced allow rules to solve the XOR problem (f). As can be seen in Fig. 2, rules without complex conditions are not able to solve this problem.

In addition to the generated synthetic sets, it was decided to present the advantages of introducing complex conditions also on the publicly available Monk synthetic sets [17] often used in the literature.

Target Concepts associated with the MONK's problem:

1. MONK-1: $(a1 = a2)\ or\ (a5 = 1)$
2. MONK-2: exactly two of $a1 = 1,\ a2 = 1,\ a3 = 1,\ a4 = 1,\ a5 = 1,\ a6 = 1$
3. MONK-3: $(a5 = 3\ and\ a4 = 1)\ or\ (a5! = 4\ and\ a2! = 3)$ [5% class noise added to the training set].

Looking at the sets of rules generated for Monk problems (Fig. 3), we can see that for Monk-1 and Monk-3 problems, the complex conditions introduced made it possible to find exactly the relationships on the basis of which the sets were created.

	Rules with simple conditions	Rules with complex conditions
Monk 1	IF a5 = {1} THEN class = {True} IF a1 = {3} AND a2 = {3} THEN class = {True} IF a4 = {1} AND a6 = {1} THEN class = {True} IF a4 = {1} THEN class = {True} IF a3 = {1} AND a6 = {2} THEN class = {True} IF a6 = {1} THEN class = {True}	IF a1 = a2 THEN class = {True} IF a5 = {1} THEN class = {True}
Monk 2	IF a2 = {2} AND a3 = {1} AND a5 = {2} THEN class = {True} IF a3 = {1} AND a5 = {2} AND a6 = {2} THEN class = {True} IF a2 = {1} AND a3 = {2} AND a5 = {2} THEN class = {True} IF a1 = {2} AND a3 = {1} AND a4 = {3} THEN class = {True} IF a3 = {1} AND a4 = {3} THEN class = {True} IF a3 = {2} AND a4 = {3} AND a6 = {1} THEN class = {True} IF a1 = {3} AND a3 = {1} AND a6 = {2} THEN class = {True} IF a2 = {1} AND a6 = {2} THEN class = {True} IF a1 = {1} AND a6 = {2} THEN class = {True} IF a2 = {2} AND a4 = {2} AND a6 = {1} THEN class = {True} IF a1 = {2} AND a6 = {1} THEN class = {True}	IF a3 != a6 AND a1 != a2 AND a1 = {1} AND a4 != {1} AND a5 = {2, 4} THEN class = {True} IF a3 != a6 AND a1 != a2 AND a2 = {1} AND a4 != {1} AND a5 = {2, 4} THEN class = {True} IF a3 != a6 AND a1 != a2 AND a1 = {1} AND a4 != {1} AND a5 != {1} THEN class = {True} IF a3 != a6 AND a1 != a2 AND a1 != {2} AND a4 != {1} THEN class = {True} IF a3 != a6 AND a2 != {1} AND a4 != {1} AND a5 != {3} THEN class = {True} IF a1 != a4 AND a3 != {2} AND a5 != {1} AND a6 != {1} THEN class = {True} IF a1 != a2 AND a1 != a4 AND a5 = {2} THEN class = {True} IF a1 != a2 AND a5 = {2, 3} THEN class = {True} IF a1 != a2 AND a4 != {1} AND a5 != {3} THEN class = {True} IF a2 != {3} AND a4 != {1} AND a5 != {1} THEN class = {True}
Monk 3	IF a2 = {2} AND a5 = {1} THEN class = {True} IF a1 = {1} AND a2 = {2} AND a5 = {2} THEN class = {True} IF a2 = {2} AND a3 = {2} AND a5 = {3} THEN class = {True} IF a2 = {2} AND a6 = {2} THEN class = {True} IF a2 = {1} AND a5 = {1} THEN class = {True} IF a2 = {1} AND a5 = {2} THEN class = {True} IF a2 = {1} AND a5 = {3} THEN class = {True} IF a3 = {1} THEN class = {True}	IF a2 != {3} AND a5 != {4} THEN class = {True}

Fig. 3 Rules generated for Monk problems

The Monk-2, on the other hand, is an example of a data set on which the conditions M of N would work. The introduction of such conditions is planned in further phases of the development of the algorithm. The conditions introduced so far do not find exactly the relationship in the Monk-2 data set on the basis of which it is created, but they allow better generalization. Rules without complex conditions obtained accuracy on the training set of 0.80% and on the test set of 0.76%, while rules with complex conditions obtained accuracy on the training set of 0.78% and on the test set of 0.80%.

4.2 *Experimental Evaluation*

The extended rule induction algorithm was evaluated and compared with the original one on the set of 32 publicly available data sets. All presented results were obtained by calculating the average of 10-fold cross-validation process. During rule induction well known C2 measure [3] was used for all tested data sets.

Table 1 presents results for all evaluated data sets. Columns denoted as *simple* contain metrics values for the original rule induction algorithm while columns denoted as *complex* relate to the authors' extended implementation. *BAcc (test)* column refers to balanced accuracy score calculated on test datasets.

Based on results from Table 1 it can be concluded that the usage of complex conditions during rule induction did not affect balanced accuracy score on most of the datasets. Wilcoxon Signed-Rank Test was conducted which confirmed that at a confidence level of 5% there is no significant difference for both distributions. The

Table 1 Experiment results

	BAcc (test)		Rules		Avg. conditions per rule	
	Simple	Complex	Simple	Complex	Simple	Complex
Anneal	0,931	0,928	70,5	51,6	2,365	2,790
Auto-mpg	0,674	0,677	125	137	3,657	4,798
Autos	0,699	0,733	100,3	72,2	3,566	3,122
Balance-scale	0,629	0,892	164	58,4	3,294	2,473
Bupa-liver-disorders	0,713	0,641	216,5	233,3	4,133	4,551
Car	0,558	0,658	232	120,5	3,421	4,151
Cleveland	0,229	0,259	231,5	214,3	5,793	5,294
Credit-a	0,848	0,861	132,8	204,7	4,740	6,251
Cylinder-bands	0,753	0,758	290,1	237,6	4,164	3,599
Echocardiogram	0,817	0,829	22,7	29,7	2,759	4,026
Ecoli	0,719	0,727	103,2	91,9	3,589	3,464
Flag	0,544	0,576	96	82,4	4,226	4,269
Glass	0,661	0,622	113,6	90,2	4,173	3,476
Hayes-roth	0,799	0,808	25,9	40,3	1,812	2,760
Heart-c	0,779	0,793	128,2	141,9	4,459	4,821
Heart-statlog	0,792	0,791	120,7	119	4,491	4,397
Hepatitis	0,653	0,610	40,1	33,5	2,888	2,724
Horse-colic	0,829	0,826	112,8	135,9	3,883	4,588
Hungarian-heart-disease	0,793	0,778	106,6	129,4	4,547	5,360
Iris	0,913	0,913	11,1	8,8	2,148	1,857
lymphography	0,792	0,815	37,9	40,9	2,448	2,802
Monk-1	0,834	1,000	34	4	2,267	1,333
Monk-2	0,738	0,757	71	91	2,630	3,370
Monk-3	0,805	0,805	19	10	1,900	2,000
Mushroom	0,967	0,999	30,7	19,9	1,953	1,990
Nursery	0,764	0,660	1414,2	508,3	4,666	4,812
Soybean	0,934	0,935	130,1	124	3,149	3,316
Tic-tac-toe	0,954	0,958	132,1	158,4	3,184	3,626
Titanic	0,711	0,705	22,4	25	1,576	1,621
Vote	0,952	0,955	9,6	28,6	1,515	3,976
Wine	0,955	0,931	15,5	16	1,886	2,025
Zoo	0,860	0,888	28,9	15,6	2,390	1,661

approach utilizing complex conditions achieved better balanced accuracy for 21 of 32 data sets and for one data set it was exactly the same value.

The complexity of achieved rule sets was measured using the number of rules in them and an average number of conditions in rules. It can be observed that using complex conditions has an effect on both of that metrics. For 18 data sets the number of rules decreased for the complex conditions approach on average by 31.26%. For the rest of the data sets the number of rules increased by 15.82% on average. The usage of complex conditions decreases the average number of conditions in the rules on 17 data sets while increasing it on the rest of them. The average improvement was 15.7% while the average deterioration was 15.62%. To evaluate the holistic complexity of rule sets authors proposed a custom metric calculated by multiplying the rules count by the number of conditions in the rule set. For this self-defined measure, improvement was observed for 19 data, on average by 46.08%. For the rest of the data sets its value was worse by 33.29% on average.

Custom implementation was measured to be around 7.2 times slower than the original RuleKit while still giving acceptable training times for all tested data sets. Turning off Alternative Conditions during induction significantly improves times making them around 37% faster compared to turning them on.

Full experiments results together with all the code are available on GitHub repository.[1]

5 Conclusion

The introduction of complex conditions into the rule induction algorithm enhances their descriptive capabilities while maintaining the accuracy of the models. It is shown on synthetic data sets that by introducing complex conditions it is possible to obtain simpler sets of rules, directly describing the relationships present in the data. Also, experiments on real sets showed that the introduction of complex conditions positively affected the complexity of the models while maintaining accuracy. However, it is important to remember that not all data sets have complex relationships that can be described more simply using complex conditions. In such situations, the models are not simpler, but it is also important that they do not become more complicated. The set of complex conditions to be checked will be expanded in the future to include M of N conditions (e.g., 2 conditions out of 3 are satisfied) and intermediate concepts that will help reduce complexity for problems like the Monk-2 problem presented, or the parity problem. Further work is also planned to computationally optimize the solution and propose Deep Discrete Learning methods for generating rules.

References

1. Beck, F., Fürnkranz, J.: An empirical investigation into deep and shallow rule learning. CoRR (2021). ArXiv:abs/2106.10254
2. Bloedorn, E., Michalsi, R.: Data-driven constructive induction. Intell. Syst. Their Appl. IEEE **13**, 30–37 (1998). https://doi.org/10.1109/5254.671089

[1] https://github.com/cezary986/complex_conditions.

3. Bruha, I.: Quality of decision rules: Definitions and classification schemes for multiple rules. Mach. Learn. Stat. Interface 107–131 (1997)
4. Clark, P., Niblett, T.: The cn2 induction algorithm. Mach. Learn. **3**(4), 261–283 (1989). https://doi.org/10.1023/A:1022641700528
5. Cohen, W.W.: Fast effective rule induction. In: Prieditis, A., Russell, S. (eds.) Machine Learning Proceedings 1995, pp. 115–123. Morgan Kaufmann, San Francisco (CA) (1995). https://doi.org/10.1016/B978-1-55860-377-6.50023-2, https://www.sciencedirect.com/science/article/pii/B9781558603776500232
6. Demšar, J., Curk, T., Erjavec, A., Gorup, v., Hočevar, T., Milutinovič, M., Možina, M., Polajnar, M., Toplak, M., Starič, A., Štajdohar, M., Umek, L., Žagar, L., Žbontar, J., Žitnik, M., Zupan, B.: Orange: Data mining toolbox in python. J. Mach. Learn. Res. **14**(1), 2349–2353 (2013)
7. Fürnkranz, J., Gamberger, D., Lavrač, N.: Foundations of Rule Learning. Cognitive Technologies. Springer, Berlin (2012). https://books.google.pl/books?id=JsTXnjWexCEC
8. Fürnkranz, J.: Separate-and-conquer rule learning. Artif. Intell. Rev. **13**, 3–54 (1999)
9. Fürnkranz, J., Kliegr, T., Paulheim, H.: On cognitive preferences and the plausibility of rule-based models. Machine Learning **109**(4), 853–898 (2019). https://doi.org/10.1007/s10994-019-05856-5
10. Gudyś, A., Sikora, M., Łukasz Wróbel: Rulekit: A comprehensive suite for rule-based learning. Knowl.-Based Syst. **194**, 105480 (2020). https://www.sciencedirect.com/science/article/pii/S0950705120300046
11. Holmes, G., Hall, M.A., Frank, E.: Generating rule sets from model trees. In: Australian Joint Conference on Artificial Intelligence (1999)
12. Katzir, L., Elidan, G., El-Yaniv, R.: Net-dnf: Effective deep modeling of tabular data. In: International Conference on Learning Representations (2021)
13. Michalak, M., Sikora, M., Wróbel, L.: Rule quality measures settings in a sequential covering rule induction algorithm—an empirical approach. In: 2015 Federated Conference on Computer Science and Information Systems (FedCSIS), pp. 109–118 (2015). https://doi.org/10.15439/2015F388
14. Michalski, R.: On the Quasi-Minimal Solution of the General Covering Problem (1969)
15. Qiao, L., Wang, W., Lin, B.: Learning accurate and interpretable decision rule sets from neural networks. CoRR (2021). ArXiv:abs/2103.02826
16. Sikora, M., Wróbel, L., Gudys, A.: Guider: a guided separate-and-conquer rule learning in classification, regression, and survival settings. CoRR (2018). ArXiv:1806.01579
17. Thrun, S., Bala, J., Bloedorn, E., Bratko, I., Cheng, J., De Jong, K., Džeroski, S., Fahlman, S., Fisher, D., Hamann, R., Kaufman, K., Keller, S., Kononenko, I., Kreuziger, J., Mitchell, T., Pachowicz, P., Reich, Y., Vafaie, H., Wnek, J.: The Monk's Problems a Performance Comparison of Different Learning Algorithms (1992)
18. Tjoa, E., Guan, C.: A survey on explainable artificial intelligence (XAI): Toward medical XAI. IEEE Trans. Neural Netw. Learn. Syst. **32**(11), 4793–4813 (2021). https://doi.org/10.1109/tnnls.2020.3027314
19. Witten, I.H., Frank, E.: Data mining: Practical machine learning tools and techniques with java implementations. SIGMOD Rec. **31**(1), 76–77 (2002). https://doi.org/10.1145/507338.507355
20. Wnek, J., Michalski, R.S.: Hypothesis-driven constructive induction in aq17-hci: A method and experiments. Mach. Learn. **14**, 139–168 (1994)
21. Wojna, A., Latkowski, R.: Rseslib 3: Open Source Library of Rough Set and Machine Learning Methods, pp. 162–176 (2018). https://doi.org/10.1007/978-3-319-99368-3_13
22. Yang, Y., Morillo, I.G., Hospedales, T.M.: Deep Neural Decision Trees (2018). ArXiv:abs/1806.06988

Softcomputing Approach to Music Generation

Jacek Mazurkiewicz

1 Introduction

The work aims to create a softcomputing system for sound generation based on data representing music in a selected style. The concept of a learning system means algorithms in the field of machine learning that allows the creation, training, and use of intelligent models to perform the sound tasks processing assigned to them. The music considered in this paper will be presented using the abstraction of a note. Information such as pitch, duration, or point in time will be processed. Data representing music will be understood as sequences of notes that make up a piece of music. The style means a set of features to create the impression of similarity both in standard music notation and in the perception of the sound material by the listener. Many applications of AI solutions are known, and the number is still growing. Despite this, the combination of music and intelligent systems is not popular. Interest in the subject is present mainly in research labs. This introduces some problems because it is impossible to talk about the quality of the music generator without defining the quality of the music itself. Although there are cases in which the researcher has a musical education, this is a rather rare situation. The work combines the fields of art and science. Particular attention will be paid to the quality of the generated music in the sense of art. The advantages and disadvantages of the intelligent solution will be defined from the point of view of the recipient who knows the features of a musical work that define it as a work of artistic value. This approach aims to combine two separate fields. The effect of such a procedure is to make another step on the way to combine the knowledge of artists with the knowledge of researchers. This way, the resulting solution will define not only technical but also artistic problems, the solution of which is necessary to create a high-quality music generation system. The

J. Mazurkiewicz (✉)
Faculty of Information and Communication Technology, Wrocław University of Science and Technology, ul. Wybrzeże Wyspiańskiego 27, 50-370 Wrocław, Poland
e-mail: jacek.mazurkiewicz@pwr.edu.pl

W. Zamojski et al. (eds.), *Dependable Computer Systems and Networks*, Lecture Notes in Networks and Systems 737, https://doi.org/10.1007/978-3-031-37720-4_14

analyzed issues include features and standards of notation of musical compositions, ways of representing music by a computer, methods of music data processing, and principles and methods of creating intelligent systems [1]. The preparation of music data is related to collecting music tracks saved in a format that allows them to be used in the created system, selecting similar tracks, creating training data sets from selected tracks, and processing data set into the form of emphasizing their use in the generator training phase [2]. Testing the generator's work will consist of the manual analysis of the generated compositions by listening to them. Generation rating will be a subjective opinion about the generated work; its strengths and weaknesses. The listener will identify the similarities and differences between the generated song and the original songs contained in the training dataset. As part of the selection of the optimal operating point, the impact of the values of available system parameters on the quality of generation will be examined. Values that maximize this quality will be retained.

2 Fundamentals of Sound Theory

Vibrations are a necessary phenomenon for the creation of sound. On the other hand, matter can vibrate; the type of this matter determines the characteristics of the sound. The basic characteristic of a sound is its pitch. It depends on the frequency of vibrations of the medium as the source of the sound. When the medium vibrates with a single frequency, we say that the so-called tone—sound with one component. Tones, as separate entities, are rare in nature. In most cases, we are dealing with complex vibrations. The vibrating medium in this case vibrates at many frequencies simultaneously. The sound created as a result of the vibration of such a medium can be presented as a harmonic series—it is possible to distinguish the component tones of such a sound [3]. Another important feature of sound is its intensity. We say that a sound is loud when the vibration amplitude of the sound source is large. Qualitative terms are used intentionally, as whether a sound is "loud" or "soft" at a given moment depends on the context. For example, along with the path traveled by a sound wave, we observe the decay of the sound. The sound wave loses energy and as the distance from the source increases, the listener will perceive the sound as getting quieter. Musicians manipulate the intensity of sound because it is one of the basic means of expression in musical works. This is related to the concept of dynamics which determines the force with which the sounds, phrases, or musical sentences should be performed. Combining sounds was systematized in the ninth century—the branch of music theory dealing with combining sounds and studying the interdependence between them was called "harmony". The Baroque period introduced significant changes in the understanding of harmony. At that time, functional harmony was defined, which to this day is the basis for the construction of musical works [4]. When we talk about even rhythmic beats, we mean regular beats, showing a certain temporal structure. The colloquial meaning of this expression is not far from the notion of musical rhythm, which can be defined as the time dependence of sounds and their sequences in a piece of music. The BPM (Beats Per Minute) unit is used to specify the time base for a given beat value [5].

3 Assumptions and Data Preprocessing

The task set for the created system is to generate music in a selected style. The generated fragments should feel similar to the pieces presented to the system in the training phase, however, creativity in the created compositions is also highly desirable. The feature of creativity means the presence of fragments unheard of in training examples, additionally not disturbing the style of the piece and fitting well into the melodic, harmonic, and rhythmic framework of the created composition. The model of the generator will accept data in the form of processed melodies exported from MIDI files [5]. The data will be a sequence stored as a list of notes. A single note will be represented by three variables: shift in time relative to the previous note, duration, and pitch. Elements of the work such as tempo, dynamics, or articulation will be omitted. The generation system should create works with the features of an artistically valuable composition. Songs generated should include proper order in terms of melody, harmony, and rhythm, well-conducted motifs, phrases, repetitions of important fragments of a piece with sensitivity, and as few false or out-of-context sounds as possible. The listener should have difficulty distinguishing the style of the generated work from the style of the examples used to train the model but should hear that the generated work is not a copy of the source work. The listener's opinion about the quality of the generated compositions will be the only criterion for evaluating the model's performance. The input point of the designed system was music data saved in the MIDI format. For their initial processing, the *music21* library was used [6]. With the help of this library, it was possible to extract the necessary data without going into the implementation details of the format itself. In the course of processing, irrelevant data was filtered out and individual MIDI tracks were sorted in memory [5]. After filtering, it was possible to iterate over the strings of sounds that make up the song. Data must represent a sequence. The project assumes that this sequence will be represented by a sequence of fixed-point numbers. On the other hand, it is assumed that the generator will treat the note as a unit consisting of three variables. This conflict has been solved by using an auxiliary data structure—a dictionary of notes. This structure was implemented as a real dictionary with a key being a significant representation of a note and a value corresponding to the index of the note in the dictionary. To allow for the note-index conversion in both directions, the key and the value have been swapped to create an additional list of notes. The dictionary contained unique words, i.e. notes with different values of their representation variables [4]. The basis for its creation: all notes were from songs loaded into the program from MIDI files. When the list of sheet music and the dictionary were created, the data was written to the file. This way we allow for the reuse of once-processed data without the need to perform the same operations again. The data prepared this way were further processed already in the training phase. Before starting the training, it was necessary to additionally prepare the processed music data due to the limitations imposed by the TensorFlow library API. This API required data to be grouped to distinguish the following dimensions in the training set: a batch of data, sequence destination, and notes in the selected sequence. The meaning of the sequence is understood as the nature of one of the two sequences provided as a pair to the network [7]. The first sequence of the pair is used as input to the model, and the second one has shifted forward one element and serves as the generation target to which the model should adapt. Individual melodic parts (voices in a piece or instrument) were processed separately. The algorithm of the sequence generator consisted in dividing the song part into sequences of a certain length and returning a pair of sequences [8]. The first one was the entrance to the network, while the second served as the model prediction target. This is a reason why, it was shifted one word forward compared to the input sequence. When a list of sequence pairs of a certain length was created, the processed data was returned for further processing. Additional functionality was to treat each musical part as a separate entity and return the sequence divided into parts [9]. The choice of such behavior was possible thanks to the appropriate parameter of the function. The default behavior of the data format conversion algorithm was to return a flat list of sequence pairs, where all created pairs were placed in the same dimension of the list, sequentially from each musical part.

4 Model of System

4.1 Structure

The neural system model is as follows: embedding layer, variable number of LSTM recursive layers, and fully connected layer [10]. The task of the embedding layer was to create a vector representation of the index processed by the network. This layer had two purposes: creating an additional dimension for the LSTM layer, and capturing spatial dependencies of vectors representing particular notes, directly linked to the indexes of the dictionary. Especially the second purpose of the embedding layer is interesting here. Captured relationships could be corrected in the training phase. The task of the LSTM layer was the generation of successive notes of the song, as it results from the assumptions of this project [11]. Thanks to the characteristics of this layer, the task of predicting the next note based on the data given so far was not a challenge. The implementation made it possible to set up many such layers in succession. Subsequent LSTM layers operated on the output of the previous layers—their operation principle was identical. The last layer of the model was fully connected. This is a classic layer where each neuron computes the weighted sum of the input and with the help of a non-linear activation, the function produces the output result. The purpose of this layer in the created model was to read the results of the recursive layer and process them to produce the size of the output vector of the model equal to the size of a dictionary of notes—each output corresponded to the right word in the dictionary [12].

4.2 Training Procedure

After creating the model, it was possible to train it with the use of previously prepared training data. This training was carried out using the *tf.keras.Model.fit()* method, setting the number of training epochs was given—the processing of the supplied data set by the network. Models achieving better and better results were systematically serialized to a file, which allowed for the later continuation of the training and independent use of the saved model for the task of generating a musical piece. In addition, a custom training algorithm was defined with a minor modification compared to the standard *tf.keras.Model.fit()*, aimed at resetting the state of the recursive layer after processing the extracted part of the song. The aim was to examine the behavior of the model, assuming that it only has the context of the previously processed part of a musical piece [12].

The operation of the functions used in the model training phase could be modified by the system user. The training phase depended on the following parameters: a selected set of training data (out of many created), the format of training data (with or without division into batches), the length of a single sequence processed by the network, size of batches of input data to the network, the dimension of the vector created by the embedding layer, number of recursive layers, size of the recursive layer, state storage mode by the recursive layer, the number of epochs that fulfill a single training session. The mode of storing the state by the LSTM layer was a logical type variable. Its value defined if the state should be preserved between batches of training data or not. This defined the context of the network when it predicted the next note. If the state was not saved, it was reset after each batch of data. The data batch usually consisted of one training pair (sequence). In this case, the network output was determined based on the data in the given sequence and did not go beyond this sequence. In the case of maintaining the state, the entire dataset became the context. The state was reset only after the processing of all data in the set was completed, i.e. after the end of the training epoch. The size of the recursive layer was the number of duplicated LSTM units in the layer [10]. Each of these units has a unified structure and operates on the same data. However, these units differed in the direction of training—they had different initial weights, and thus obtained different final weights. Each of these units performed the same task. The results of individual units were aggregated by the fully connected layer, which made the model more confident in proposing the next notes during generation. Using the presented parameters, it was possible to experimentally determine the optimal operating point of the created system.

4.3 Music Generation

For the music generation task, a set of training data created in the data preparation phase and an already trained model were in use. The set and the model have not been related, but in practice, they turned out to be of little importance. The first phase of generation asks the model for the next notes; a sequence was generated based on the data acquired by the model from the training sets. The initial fragment from the set was selected for processing by the network. The use of a fragment of existing work as the model input in the initial generation phase was to initialize the state of the recursive layer. The output of the model was the basis for building a distribution to point to the next note. This way we try to reduce the probability of the model output looping–stalling in the generation dead point, where only a sequence of the same notes is generated in the extreme case. The note selected by the distribution was prepared for processing by the model in the next step, to generate the next part of the song. All generated notes, or rather indexes pointing to notes in the dictionary, were stored until the required number of notes was generated. The output from the stage of generation as the sequence representing the song by the TensorFlow model was a list of strings (dictionary keys) representing notes. Creating a MIDI file from this kind of data required an additional step to convert the string representation of the note into an object form supported by the *music21* library [3]. After creating note objects, it was possible to create a MIDI file with the generated song. Using the *music21* API, a MIDI file was created and then saved to disk, completing the song generation process. The user could parameterize the stage of music generation using a trained model. Modifiable parameters were available to him: an indication of the data set initializing the state of the recursive layer, an indication of the selected TensorFlow model, initialization sequence length, number of notes to generate, and temperature model. The temperature modifies the distributions used for the next note sequences pointed for the sample. Reducing the temperature resulted in the generation of fewer uncertain sounds—the generation was more predictable and potentially less interesting. Increasing the temperature made subsequent notes less predictable. This increased the "creativity" of the model and thus increased the number of sounds that did not fit the generation context. In practice, it was assumed that the length of the initializing sequence is always one to force the model to use the self-generated context of the song to the maximum extent to generate subsequent notes.

5 Results Analysis

In this chapter, training datasets used in the music generation task will be presented. Next, examples of songs included in the dataset and examples of songs generated with generation problems will be shown. Next, the impact of the parameters available in the system for the user on the generation quality will be discussed. The analysis will be summarized by a study with the respondents, consisting in identifying the

Table 1 Dataset based on variations by JS Bach

Parameter	Value
Number of MIDI files	30
Number of notes in the file	22,958
Number of notes in the dictionary	1940

type of work (original or generated work). The generation of tracks was based on training data sets previously prepared and later delivered to the TensorFlow model in the training phase. Such a set had the structure of a list of musical parts (voices or instrumental parts), which was built of notes represented by a sequence of characters. The dataset was also accompanied by a dictionary that contained unique notes. Each element of the dictionary was identified by a unique index. The created collections were characterized by a coherent style. However, the melodic material did not always fit the works contained in the same collection. This means that by compiling random fragments of songs contained in the collection, there was a possibility that the song created in this way would be of low quality. This will prove important in the analysis of the generated tracks. In the project, five music datasets were created and used: JS Bach variations, JS Bach fugues, jazz pieces (large collection), jazz pieces (small collection), and jazz pieces (single-element set). Large collections of variations, fugues, and jazz pieces were created based on 20–30 MIDI files, several minutes long. Two representative above-mentioned collections will be presented in detail. JS Bach's set of variations is described in Table 1 and histograms (Fig. 1). Table 1 lists the basic features of the set. It is worth paying attention to the ratio of unique words to all words (notes) of the set, which in this case is approximately one to twelve. Histogram (Fig. 1) shows the distribution of the notes in the collection. The x-axis of this graph represents the index of the unique note in the dictionary. This is the basis for this histogram because this is how the TensorFlow model considers the supplied data; after all, it processes sequences of fixed-point numbers, not audio characteristics such as pitch or duration.

From the histogram, it can be concluded that the distribution of notes in Bach's works is uneven. Some notes appear much more often than others. The cumulative

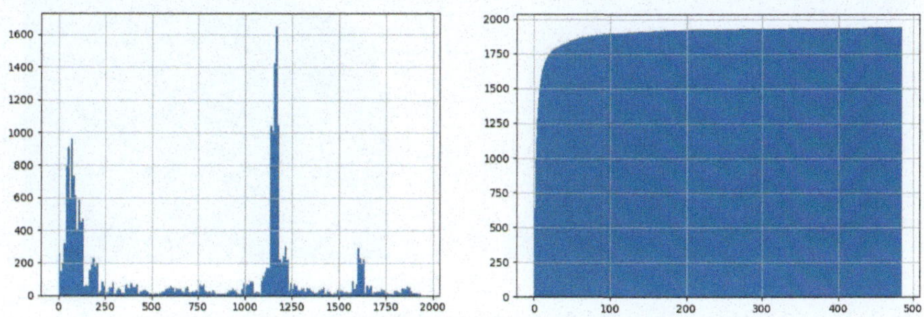

Fig. 1 Dataset histogram (left) and dataset dictionary histogram (right)—variations by JS Bach

histogram (Fig. 1) characterizes the vocabulary of the dataset. The x-axis of this histogram describes the number of occurrences of a unique dictionary note in the dataset. The y-axis represents the cumulative number of occurrences of the x-axis data. On the y-axis, you can read the number of dictionary elements that meet the criteria imposed by the data on the x-axis. The interesting points of this histogram are located on the envelope of the shaded figure. Any point selected from the envelope shows that there are y dictionary elements that have occurred x or fewer times in the training data set. Trying to understand the meaning of the histogram, it can be read that in the set based on Bach variations, one can distinguish many notes that appear rarely. There are about 100 notes that occur more than 100 times. In these 100 notes, some notes occur even less than 500 times. This plot confirms the non-uniformity of the set noted in the histogram (Fig. 1). Using the same methodology of histograms and tables, a dataset based on three jazz pieces is described. This set is described in Table 2 and histograms (Fig. 2). We can see that the ratio of words unique to all words in this set is approximately one to two. So, it differs significantly from the set of Bach variations.

Analyzing the histogram (Fig. 2), it can be concluded that the set of jazz pieces does not distinguish the notes contained in it. The dictionary histogram (Fig. 2), however, shows that more than half of the unique words appear in the data set up to 2–3 times. It is also evident that only about 100 dictionary words occurred more than 10 times in the dataset. This proves the high dynamics of this set.

TensorFlow models were trained with the presented datasets. Each model was trained on a single dataset. The end of the training was defined by the point where the value of the model's loss function did not change significantly. After training the

Table 2 Dataset based on jazz pieces

Parameter	Value
Number of MIDI files	3
Number of notes in the file	5545
Number of notes in the dictionary	2593

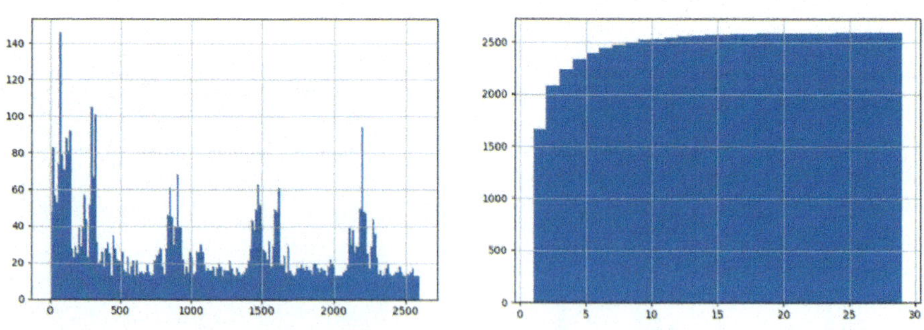

Fig. 2 Dataset histogram (left) and dataset dictionary histogram (right)—jazz pieces

Fig. 3 Fragment of an original piece by Bach (left) and generated piece in Bach style (right)

model, tracks were generated. On average, 3 pieces of 1000 notes were generated for each trained model.

Two generated musical examples will be presented. The first of the analyzed pieces were created using a model trained on a data set based on Bach variations. A fragment of the original composition contained in this collection is shown in Fig. 3. This fragment has a flowing melody, interspersed between two voices, accompaniment in the form of a harmonic complement in the accompanying voice—"lonely" notes visible, Bach's pure harmony, material free of false notes. The piece by Bach, although difficult to read from the sheet music alone, is present here.

Figure 3 shows a fragment of a generated piece based on pieces similar to the original by JS Bach. Not all notes follow the rules of rhythm and meter. This is most likely due to non-standard note lengths caused by rounding errors. In the audio material, this fact is not noticeable.

Main features: smooth melody passed between voices, accompaniment that well complements the harmony of the composition, false sounds that do not fit the context, repetition of short fragments contained in it. We can notice that the errors here are a distinct difference compared to the original work. The imperfections are caused by the uncertainty of the next note selection during generation. It is selected using a fixed distribution, new for each successively generated note. The repetition of the material is also noticeable. This may be due to the repetition of selected fragments between the tracks of the data set. The built TensorFlow model was characterized by a high memory capacity. Repeated fragments had a greater chance of appearing in the generated work because the model perceived such fragments as easier to match with other possibilities, in the current context of the work. Both the number of errors and the creativity of the model could be controlled by the temperature parameter present in the generation phase. In practice, however, this did not bring good results. The modifications of this parameter were difficult because the change in the behavior of the model resulting from the temperature change was abrupt. Adding other parameters controlling the discussed behavior of the model could significantly improve the obtained results. The bags generated on the Bach data sets were qualitatively good. It was hard to determine whether the piece we listening to came from a dataset or was a generator composition. The prepared TensorFlow model captured the dependencies of Bach's pieces well and successfully remembered them. The similar musical material of the tracks in the data set was the reason for a high-quality generation. Although the creativity of the model was limited and fragments of the generated work were often copied from the data set. The size of the copy and the size of the set did not allow us to easily notice these facts, which effectively confused the recipient in assessing whether he or she is listening to the original or generated work.

The second composition was generated based on a set of three jazz pieces characterized by a similar style. Figure 4 shows an excerpt from one of the three tracks in the training dataset. This piece has: visible at first glance, complicated melodic relationships and rhythmic, hard to notice in the notes, but audible in the listening session, proper control of the tension, and correct phrase guidance. Jazz bags are characterized by a complicated formal structure. The musical material of such works is rarely repetitive concerning the main melody. The accompaniment is often distinguished by the fact that it is constant throughout the piece. Cutting out fragments of a jazz piece and mixing them, and then putting them back together in a different order has a low chance of creating an interesting and valuable composition in this case. Figure 4 shows a fragment of a generated jazz piece. At first glance, it has a lot in common with the original. It is characterized by: drawn from examples, attempts at conducting melodies, phrases, rhythmic and harmonic relationships built based on examples, low repetition of its fragments, and a large number of notes that do not fit the context.

The greater variety of sounds in the jazz set compared to the Bach set makes it difficult for the model to combine successive elements. There was noticeable confusion in the generated jazz compositions, which were punctuated by larger or smaller fragments known from the set of training examples. This was possible due to the small number of tracks in the dataset. However, sometimes the taken constructions were modified, with a good impression when the sound material was analyzed. Large the number of false sounds was less noticeable in jazz pieces due to their complex structure. Nevertheless, the presence of sounds that did not match could be noticed. This is most likely due to a greater variety of sounds than in the case of Bach's works, and thus less certainty as to the choice of the next generated note. The cost of the jazz generation was moderate. On the one hand, they had a jazz character, on the other hand, they were very chaotic, which made it easy to see that they

Fig. 4 Fragment of original jazz song (left) and generated jazz-style song (right)

were generated artificially. It is easy to notice that the generation quality was highly dependent on the content of the training dataset. Generation using a model trained on a data set with similar track characteristics was performed with much higher quality compared to generation on a set with greater diversity. This is not a desirable feature of the system, but unfortunately unavoidable. To achieve a higher quality of generation, it was necessary to get acquainted with the training material in-depth and to reject examples that significantly disrupt the uniformity of the collection style. The important problem encountered in the generated compositions was the presence of long, uniform musical fragments. This type of part stretched for a long time. This was especially noticeable in jazz pieces. The presence of such musical material in the generated compositions significantly hindered their reception. The generation quality decreased drastically, which was easily noticeable when listening to the generated track. The increasing number of LSTM layers and the increasing number of cells per LSTM layer showed better and better memory capacity. It was hard to determine whether the generation quality with more accurate memory was always better. A higher degree of remembering the characteristics of the training data set is associated with their better mapping during generation, but it carries the risk of reducing the creativity of the model, which may result in generating uninteresting compositions. To examine the quality of the generated works, a survey was conducted with the participation of the respondents. The task presented in the study was to classify work into one of two categories: original song, and track generated. The survey was divided into two parts: Baroque and jazz pieces were examined separately. From the point of view of the created system, the most important is the number of incorrect classifications made by the respondents. Its high value means that the generated works seemed to be original, and the original ones seemed to be generated; the system tends to a high degree of uncertainty in user classification. The study shows—Table 3—a significant advantage of incorrect classifications over correct classifications. This means that the respondents had problems recognizing the type of work. The second part of the study concerned jazz works and was carried out in the same way as the first. Basic test data are presented in Table 3. It was decided to distribute the generated and original jazz pieces unevenly to distort the respondents' answers and thus force them to answer only according to their convictions. One can notice the opposite characteristics of the answers given about Bach's works. Correctly classified works predominate here, which gives grounds to conclude that the generator does not cope well with the generation of jazz.

6 Conclusions

Understanding both musical and machine learning issues was a demanding task. Music data in the form of MIDI files of Bach and jazz works were collected as sources of created datasets with a consistent style. Such sets were processed necessary from the point of view of teaching the TensorFlow model. In this process, the main emphasis was placed on creating an appropriate data format compatible with

Table 3 Reception on JS Bach and jazz pieces

Parameter	JS Bach	Jazz
Number of total tracks	8	8
Number of original tracks	4	2
Number of total tests	4	6
Number of correctly recognized	40	40
Number of correctly recognized	15	26
Number of incorrectly recognized	25	14

the TensorFlow API supporting the network training process. The architecture of the neural network, which is the heart of the system, was designed. The trained model was successfully used to generate music. Despite some problems, such as the looping of short fragments of the song, it was possible to generate high-quality music. The created model was tested in terms of generating quality. Using the available parameters, the optimal working point was adjusted through manual experiments. The assessment of the music generator's work consisted in subjectively determining its positive and negative features from the artistic point of view. The level represented by the generated music was compared to the quality of the music used in the training phase of the model. The music generator of acceptable generation quality is achievable without much time and money. The proposed solution is based on open-source technology. Thanks to this, potential further research can be conducted without artificial constraints. The problems that arose during the development of the system do not seem insurmountable. The largest number of them was related to the quality of the work generated by the system. The problem of composition looping was not common enough to occur in every generated song. A more serious impact on the quality had the level of creativity of the model in generating successive musical fragments. Experiments with the current parameters of the system allowed us to conclude that to solve the problem of low creativity it will probably be necessary to use an additional solution in the form of an algorithm forcing the generator to creative generation. Despite the difficulties encountered, it can be stated that the quality of generation of the complexity of the created system was satisfactory.

References

1. Briot, J.P., Pachet, F.: Deep learning for music generation: challenges and directions. Neural Comput. Appl. **32**(4), 981–993 (2020)
2. Briot, J.P., Hadjeres, G., Pachet, F.D.: Deep Learning Techniques for Music Generation. Springer Nature Switzerland AG (2020)
3. Herremans, D., Chuan, C.H., Chew, E.: A functional taxonomy of music generation systems. ACM Comput. Surv. (CSUR) **50**(5), 1–30 (2017)
4. Huang, C.F., Huang, C.Y.: Emotion-based AI music generation system with CVAE-GAN. In: 2020 IEEE Eurasia Conference on IOT, Communication and Engineering (ECICE), pp. 220–222 (2020)

5. Association, T.M.: Standard MIDI Files (SMF) specification. https://www.midi.org/specifications-old/item/standard-midi-files-smf (2020)
6. Bengio, Y., Simard, P., Frasconi, P.: Learning long-term dependencies with gradient descent is difficult. IEEE Trans. Neural Networks **5**(2), 157–166 (1994)
7. Zhao, K., Li, S., Cai, J., Wang, H., Wang, J.: An emotional symbolic music generation system based on LSTM networks. In: 2019 IEEE 3rd Information Technology, Networking, Electronic and Automation Control Conference (ITNEC), pp. 2039–2043 (2019)
8. Karpathy, A.: The unreasonable effectiveness of recurrent neural network. https://karpathy.github.io/2015/05/21/rnn-effectiveness/ (2015)
9. Zimmermann, H.G., Neuneier, R.: Neural network architectures for the modeling of dynamical systems. In: A Field Guide to Dynamical Recurrent Networks, pp. 311–350. IEEE Press, Los Alamitos (2001)
10. Mangal, S., Modak, R., Joshi, P.: LSTM based music generation system. arXiv https://doi.org/10.17148/IARJSET.2019.6508 (2019)
11. Hochreiter, S., Schmidhuber, J.: Long short-term memory. Neural Comput. **9**(8), 1735–1780 (1997)
12. Greff, K., Srivastava, R.K., Koutnik, J., Steunebrink, B.R., Schmidhuber, J.: LSTM: a search space odyssey. IEEE Trans. Neural Netw. Learn. Syst. **28**, 2222–2232 (2017)

Identification of the Language Using Statistical and Neural Approaches

Szymon Nagel, Magdalena Nagel, Rozalia Solecka, Julian Szymański, David Gil, and Higinio Mora

1 Introduction

Language Identification is a task in Natural Language Processing domain. For the last few years, the interest in research in the field of identifying languages has been rising due to high demand among new systems using textual data manipulation. The primary purpose of this research is to test the quality of the neural network approach in comparison with the statistical one for identifying what language the given text was written in. The statistical approach is based on comparing letter frequency distributions in a training set composed of texts that are written in different languages. We also test the application of a LSTM network that has been trained to learn the language representation and make a classification using a dense layer. As a data set, we use varied texts in 24 European languages written only in the Latin alphabet to limit the influence of characteristics like diacritic letters for a particular language.

S. Nagel · M. Nagel · R. Solecka · J. Szymański (✉)
Gdansk University of Technology, Narutowicza 11/12, 80-233 Gdansk, Poland
e-mail: julian.szymanski@eti.pg.edu.pl

S. Nagel
e-mail: s175818@student.pg.edu.pl

M. Nagel
e-mail: s175741@student.pg.edu.pl

R. Solecka
e-mail: s175735@student.pg.edu.pl

D. Gil · H. Mora
University of Alicante, 03690, San Vicente del Raspeig Alicante, Spain
e-mail: dgil@dtic.ua.es

H. Mora
e-mail: hmora@dtic.ua.es

W. Zamojski et al. (eds.), *Dependable Computer Systems and Networks*, Lecture Notes in Networks and Systems 737, https://doi.org/10.1007/978-3-031-37720-4_15

2 Current Approaches for Language Identification

Much research has already been led in the field of language classification. Researchers have carried out works under processing sign [4], speech [8], and also text [14] language using different approaches. For example in the domain of text classification, [10] describe n-gram methods for language identification. In this paper, the authors propose a method based on N-gram cumulative frequency addition (CFA). The proposed method is simpler than other mentioned techniques like Naive Bayesian, but reaches similar accuracy. The proposed CFA method works by summing up the frequencies of all N-grams in the text and usage of cumulative frequency built on the training set as features for classification.

The research on automatic language classification, using a computer, is inspired by the cognitive theories of how language is recognized by humans [5]. The usage of a language requires humans to learn it over a longer time The level of knowledge can differ if it is a native or second language. In order to identify language, knowledge of vocabulary, syntax, grammar, and structure is used by humans. What is interesting it was observed is without detailed knowledge about the language, people are able to properly classify language only based on specific patterns directly from the data [2].

Based on these observations, researchers became interested in mapping human mechanisms to automatic solutions. Various approaches to this problem have been tested over the years, including methods based on pure statistics and artificial intelligence, eg.: Gaussian Mixture Models [16], N-gram Models [2], Decision Tree Models [7], Artificial Neural Network Models [11], Support Vector Machines Models [17].

Gaussian Mixture Models (GMMs) are known to be a successful method of classification language for its identification. In GMM, each language unit is modeled as a probabilistic source with an unknown but fixed probability density function. In the training phase, the parameters of the probability density function are estimated from the training samples. For classification, the likelihood of the test with respect to the model is computed [12].

The n-gram method uses letter n-grams distributions, representing the frequency of occurrence of various n-letter combinations in a particular language. The language identification process can be divided into two phases. The purpose of the first one is to create a frequency distribution knowledge base for each language using the training set. In the second phase in order to classify language, distributions from two texts are compared using particular metrics. The classification is performed based on the most similar distribution [7].

Decision trees are used to determine the most likely language for each letter in the input word. The language is obtained by asking a series of questions about the context of the current letter, as determined by the corresponding decision tree. The data set, before selecting the most appropriate context for an input letter, should be partitioned to minimize entropy. Due to storing only the letter context, the representation is in a very compact form. Each node of the tree contains an attribute (information about the most appropriate context) and a tag representing the most likely language in the current context. The tree consists of a root node and internal nodes, each containing an attribute and a language tag, and lists containing only language tags. Once trained, decision trees can be used to identify the language of any word [7].

Support Vector Machines (SVMs) in the domain of language classification, were also used to classify the language of a given text. SVMs are trained on a dataset of texts from different languages, and they are then used to classify a given text into one of the languages in the dataset. The features used for language classification include character n-grams, word unigrams, and bigrams. The state-of-the-art SVM models used in language classification tasks are usually based on Support Vector Regression (SVR) and Support Vector Clustering (SVC). SVR is used to capture the relationship between the inputs and the output, while SVC is used to group the input data points into clusters. Both SVR and SVC can be used to train the classifier with the input data [9].

Deep Neural Network (DNN) models are a powerful tool for Natural Language Processing (NLP). These models in the convolutional setting have been successfully applied to language identification [6]. The other approach employed Recurrent Neural Network (RNN) models [3] that allows analyse of sequential data. In the context of language processing, RNNs are used to process letter sequences and thus capture contextual information. RNNs have been used for language identification by learning the character-level distributions of different languages.

DNNs have been used to create language classifiers that are capable of distinguishing between languages with high accuracy [11]. These models are trained on large datasets consisting of texts written in different languages. The model is then able to predict the language of an unseen text by analyzing its character-level distribution, but as it was reported requires sufficient pieces of a text to properly detect a language.

The performance of a language classifier is typically measured by its accuracy, which in our case is a measure of how many languages the model correctly identifies out of a set of test data. The accuracy of a language classifier is typically between 80 and 95% [13], depending on the degree of difficulty of the task resulting from the number of languages, their similarity, and the amount of data: both training and test, given at the input of the model.

Yet much research has been done in the field of language identification there is a lack of studies that compare the statistical and neural approach for the strict domain, e.g.: of the Latin alphabet as well as identifying the language families based on classification models. Also, there is a lack of studies on the influence of the input data size on the quality of the results.

3 The Data

It was necessary to ensure that the language classification task was sufficiently challenging. Therefore, we focus on European languages, that share the same alphabet and thus have more similarities between them than languages having specific letters that can characterise clearly the language.

Input data comes from randomly selected Wikipedia articles [15]. However, the prepared dataset contains only a small part of it—24 manually selected European languages divided into groups related to language families shown in Fig. 2. The data set is divided into two files:

- 'train.csv' contains 21600 lines of text, where each line belongs to one language in the format: 'id, text, lang'
- 'test.csv' contains 2400 lines of text, where each line belongs to one language in the format: 'id, text, lang'.

Finally, each language is represented by corpora consisting of 30,000 lines, which gives about 7,680,000 symbols for the data set. The codes of our experiments are available at https://kask.eti.pg.gda.pl/gitlab/julian/nlc.

4 Statistical Approach

In order to implement this step, a distribution of letter frequencies in the text was generated for each language. The frequencies of occurrence of letters were sum-up, and then the percentage distribution of a given letter in relation to the rest of the alphabet was determined.

The same approach was repeated for bigrams, which means each combination of the letter's pair, as well as trigrams—a combination of letter's triples. It should be noted increasing n in n-grams, produces very high and spare representation, which may be problematic for some models, but should allow the automatic detection of more characteristic letter sequences for a particular language.

An example unigram distribution chart for the Italian and English languages is given in Fig. 1.

Kullback-Leibler divergence given by Formula 1 was used to determine the similarities between the two probability distributions p and q [10].

$$KL(p, q) = \sum_{i=1} p(i) \log_2 \frac{p(i)}{q(i)} \tag{1}$$

The formula produces non-negative values and 0 value only if the compared distributions are identical. This measure is not symmetrical. When the distribution contains an element with a value of zero, normalization is applied so as not to perform a division by zero operation. Therefore, Kullback-Leibler divergence method was very suitable to compare statistical distributions of the letters.

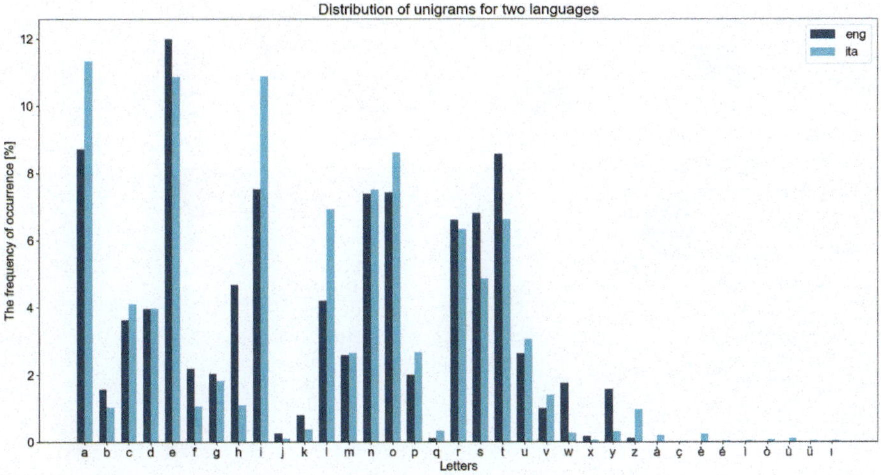

Fig. 1 Comparison of unigram distribution for Italian and English languages

4.1 Similarity Analysis

Graph Fig. 2 presents language families divided according to a linguistic point of view on the basis of their common characteristics, geographic location, or detected relationship. There are seven color-coded word families in 24 languages: Germanic, Balto-Slavic, Uralic, Italic, Hellenic, Semitic, Celtic.

On the other hand, similarities between languages based on the distribution of unigrams have been presented by the graph in Fig. 3. The numbers on the edges represent the degree of similarity (based on Kullback-Leibler divergence) between the languages represented by the vertices. The lower the number is, the more similar the languages are to each other and the line is stronger. If there is no edge between vertices, it means that similarity is insignificantly small and overlooked. The nodes are colored in pursuance of the language family.

From the graph, it can be seen that in languages that are similar in terms of statistical method, the majority belong to the same language families. For instance, Croatian and Slovenian as well as Portuguese and Spanish—languages that are known to be similar in general without detailed knowledge of them, have higher similarity than English and Spanish which differ more even at the first glance of an eye. It leaves no doubt that the similarities determined by the statistical method relates with the ones formed by the linguists.

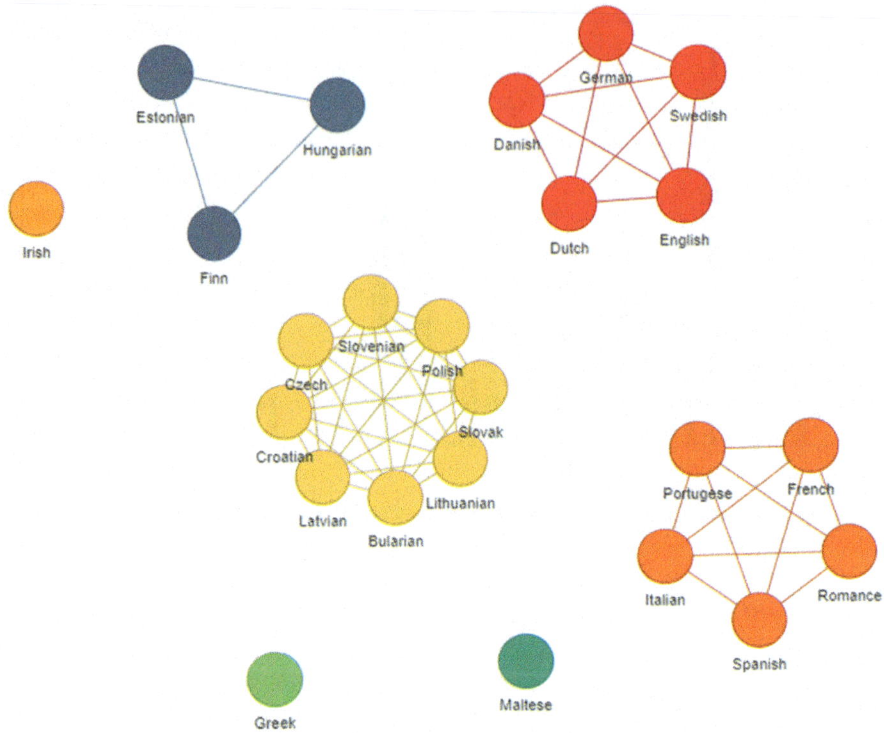

Fig. 2 Graphs of groups related to Language families used in our experiments

5 Neural Classification Approach

The development of neural networks allowes to apply this method to a variety of problems, including the task of language classification. In our approach, it was necessary to adjust the data so that it could be used as training data for the neural network. The concept behind recurrent neural networks is to employ sequential information. To create a sequence from a set of data, Tokenizer was used. This module creates a database of letters and assigns them a corresponding token, which is a numerical value. Unrecognized symbols receive the token <OVV>.

The neural network model used in our approach has the following architecture:

- Embedding layer (Length of registered tokens, 64 output dimension,400 input length)
- LSTM layer (32 neuron input, 32 neuron output)
- LSTM layer (32 neuron input, 32 neuron output)
- Dense layer (Number of languages for classification—24 neurons).

In order to preprocess the training and validating datasets, a Tokenizer was initialized, and the text was padded with a maximum length of 400. Using this Tokenizer,

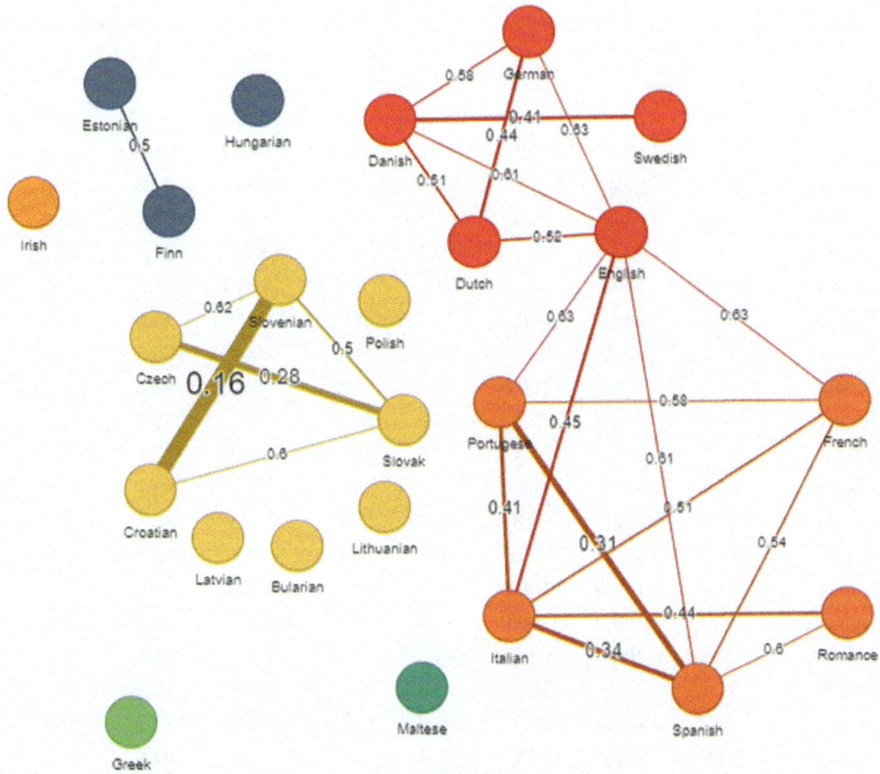

Fig. 3 Graph of language groups identified with similarity measures

the text was transformed into sequences, which were then padded. After the preprocessing stage, an Embedding layer with 64 nodes was added to the model in order to represent the words with a dense vector. Following this, two LSTM layers of 32 nodes each were added, followed by a Dense layer with the number of nodes equal to the number of languages in the dataset. Finally, the model was compiled using the sparse categorical cross entropy loss function and the Adam optimizer. The models were trained for 100 epochs, with batch size 32. Subsequently, the model achieved accuracy at level 0.96.

6 Comparison of Approaches and Result Analysis

By implementing both approaches, it was possible to compare the effectiveness, as shown in Table 1.

The statistical approach, as well as a neural network, reach similar efficiency at the level of 95–97%. Nevertheless, using n-grams, better results were achieved using

Table 1 Results of statistical and neural network models for 400-letter and 40-letter input sentences

	400-letter				40-letter			
Language	1 g (%)	2 g (%)	3 g (%)	NN (%)	1 g (%)	2 g (%)	3 g (%)	NN (%)
Bulgarian	**97,1**	**97,1**	**97,1**	95,1	91,2	91,2	91,2	**94,6**
Czech	97,2	**97,4**	97,2	96,6	58,3	75,0	**77,8**	62,0
Danish	**94,4**	**94,4**	**94,4**	93,3	61,1	66,7	72,2	**80,7**
German	90,0	**92,5**	**92,5**	90,2	70,0	85,0	**87,5**	40,7
Greek	**100,0**	**100,0**	**100,0**	98,8	**100,0**	**100,0**	96,0	98,0
English	97,8	**100,0**	**100,0**	90,7%	75,6	88,9	**95,6**	23,2
Estonian	92,7	92,7	92,7	**96,7**	53,7	73,2	**80,5**	57,8
Finnish	98,9	**100,0**	**100,0**	98,4	69,4	88,9	**97,2**	46,7
French	90,3	93,6	96,8	**96,9**	67,7	90,3	**93,5**	54,2
Irish	97,4	97,4	97,4	**97,5**	89,5	**94,7**	92,1	67,5
Croatian	90,7	97,7	**100,0**	88,4	69,8	76,7	**83,7**	32,4
Hungarian	97,8	97,8	97,8	**98,3**	80,0	**93,3**	86,7	80,3
Italian	93,0	90,7	93,0	**96,1**	60,5	74,4	**83,7**	43,4
Latvian	99,8	**100,0**	**100,0**	99,5	91,7	**94,4**	88,9	84,5
Lithuanian	99,9	**100,0**	**100,0**	99,8	79,5	**89,7**	84,6	60,6
Maltese	99,5	**100,0**	99,9	98,2	80,4	91,3	**93,5**	65,6
Dutch	91,7	99,1	**100,0**	93,6	52,1	66,7	**79,2**	18,2
Polish	99,2	**100,0**	99,7	99,6	87,5	**92,9**	89,3	67,2
Portuguese	**94,9**	**94,9**	**94,9**	91,4	64,1	79,5	**84,6**	69,3
Roman	97,6	97,6	97,6	**98,7**	63,4	73,2	**80,5**	68,3
Slovak	95,5	97,7	**100,0**	95,9	68,2	**68,2**	65,9	56,4
Slovenian	84,4	91,1	**93,3**	89,2	55,6	**84,4**	80,0	64,7
Spanish	97,7	98,7	**100,0**	97,5	65,1	83,7	**86,0**	31,6
Swedish	97,7	98,3	97,7	**98,7**	**79,5**	77,3	72,7	76,0
Average	95,6	97,0	**97,7**	95,8	72,2	83,3	**85,1**	60,2

a statistical approach. However, the second approach also achieved high language classification results, for some languages: Estonian, French, Irish, Hungarian, Italian, Roman, and Swedish even higher than the statistical approach.

It is worth mention that using unigrams, average efficiency is lower than using bigrams. For trigrams, the average efficiency result is the greatest out of the three. Only Czech, Polish, and Maltese trigrams caused the loss of efficiency in comparison to bigrams. We may conclude that there are language-specific combinations of letters (e.g. the ending *-ing*, *-ed* for English) on the basis of which a better classification can be conducted.

When both approaches were tested on shortened sentences, on average, the best results were also achieved for trigrams. However, both methods lost classification efficiency—average results are between 60,2 and 85,1%. For only two languages, the use of a neural network improved results—Bulgarian and Danish. An interesting observation is a negative effect of increasing the n-grams to 3 g for some languages, e.g. Greek, Irish. The negative effect can be especially observed when tested on short sentences.

7 Summary and Future Works

In this paper, a comparison between a statistical and a neural network approach to language identification was presented. The results obtained demonstrate that both approaches have similar efficiency when applied to text of a large size. However, when the text is shorter, the statistical approach outperforms the neural network approach. This is because the statistical approach is better suited to capturing the subtle differences between languages, whereas the neural network approach is better suited to capturing the overall patterns of language. It is worth mentioning that usage of trigrams gives the best results in terms of accuracy, suggesting that more complex patterns of language are better captured using this approach. It can be concluded that both approaches are effective for language classification, and each has its advantages and disadvantages.

Based on the obtained results and conducted analysis, we argue there are still some fields that may be more deeply investigated as future work on the topic of language classification. It is worth taking into consideration characteristics of languages represented with n-grams e.g.: like *-ing, -ed, -able -acy, -ance, -ate, -fy* in the case of English. Such knowledge appears to be promising and can contribute to the improvement of the classification system. In the model based on a statistical approach, it can be achieved using feature selection approaches, which can be compared with a selection that the neural network reaches by assigning the weights to particular features.

Another issue is to deeper study the dependencies between linguistic and computer science points of view. Partitioning languages into smaller and more detailed family groups could have a positive impact on overlapping mentioned similarity graphs, also, the reconstruction of the hierarchical structure may be tested.

We also plan to evaluate more sophisticated models for language identification, especially transformer-based neural networks, eg.: [1]. In that case, a detailed analysis of the influence of dataset size, used in the training phase, must be carried out to answer what and how the model improves in the task. In that study influence of large amounts of data should be analyzed as well as this approach allows to usage of additional information, that is outside of the training dataset especially transfer learning should be analyzed in detail. Also, in the test phase, the analysis of the input data size should be carefully investigated, as the results of our experiments show that statistical models outperform neural ones while there is a small input probe. As the size of the input data, during the test, is challenging for neural models, besides comparison to the statistical models, it is also worth considering dictionary-based approaches.

Acknowledgements The work was supported partially by funds of the Department of Computer Architecture, Faculty of Electronics, Telecommunications and Informatics, Gdańsk University of Technology.

References

1. Xlm-roberta transformer model for language detection (2020)
2. Ahmed, B., Cha, S.H., Tappert, C.: Language identification from text using n-gram based cumulative frequency addition. Proc. Stud./Fac. Res. Day, CSIS, Pace Univ. **12**(1) (2004)
3. Bartz, C., Herold, T., Yang, H., Meinel, C.: Language identification using deep convolutional recurrent neural networks. In: International Conference on Neural Information Processing, pp. 880–889. Springer, Berlin (2017)
4. Dziubich, T., Szymański, J.: Automatic classification of polish sign language words. Przegląd Elektrotechniczny **90**(3), 191–194 (2014)
5. Ellis, N.C.: Meta-analysis, human cognition, and language learning. In: Synthesizing Research on Language Learning and Teaching, pp. 301–322 (2006)
6. Ganapathy, S., Han, K., Thomas, S., Omar, M., Segbroeck, M.V., Narayanan, S.S.: Robust language identification using convolutional neural network features. In: Fifteenth Annual Conference of the International Speech Communication Association (2014)
7. Hakkinen, J., Tian, J.: N-gram and decision tree based language identification for written words. In: IEEE Workshop on Automatic Speech Recognition and Understanding, 2001. ASRU'01, pp. 335–338. IEEE (2001)
8. Huang, J.T., Li, J., Gong, Y.: An analysis of convolutional neural networks for speech recognition. In: 2015 IEEE International Conference on Acoustics, Speech and Signal Processing (ICASSP), pp. 4989–4993. IEEE (2015)
9. Joachims, T.: Learning to Classify Text Using Support Vector Machines, vol. 668. Springer Science & Business Media (2002)
10. Kim, T., Oh, J., Kim, N., Cho, S., Yun, S.Y.: Comparing Kullback-Leibler Divergence and Mean Squared Error Loss in Knowledge Distillation (2021). arXiv:2105.08919
11. Lopez-Moreno, I., Gonzalez-Dominguez, J., Plchot, O., Martinez, D., Gonzalez-Rodriguez, J., Moreno, P.: Automatic language identification using deep neural networks. In: 2014 IEEE International Conference on Acoustics, Speech and Signal Processing (ICASSP), pp. 5337–5341. IEEE (2014)
12. Mary, L.: Extraction and Representation of Prosody for Speaker, Speech and Language Recognition. Springer Science & Business Media (2011)
13. Simoes, A., Almeida, J.J., Byers, S.D.: Language Identification: A Neural Network Approach (2014)
14. Szymański, J.: Comparative analysis of text representation methods using classification. Cybern. Syst. **45**(2), 180–199 (2014)
15. Thoma, M.: Wili-2018-Wikipedia Language Identification Database (2018)
16. Torres-Carrasquillo, P.A., Reynolds, D.A., Deller, J.R.: Language identification using gaussian mixture model tokenization. In: 2002 IEEE International Conference on Acoustics, Speech, and Signal Processing, vol. 1, pp. I–757. IEEE (2002)
17. Zhang, W., Li, B., Qu, D., Wang, B.: Automatic language identification using support vector machines. In: 2006 8th International Conference on Signal Processing, vol. 1. IEEE (2006)

Smart Data Logger with Continuous ECG Signal Monitoring

Jan Nikodem, Ryszard Klempous, Konrad Kluwak, Dariusz Jagielski, Dorota Zyśko, Bruno Hrymniak, Jerzy Rozenblit, Thomas A. Zelniker, and Andrzej Wytyczak-Partyka

1 Introduction

Increasing workloads and economic pressures are forcing healthcare providers worldwide to improve operational efficiencies and introduce innovative healthcare models. To help address these challenges, we propose innovative technologies that will enhance healthcare as we expect them to become even more popular soon [1].

Wearable and IoT technologies adoption in healthcare has lagged. However, in recent years the fast-growing acceptance and adoption of IT solutions in many fields of medicine have been observed. This raises expectations that such trends will continue allowing for patient monitoring beyond hospital walls.

Data acquisition and storage is one of the critical technological enabler for creating genuinely connected and integrated IT infrastructures in healthcare. In tandem, we see a further proliferation of significant challenges caused by big-data problems such

J. Nikodem (✉) · R. Klempous · K. Kluwak · A. Wytyczak-Partyka
Wrocław University of Science and Technology, Wrocław, Poland
e-mail: jan.nikodem@pwr.edu.pl

R. Klempous
e-mail: ryszard.klempous@pwr.edu.pl

K. Kluwak
e-mail: konrad.kluwak@pwr.edu.pl

D. Jagielski · B. Hrymniak
Department of Cardiology, Center for Heart Diseases, 4th Military Hospital, Wrocław, Poland

J. Rozenblit
Department of Electrical and Computer Engineering, University of Arizona Tucson, Tucson, Arizona, USA

T. A. Zelniker
Division of Cardiology, Medical University of Vienna, Vienna, Austria

D. Zyśko
Department of Emergency Medicine, Wrocław Medical University, Wrocław, Poland

W. Zamojski et al. (eds.), *Dependable Computer Systems and Networks*, Lecture Notes in Networks and Systems 737, https://doi.org/10.1007/978-3-031-37720-4_16

as energy consumption, cooling and localization issues, reliable communication, and finally fast and effective data processing.

It is desirable to empower healthcare professionals with new technology, but we do not want to flood them by an ocean of bytes. Therefore, already at the stage of data acquisition, we propose innovative actions that allow replacement of raw data files with structured sets of aggregated information with annotations.

We consider a smart data logger collecting samples of slow-changing signals provided from sensors placed on the human body. The typical ones include physiological signals ECG, EMG, EEG, PPG, and sensors detecting body fluids such as blood, saliva, sweat, tear, or interstitial fluid sensors, as well as position sensors and accelerometers.

The biomedical sensors used by data loggers belong to the group of wearable devices. They are of very different types, ranging from electrodes collecting electrical signals, as well as electrochemical, optical (photoplethysmogram colorimetry, fluorescence). Such sensors provide an analog conditioned waveform signal that is sampled by the data logger. In the context of diagnostic requirements, conditioning allows for a sufficiently effective and satisfactorily precise signal sampling. Still, it requires several functionalities, depending on the type of measurement and sensor used. Therefore, there is no universal system for conditioning analog signals obtained from different kinds of sensors. For this reason, apart from sensors and conditioning systems, we will deal with a digital signal obtained from a 16-bit analog-to-digital converter (ADC). In this work, we focus not on conditioning/converting biomedical signals, but on their storage, pre-processing, and making them available for further analysis.

Aggregation of information and its annotation requires some functionalities, depending on the type of measurement and sensor used. Therefore, because there is no universal way to solve these problems for different types of biomedical signals, in this work we focused on the aggregation and annotation of ECG signals [2, 3, 5].

2 Smart Acquisition of ECG Data

Over the years of practice and experience, many different ECG electrode constellations been developed, differing in the number and location of electrodes on the patient's body. In this work, we will deal with a single bipolar lead corresponding to the first limb lead of Einthoven's triangle, assuming that proposed solution can also be applied to any number of channels.

The smartness of the proposed data logger consists in adaptive and selective processing of the collected data presented in the Fig. 1.

To reduce noised data the use of stochastic filters is effective enough. Moreover, due to the signal conditioning on the analog side, we do not have many of noise in the collected ECG signal samples [13]. For data with deterministic disturbances, we used long impulse response recursive integer filters because they do not require performing a long convolution. In addition, integer filters can run on simple and

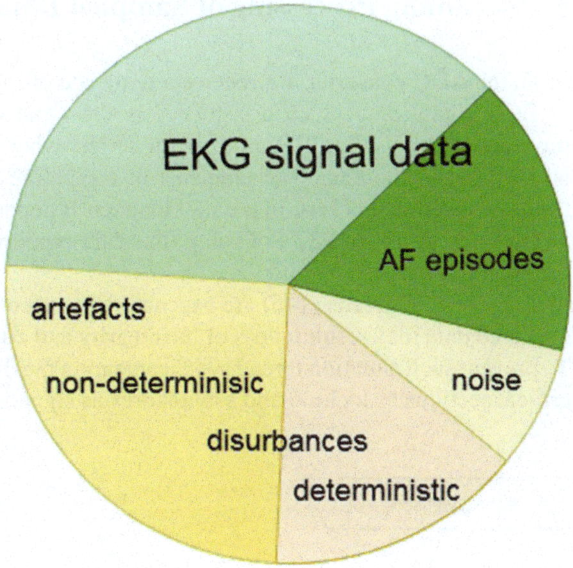

Fig. 1 The structure of data sampled by the data logger

slow microprocessors and in power-saving sleep modes. Detailed information can be found in the following chapters.

Adaptability comes down to the parameterized operation of algorithms, especially in terms of detecting non-deterministic disturbances. The described device samples ECG signal during the patient's everyday life activity, which makes it significantly different from those sampled in laboratory or hospital conditions. So, there are many more physiological artefacts in sampled ECG signal (i.e., spontaneous electrical activity of peripheral muscles) and day activity artifacts (i.e., electrical activity of patient mobility muscles, changes of skin-electrode impedance or ESD discharges), which practically do not occur in laboratory or hospital ECG records.

Selectivity, on the other hand, means that we focus on data that allows us to catch atrial fibrillation episodes. For this purpose, firstly in the sampled ECG signal, we annotate the R wave as a reference point, then we determine the Q wave and the S wave, resulting in QRS complexes. We extract from the supplied record what is important, i.e., the so-called fragments of the record, which are continuous sequences of multiple QRS complexes (i.e., from the circle in Fig. 1 we focus only on the green shades one). Secondly, using simple morphological and spectral discriminators, we determine the set of atrial fibrillation (AF) episodes (i.e., concerning the green shades in Fig. 1 we focus only on the darkest one).

3 Streaming Processing of Sampled Data

From the ADC converter, we receive a sequence of byte values represented samples of ECG signal data. Such a sequence is the input element for the data pipeline procedures (1–5) (Fig. 2) connected in series, where the output of one procedure is the input of the next. The elements of a pipeline can be executed concurrently (in time-sliced manner) or in parallel manner. It depends on the hardware used, and in our case, using ESP32 SoC chip, the data pipeline is executed concurrently in different threads.

Because the threads (1–5) are executed concurrently, hence simultaneous access to shared data may result in loss of its integrity and consequently in program failure. To prevent such situation thread synchronization mechanisms are used. Mutexex and critical sections as locking and semaphores as signaling mechanisms.

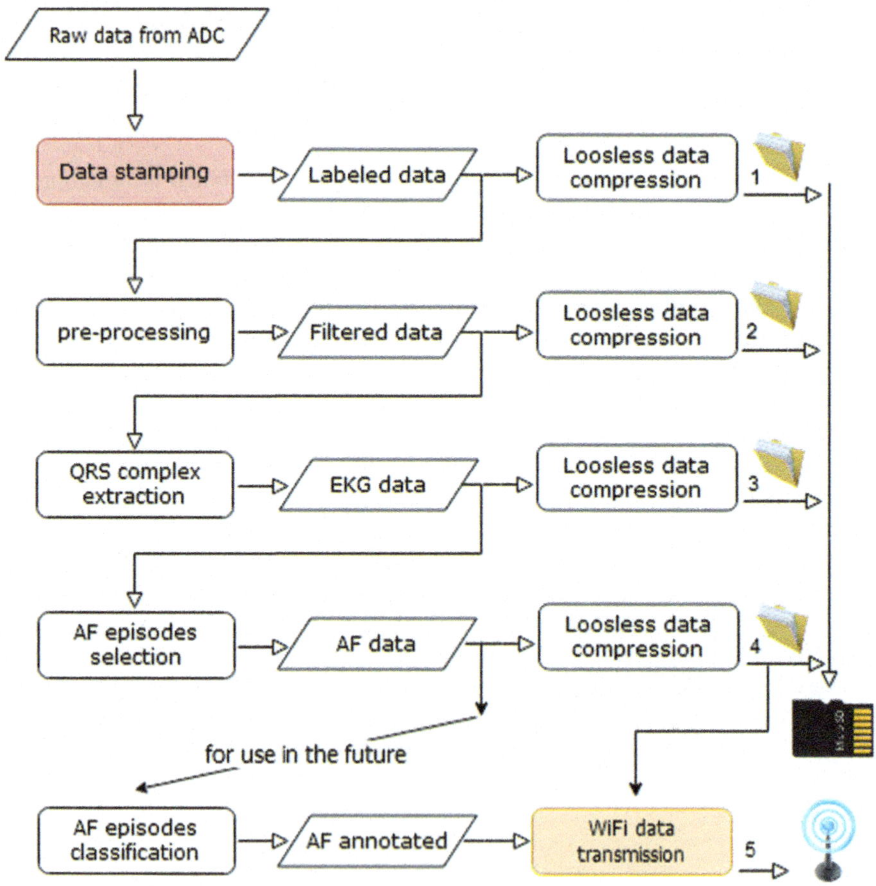

Fig. 2 The sampled data streaming procedures

The execution of threads (1–5) within the data pipeline procedures is conditioned by the readiness of the input data and the availability of resources necessary to perform the calculations. Moreover, two of the treads (1 and 5, marked in color in Fig. 2) are additionally time-dependent. The thread (1)—data sampling and stamping is (due to constant sampling frequency) strongly time-depended, but the thread (5)—WiFi data transmission is (due to WiFi transmission requirements) weakly time-dependent. Therefore, they both work on interrupts. The first (1) with the highest priority and the fifth (5) with the medium. Other (2–4) threads are assigned the lowest priority.

The raw data we receive from the ADC is a sequence of byte value samples, which should be supplemented with organizational information, such as the identifier of the measurement channel or the time stamp of the sample. This process (data stamping) provides additional context for sampled data, and as a drawback, it increases the collected file size by a small amount. In this way, we get labeled data that we make available as input to the next procedure in the stream and save it on the micro SD card. The saved data serve as a reference point for evaluating the performance of individual procedures during pipeline processing.

3.1 Lossless Data Compression

The execution of an algorithm in each thread (1–4) within the data pipeline routine results in output data, which are passed on to the next thread to perform further computation. For the purpose of debugging process and to ensure the ability to evaluate the calculations efficiency in each thread, the results of its operation (output data) are saved in non-volatile memory (micro SD card). However, since we perform these activities often and on big data sets, we use selected source coding methods before saving them.

From the wide range of sources coding methods, we choose lossless methods of data compression that ensure a reduction in the number of bits necessary for the binary representation of this data. As a criteria for the encoding efficiency, we applied both compression factor and speeding up the implementation or transmission. One of the tested algorithms is arithmetic coding (AS), the other is the lossless compression algorithm—Asymmetric Binary System (ABS), based on the family of entropy coding methods—Asymmetric Numeral Systems (ANS). The AS algorithm, as well known and relatively simple to implement, was chosen as the benchmark. The ABS algorithm turned out to be more demanding to perform but allowed faster implementations (in Phyton language) with a similar compression factor.

The projection used in the AS and ABS algorithms only modifies the form of the record without changing the set of data contained in this form. The data has therefore remained unchanged, but its storage on the micro SD card requires less memory space.

The second property of the coding used in AS and ABS is a bijection, thanks to which it is possible to unambiguously reverse the whole process (decoding)—to decode the final state back to the initial sequence of symbols.

As a result of lossless data compression, they are stored in non-volatile memory. This enables their use for debugging and further evaluation of processing results, which will be discussed below.

3.2 Deterministic Disturbances Filtering

During continuous monitoring, ECG signal acquisition is very often corrupted by different disturbances (Fig. 1). Non-deterministic disturbances result mainly from the patient's daily activity and are difficult to eliminate. We will deal with this issue in the next chapter.

It is simpler with the elimination of deterministic disturbances. Using adequate filtration we can neutralize the effects of such phenomena as baseline wandering or power-line interference. The filtering is realized with software procedures that should preserve the most significant ECG frequency components and eliminate the disturbance frequencies. In order to reduce the deterministic disturbances influence as much as possible, we choose long impulse response recursive integer filters which are conformed to requirements for real-time operation at simple and slow microprocessors and in power-saving sleep modes.

Recursive filters are very fast, but have less efficiency and flexibility than other digital filters. One of the most effective filtering mechanisms are convolutional techniques. However, this efficiency is usually associated with the need to perform long convolutions. This, in turn, is computationally intensive and slows down real-time performance, which we want to avoid. Therefore, the filtering method we use is based on simple convolutions (the set of coefficients $a_0, a_1, \ldots, a_k$) forming the convolution kernel is composed of several elements) and is combined with the use of additional information from the filtered signal. This additional information constitutes a feedback loop responsible for a specific aggregation of information. In this way, our filtering algorithm will be determined by the difference equation (1)

$$y[n] = a_0 x[n] + a_1 x[n+1] + a_2 x[n+2] \cdots + a_k x[n+k] - \\ b_1 y[n-1] - b_2 y[n-2] \cdots - b_j y[n-j] \quad (1)$$

To describe the behavior of proposed filters (1), i.e., to describe the relationship between the coefficients a, b of the recursion, and the response of the filter y, we use z-transform, which is an equivalent of the Laplace s-transform but used to describe discrete signals. By transforming equation (1) we get:

$$H(z) = \frac{Y(z)}{X(z)} = \frac{a_0 + a_1 z^{-1} + a_2 z^{-2} \ldots + a_k z^{-k}}{1 - b_1 z^{-1} - b_2 z^{-2} \ldots - b_j z^{-j}} \quad (2)$$

Based on zeros and poles of a transfer function (2), we determined the difference equation for the high-pass, single pole filter, which eliminates baseline wandering. For the cutoff frequency requirements of –3 dB and 2.44 Hz, we obtain:

$$a_0 = (1 - x)/2 = 0.985 \qquad a_1 = -a_0 = -0.985$$
$$b_1 = x = 0.970 \tag{3}$$

Comparison to the Pan-Tompkins filter [8] known from the literature with the following parameters:

$$a_0 = -0.03125 \qquad a_{16} = 1 \qquad a_{17} = -1 \qquad a_{32} = a_0$$
$$b_1 = 1 \tag{4}$$

indicates that the proposed filter (3) requires less computation (3 non-zero coefficients versus 5 coefficients in Pan-Tompkins filter) and requires a smaller buffer capacity with sampled data (range of a requires only one sample, while Pan-Tompkins requires 32 consecutive samples).

3.3 Identification of QRS Complexes

The operating conditions of the developed data logger which online and continuously monitoring the ECG signal, are radically different from ECG records obtained in outpatient or hospital conditions. Three aspects are crucial: significant occurrence of non-deterministic disturbances and considerable computational and energy limitations. By definition, software based detection of non-deterministic disturbances is difficult. Therefore, among all collected ECG samples, we are looking for QRS complexes, and finally we treat a set of them as ECG signal data (Fig. 1). Other elements not identified as QRS complexes are treated as non-deterministic disturbances and are not subject to further analysis.

Many effective algorithms for identification of QRS complexes can be found in the literature [4, 10–12]. We deployed the solution proposed in [12]. Deployed solution meets second and third constraints satisfactorily. The effectiveness of identifying the QRS complex among artifacts and non-deterministic disturbances (verified on the basis of ECG records [6]) has been confirmed. Now, we are waiting for ECG records from a real ECG data logger to assess the effectiveness also on ECG records containing much more non-determined disturbances than these in outpatient or hospital conditions.

4 Conclusions

The discussed results indicate that with the current technological offer of System on Chip systems, data loggers can not only collect data but also process this data on an ongoing basis. However, the computational possibilities offered by new technologies should be used wisely. It's not about collecting faster and more data, drowning us in an ocean of bits. Data should be collected wisely. First, only data containing an information, not noise, disturbances and garbage. Only those data that are relevant and only as much as is necessary (*Entia non sunt multiplicanda praeter necessitatem*).

Data selection/validation and information aggregation may (current hardware and software technology allow for this) take place up to date, in the place of their acquisition, because the site and conditions of data acquisition determine the best methods of their further processing.

Parameterized procedures of lossless data compression, filtering of deterministic disturbances are universal methods that can improve the efficiency of any data logger [7]. However, each data logger that is to aggregate the collected information must have specific procedures reflecting the type and scope of information collected. For ECG signal monitoring, these are modules of identification of QRS complexes, atrial fibrillation episodes selector, and (on which we will work in the future) atrial fibrillation episodes classificator [14].

References

1. Hadjem, M., Salem, O., Naït-Abdesselam, F.: An ECG monitoring system for prediction of cardiac anomalies using WBAN. In: 2014 IEEE 16th International Conference on e-Health Networking, Applications and Services, pp. 441–446. Natal, Brazil. https://doi.org/10.1109/HealthCom.2014.7001883
2. Ingole, M.D., Alaspure, S.V., Ingole, D.T.: Electrocardiogram (ECG) signals feature extraction and classification using various signal analysis techniques. Int. J. Eng. Sci. Res. Technol. **3**(1), 39–44. ISSN: 2277-9655
3. Joshi, S.L., Vatti, R.A., Tornekar, R.V.: A survey on ECG signal denoising techniques. In: IEEE International Conference on Communication Systems and Network Technologies, pp. 60-64. IEEE. https://doi.org/10.1109/CSNT.2013.22. ISBN 978-0-7695-4958-3/13
4. Köhler B. U., Hennig C., Orglmeister R.: The Principals of Software QRS Detection, IEEE Engineering in Medicine and Biology (2002)
5. Martis, R.J., Rama, K.M., Chakraborty, C.H., Pal, S., Sarkar, D., Mandana, K.M., Kumar, R.A.: Automated Screening of Arrhythmia Using Wavelet Based Machine Learning Techniques. J. Med. Syst. **36**, 677–688 (2012). https://doi.org/10.1007/s10916-010-9535-7
6. Massachusetts Institute of Technology. MIT-BIH ECG database. Available at, http://physionet.org/physiobank/database/mitdb/
7. da, S., Luza, E.J., Schwartzb, W.R., Càmara-Cháveza, G., Menottia, D.: ECG-based heartbeat classification for arrhythmia detection: A survey. In: Computer Methods and Programs in Biomedicine 127, pp. 144–164. Elsevier Ireland Ltd. https://doi.org/10.1016/j.cmpb.2015.12.008
8. Pan, J., Tompkins, W.J.: A real-time QRS detection algorithm. IEEE Trans. Biomed. Eng. BME-**32**(3), 230–236 (1985). https://doi.org/10.1109/TBME.1985.325532. PMID 3997178

9. Sametinger, J., Rozenblit, J.W., Lysecky, R., Ott, P.: Security challenges for medical devices. Commun. ACM **58**(4), 74–82 (2015). https://doi.org/10.1145/2667218
10. Sansone, M., Fusco, R., Pepino, A., Sansone, C.: Electrocardiogram pattern recognition and analysis based on artificial neural networks and support vector machines: a review. J. Healthc. Eng. **4**(4). Multi-Science Publishing 2016. ISSN 2040-2295
11. Singh, S., Gandhi, N.: Pattern analysis of different ECG signal using Pan-Tompkin's algorithm. Int. J. Comput. Sci. Eng. **02**(07), 2502–2505 (2010). ISSN : 0975-3397
12. Valtino, X.A.: ECG QRS detection. In: Biomedical Digital Signal Processing, pp. 236–264 (1993)
13. Villegas, A., McEneaney, D., Escalona, O.: Arm-ECG wireless sensor system for wearable long-term surveillance of heart arrhythmias. Electronics **8**(11), 1300 (2019). https://doi.org/10.3390/electronics8111300
14. Xing, D., Rozenblit, J.W., Bernau, S., Ott, P.: Cardiac arrhythmia visualization in a virtual heart for electrophysiology education. In: Proceedings of the 2014 Summer Simulation Conference, Modeling and Simulation in Medicine, pp. 152–157. CA (2014)

Movement Tracking in Augmented and Mixed Realities Impacting the User Activity in Medicine and Healthcare

Jan Nikodem, Ryszard Klempous, Jakub Segen, Marek Kulbacki, and Artur Bąk

1 Introduction

Nowadays, it is difficult to find in medicine and healthcare, areas not supported by computers. In addition to the management and administration activities, one of the largest areas is visualization, i.e. enhancing the visual message. Computer visualization helps the implementation of medical procedures (e.g. in physical therapy), and supports surgeons before and during operations, as well as raises the level of medical education in this area.

In diagnostic, magnetic resonance imaging (MRI), computer tomography (CT), positron emission tomography (PET), Roentgen visualization or coronary angiography is based on the image presented to a person on a monitor.

Similarly, inserting the camera into the patient's body (in laparoscopy, endoscopy, arthroscopy etc.) [3] gives a view of its internal organs on the monitor screen. Thanks to that the surgeon sees in high magnification what in nature is sometimes difficult to see with the naked eye.

Augmented Reality (AR) and Mixed Reality (MR) offer a lot of possibilities for healthcare [1, 12]. Many of valid solutions in medical education, surgical visual-

J. Nikodem (✉) · R. Klempous
Faculty of Information and Communication Technology, Wrocław University of Science and Technology, Wrocław, Poland
e-mail: jan.nikodem@pwr.edu.pl

R. Klempous
e-mail: ryszard.klempous@pwr.edu.pl

J. Segen · M. Kulbacki · A. Bąk
Polish-Japanese Academy of Information Technology, Warsaw, Poland
e-mail: segen@ieee.org

M. Kulbacki
e-mail: mk@pja.edu.pl

A. Bąk
e-mail: abak@pja.edu.pl

M. Kulbacki
DIVE IN AI, Wrocław, Poland

W. Zamojski et al. (eds.), *Dependable Computer Systems and Networks*, Lecture Notes in Networks and Systems 737, https://doi.org/10.1007/978-3-031-37720-4_17

isation [9], speeding up recovery in physical therapy or post-stroke rehabilitation [6] use AR and MR increasingly. Augmented Reality can assists medical staff in a operating rooms, outpatient departments and rehabilitation centers [8]. In simulators Virtual Reality (VR) is primarily used, however it is progressively being replaced by AR and MR.

Virtual Reality, AR and MR techniques has undergone significant advances and has changed the way medical education, healthcare, and medical operations are performed [1]. Technological advances have produced progressively better visualisation instruments and higher-quality imaging [10]. What's more, technology has provided us with new personal visualization devices (smart glasses), which has clearly moved the scope of the image's impact closer to the user.

Medical devices that incorporate augmented and mixed reality technologies need a new approach adequate to the latest challenges. We notice three major challenges: human-computer interface, visualization techniques and movement tracking [10, 11]. In this paper we focus on latter one. We focus on movement tracking as a means for keeping track of the point to make virtual objects visible in a real scene on smart glasses which combine the virtual and real world with the user's eye.

2 The Problem Formulation

Today technology provides us with new personal visualization devices like Head Up Display (HUD), which has definitely moved the scope of the image's impact closer to the user. Moreover, just visualizing and showing the user what is sometimes difficult to see in nature with the naked eye, is no longer satisfactory, we expect more from AR, MR. We want it, as equipped with artificial intelligence to have the features of a digital assistant that suggests the user a desired (in accordance with medical procedures) solutions.

The above trends can be observed most clearly in diagnostic, where computer-aided analysis of MRI, CT, PET images is already a fact. Equally clear are these trends in medical education where computers suggest optimal paths for medical instruments in simulators of minimally invasive techniques.

Current miniaturization and cutting-edge technology make personalised displays (smart glasses) equipped with System on Chip (SoC). These devices have low power consumption and effective computing possibilities. Therefore, these smart glasses are no longer just displays but are slowly becoming intelligent digital assistants enriching the real image with additional, computed information.

The additionally displayed information is a kind of front-end of the working software whereas the intelligence of the digital assistant is contained in the back-end, where certain features, classifiers to recognize and classify some objects or actions are incorporated. One of the main scope, from which the results obtained feed the artificial intelligence of the digital assistant, is visual movement tracking. The implemented software allows the HUD to display the optimal path for the movement of medical instruments (usually in simulators) and to signal the entering of the instrument tip in the no-fly-zone (supporting surgeons during operations). However, the appearing errors committed by the device in positioning and overlapping the virtual object with the real object reproduction cause disorientation of the brain and thus disturb the coordination of actions in the eye-brain-hand feedback. Thus, in the further chapters we cope with this issue.

3 Intelligent Movement Tracking

A trajectory is a set of points in space-time which combines the three dimensions of space and one dimension of time.

$$\mathcal{T} = \{\overline{a}_1, \overline{a}_2, \ldots, \overline{a}_k \mid \overline{a}_i \in \mathbb{R}^4,\ i \in \{1, 2, \ldots, k\},\ k \in \mathbb{N}\} \tag{1}$$

$$where\ \overline{a}_i = [x_i, y_i, z_i, t_i]; \quad \text{and } t_i \text{ is a time} \tag{2}$$

$$and\ q_i = (x_i, y_i, z_i); \quad \text{where } q_i \text{ is a position vector} \tag{3}$$

An important feature of the time concept is its strict monotonic property. And as such, it is suitable for introducing a strict order (>) which determine the sequence in which events unfold in time, creating the *path* as time-dependent sequence of points $P_i = (x_i, y_i, z_i)$ in configuration space. Each P_i is represented by position vector (3) that uniquely determine the position of the point P_i in Euclidean space.

$$\mathcal{T}_e = \{\overline{a}_1, \overline{a}_2, \ldots, \overline{a}_k \mid \overline{a}_i \in \mathbb{R}^4,\ i \in \{1, 2, \ldots, k\},\ k \in \mathbb{N}\} \tag{4}$$

$$\mathcal{T}_1 = \{\overline{b}_1, \overline{b}_2, \ldots, \overline{b}_l \mid \overline{b}_i \in \mathbb{R}^4,\ i \in \{1, 2, \ldots, l\},\ l \in \mathbb{N}\} \tag{5}$$

Movement tracking is an operation of comparing two trajectories (1), the first of which (4) is the reference trajectory $\mathcal{T}_e$ (a movement pattern). Therefore, in each comparison, we have a trajectory prepared by an expert corresponding to the desired global behavior and rules specifying the conditions for accepting local actions. Due to realize an intelligent movement tracking, an expert defines the features vector structure and the feasible ranges of its components. He also determines the classification rules to recognize and classify some situations on the compared trajectory (5).

3.1 Matching up a Samples of Two Trajectories

Let's choose a norm that enable us to compare two trajectories occurring in the $\mathbb{R}^4$ space (1). Various choices are found in the literature [2], but since the space $\mathbb{R}^4$ is described by Euclidean geometry, the Euclidean norm seems to be the most natural. Now, the comparison of trajectory $\mathcal{T}_1$ with $\mathcal{T}_e$ requires a mapping function that match up pairs, and then (according to the accepted norm) determines the distance between it elements and sum up the total distance of all component.

$$\mathcal{M}_{\mathcal{T}_e,\mathcal{T}_1} = \{ < \overline{a}_i, \overline{b}_j > \ | \overline{a}_i \in \mathcal{T}_e, \ \overline{b}_j \in \mathcal{T}_1 \} \tag{6}$$

One of the obvious match up functions is a bijection that guarantees one-to-one correspondence, but it requires the following condition: $card(\mathcal{T}_1) = card(\mathcal{T}_e)$. Most often, however, the compared trajectories have different lengths i.e., $i \neq j$. In such case, we search a match up function (6) for which the required minimum is that to be a surjective or we use relations.

In the movement tracking scope, we come across two types of algorithms:

(I) *passive movement tracking (a posteriori)*, in medical simulators when after completing the medical procedure, the system signals errors and gives a grade,
(II) *active movement tracking (in real time)*, when the result of trajectory comparison produces feedback in real time manner.

The basic, well known method used to solve the match up problem (3.1). is Dynamic Time Warping (DTW), which automatically cope with time deformations and different speeds associated with time-series samples [5]. Preserving the strict monotonic property of time, this method minimizes the effects of shifts and distortions of the trajectory by *flexible* transformation of the time signal. This *flexibility* deforms trajectories in order to achieve an optimal alignment between them. Dynamic Time Warping tries to make the paths similar and does not judge whether and to what extent they are comparable. However, guaranteeing the minimum costs of proposed match up process implicates that there is no better much up result. A detailed description of DTW algorithm can be found in [4].

The DTW method works a posteriori, i.e. matches up sample indexes on two already taken trajectories. Therefore, it is difficult to implement such trajectory comparison in real time mode. Performing trajectory comparison in real time we can realize match up procedure in time as well as space scopes.

(A) Time-domain mapping "*you are here*" on the $\mathcal{T}_1$ realized trajectory (5), "*but at this moment you should be there*" according the $\mathcal{T}_e$ reference trajectory (4).
(B) Projection in space "*you are here*" on the $\mathcal{T}_1$ realized trajectory (5), "*but the nearest position is there*" on the $\mathcal{T}_e$ reference trajectory 4).

The visual movement tracking in strong time-dependent medical procedures prefers time-domain mapping (A). However, when precision is crucial, then projection (B) is better choice. Both, the mapping (A) and projection (B) reflect the real time situations and allow system for immediate feedback.

3.2 Comparison of Trajectories

Possess set (6) we can proceed to compare the realized $\mathcal{T}_1$ trajectory (5) with the reference $\mathcal{T}_e$ trajectory (4). While creating a reference trajectory, the expert built a feature vector and on the basis of it he defined the rules for the classifier of points position on the trajectory (objects). These rules, presented as decision table form, constitute a classifier.

The feature vector is not required to be homogeneous. Its components may be global features (trajectory length L, time of passage Tp) and local features (point location) as well as static and dynamic features (e.g. speed of point to point movement). The most common used feature vector (7) consists of six components: four local features and two global.

$$\overline{\mathcal{FV}} = [x, y, z, t, L, Tp] \tag{7}$$

The vector (7) consists of three spatial (x, y, z) features, time component t, two global L, Tp features mentioned above. Sometimes it is extended with two dynamic components, $\overline{v}$ as a velocity and $\overline{P}$ as momentum vector but we won't be dealing with that in this paper.

The classification involves the knowledge of the classes to which individual objects should be assigned. These classes must be non-overlapping and mutually exclusive. Let's consider a multi-valued classifier for which four following classes are defined:

<1> red class—to far from reference trajectory,
<2> green class—accurately on reference trajectory,
<3> yellow class—verging on the reference trajectory,
<4> white class—no sufficient data for classification.

The classification rules presented in the form of a decision tree and written in pseudo-code are as follows:

$$\begin{array}{lrlcc} \text{for any pair } < \overline{a}_i, \overline{b}_j > \text{ compute } \mathcal{S}(\overline{a}_i, \overline{b}_j) & & & & \\ & \text{if } \mathcal{S}(\overline{a}_i, \overline{b}_j) & = NAN & \longmapsto & <4> \\ & \text{else if } \mathcal{S}(\overline{a}_i, \overline{b}_j) & < 0.8 \cdot B & \longmapsto & <2> \\ & \text{else if } \mathcal{S}(\overline{a}_i, \overline{b}_j) & < B & \longmapsto & <3> \\ & \text{else} & & \longmapsto & <1> \end{array} \tag{8}$$

where $B \in \mathbb{R}$ is a value determined boarder between fly- and non-fly zones, and $\mathcal{S}(\overline{a}_i, \overline{b}_j)$ means similarity relation between two vectors $\overline{a}$ and $\overline{b}$. In normed vector space the vector similarity is usually expressed as the distance between two vectors.

Comparison of the realized $\mathcal{T}_1$ trajectory (5) with the reference $\mathcal{T}_e$ trajectory (4) is performed as the series of comparisons previously matched up pairs of points (6). These comparisons incorporate feature vector (7), classification rules (8) and the idea of coordination space. On Fig. 1 a coordination space is traditional Euclidean 3D-space where the points forming the trajectory are uniquely determined by the position vector (3). We choose 3D space to simplify the visualisation, but generally a number of feature vector (7) components determine the dimension of that space.

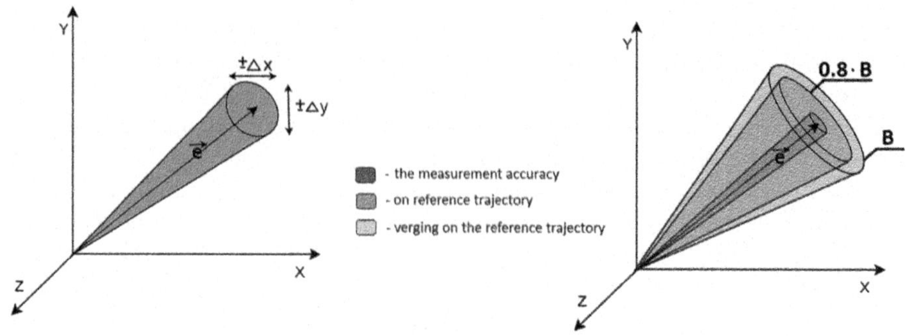

Fig. 1 Graphical representation of classification rules (8)

f not time but the precision of execution is important in the implementation of a medical procedure (see maching up procedure (B) described in Sect. 3.1), we reduce the feature vector by eliminating two components t, T_p from it. Then the considered space will be 4D.

In such, described above space, we propose two rules of vectors similarity $\mathcal{S}(\overline{a}, \overline{b})$. The first of its defines the similarity of two vectors $\overline{a}$, $\overline{b}$ as the length of vector resulting from it subtraction.

$$\mathcal{S}(\overline{a}, \overline{b}) = \|\overline{a} - \overline{b}\| = \sqrt{\sum_{i=1}^{7}(a_i - b_i)^2} \tag{9}$$

Two vectors are the more similar to each other, the more their substract is close to zero. The measure (9) is non-negative and if it is equal to zero, the compared vectors are identical.

The another way of defining the similarity of two vectors is based on their dot product.

$$\mathcal{S}(\overline{a}, \overline{b}) = \cos(\alpha) = \frac{\overline{a} \otimes \overline{b}}{\|\overline{a}\| \cdot \|\overline{b}\|} = \frac{\sum_{i=1}^{7}(a_i \cdot b_i)}{\|\overline{a}\| \cdot \|\overline{b}\|} \tag{10}$$

where α is angel between vectors, $\otimes$ is a dot product, and $\|\overline{a}\|$ means the length of vector $\overline{a}$.

Two vectors are the more similar to each other, the more their orientation and sense are the same (but not necessarily the length). Orientation and sense together determine the direction of a vector. The value of measure (10) is between $<-1, 1>$ and if it is equal to one, the directions of compared vectors are identical.

4 Discussion

Virtual Reality used in medical simulators has undergone a significant changes during the last 30 years evolving towards Augmented Reality and more recently Mixed Reality technologies. These changes were caused primarily by the development of technology and the new software that followed it. However, the eye-brain-hand coordination is still a basis of these systems.

The AR achievements extended the application area to the other issues of medical and healthcare needs. Physicians, patients, and caregivers can incorporate AR/VR to help them prepare for, or perform, certain treatments or procedures. AR technology, showing the user a real image enhanced by a virtual, imposes on the software the requirement of coordinating these two worlds, but the required precision is not exacting. As in the case of using a computer mouse, the correction of acceptable inaccuracies in the position and velocity of the cursor on the screen is corrected by a human (eye-brain-hand coordination is efficient enough for this task) for whom stability of these quantities is more important.

The Head Up Display (HUD)-based MR technology being more expansive has reinforced AR requirements. The user receives a virtual image on the HUD even when he focuses on the real image. Therefore the coordination of the virtual and real take on special importance and its lack or imperfections have a negative impact on user activity.

In the work, we focused on the movement tracking, which is one of the basic ones when enhance visual message by additional information about the desired behavior (movement). The described in chapters () methods developed for AR indicate that the software is not a problem. It can ensure coordination of the virtual and real world with an accuracy acceptable for many applications. However, these methods require one coordinate system as reference to work well, and we have problem with this in mixed reality

With the expansion of AR to MR technology, we opened Pandora's box. To make virtual objects visible in a real scene and to reflect the position and distance of the object under the user's gaze we expand software methods by spatial anchors, frames of reference, spatial mapping, etc. In this scope, even for one point perspective, the virtual world scaling (naturally to user intuition) methods require improvement. We have to do it because information about desired and performed movement must be coherent. Any discrepancy forces the user brain to correct the visual information. As we know, in such a case, the most common brain scenario is to turn off one of the messages from the virtual or real world.

5 Conclusions

In this paper we focused on Augmented and Mixed Realities technology issues use in medicine and healthcare. Appropriate technology, hardware and software for better accuracy and stable tracking with less costs are the prerequisites for a good tool.

However, user experience matters and cannot be forgotten because it shapes the user activity. The undoubted benefits of introducing the AR and MR technologies in medicine and healthcare are noticeable in the societies of physicians, surgeons, caregivers and patients, however adverse phenomena inhibiting development in these areas are also signaled.

Users from clinical-grade, education and training, post-operative and other therapies rehabilitation areas report about risks related to the usability of the device as eyes, neck and shoulders pain. Furthermore risks related to the quality and images that the AR/MR technologies provide, are reported. It's things like a low contrast, dizziness, fatigue, negative effects on vision, perspective errors such as location or depth of anatomy, trajectory and information overload.

We can deal with information overload by building balanced, intelligent interfaces with the ability to adapt them to the individual user and work environment. The key to increasing the interest of devices that incorporate AR/MR is to guarantee feedback which improved quality of the offered solutions.

Visual interface problem is more sophisticated and difficult to solve. Head Up Display (HUD) forces frequent changes of focal length, staring alternately at strongly illuminated objects located close, and hidden in the dark when they are in the distance. With HUD appliance, the crucial is not only a problem of overloading the brain, but mainly the limitations of a channel that provides this information, i.e. the human eyes.

Mixed Reality display technologies which combine the virtual and real world with the user's eye should ignore neither the physiology of looking (eye) nor the physiology of seeing (brain). If not, the list of complaints is long started from Sicca syndrome, transient myopia, image blurring, reduced visual contrast, burning eyes, feeling oppressed, dull eye pain, tearing eyes, conjunctival redness etc.).

Concerning eye accommodation, i.e. the adjusting ability of the optical system of the eye enabling sharp vision at different distances. Eye accommodation is affected by fatigue, illuminance and age. The far point is getting closer but the near point is getting away. Typical impact of age to the near point distance varies from 8 cm for 16 yr. to 100 cm for 60 age. However, a Head-Mounted Displays (HMD) typically perform holograms on 3–5 cm distance from eye. How far is this situation from healthy physiology and ergonomics of looking?

The omission during the Mixed Reality systems presentation of phenomena such as near point of accommodation, the optic chiasm, eyes convergence, stereopsis and depth perception in visual cortex, do not imply positive connotation in medical societies. The situation is getting even worse by the prevalence and impact of FUD marketing. In our opinion this is the one of main reasons for a slow (considering the powerful capabilities of MR tools) development of the use of these technologies in medicine and healthcare. The second is that while Augmented Reality over the years of development has already reached its mature level, Mixed Reality still requires improvements to approach a similar level.

References

1. Greenleaf, W.: Virtual Reality & Augmented Reality Systems. The Impact on Clinical Care. Patient Engagement and Advisory Committee Meeting on Augmented Reality and Virtual Reality Medical Devices. FDA 2022. https://www.fda.gov/advisory-committees/advisory-committee-calendar/july-12-13-2022-patient-engagement-advisory-committee-meeting-announcement-07122022
2. Han Su, H., Liu, S., Zheng, B., Zhou, X., Zheng, K.: A survey of trajectory distance measures and performance evaluation. VLDB J. **29**, 3–32 (2020). Springer Nature 2020. https://doi.org/10.1007/s00778-019-00574-9
3. Jiang, H., Xu, S., State, A., Feng, F., Fuchs, H., Hong, M., Rozenblit, J.: Enhancing a laparoscopy training system with augmented reality visualization. In: Society for Computer Simulation International, SpringSim-MSM, 2019, Tucson, AZ, 2019. Society for Modeling & Simulation International (SCS) (2019). https://doi.org/10.5555/3338264.3338280
4. Keogh, E.J., Pazzani, M.J.: Derivative dynamic time warping. In: First SIAM International Conference on Data Mining, Chicago (2001)
5. Kulbacki, M., Bąk, A.: Unsupervised learning motion models using dynamic time warping. In: Intelligent Information Systems, pp. 217–226. Sopot, Poland (2002)
6. Marzec, M., Olech, M., Klempous, R., Nikodem, J., Kluwak, K.J., Chiu, C., Kołcz, A.: Virtual reality post stroke rehabilitation with localization algorithm enhancement. In: Bruzzone, A.G., et al. (eds.) 5th International Conference of the Virtual and Augmented Reality in Education, VARE, pp. 28–35. DIME Universitádi Genova (2019). ISBN: 978-88-85741-42-3; 978-88-85741-41-6
7. Melińska, A., Czamara, A., Szuba, Ł, Będziński, R., Klempous, R.: Balance assessment during the landing phase of jump-down in healthy men and male patients after anterior cruciate ligament reconstruction. Acta Polytech. Hung. **12**(6), 77–91 (2015). https://doi.org/10.12700/APH.12.6.2015.6.5
8. Pang, J.C.Y., Tsang, R.S.W.: Reliability of three-dimensional motion analysis during single-leg side drop landing test after anterior cruciate ligament reconstruction: an in vivo motion analysis study. Hong Kong Physiother J. **4**2(1), 65–73 (2022). https://doi.org/10.1142/S1013702522500081. Epub 2022 May 30. PMID: 35782700; PMCID: PMC9244601
9. Sánchez-Margallo, J.A., Plaza de Miguel, C., Fernández Anzules, R.A., Sánchez-Margallo, F.M.: Application of Mixed Reality in Medical Training and Surgical Planning Focused on Minimally Invasive Surgery, Frontiers in Virtual Reality, vol. 2 (2021). https://doi.org/10.3389/frvir.2021.692641. ISSN:2673-4192
10. Smith, R.T., Clarke, T.J., Mayer, W., Cunningham, A., Matthews, B., Zucco, J.E.: Mixed reality interaction and presentation techniques for medical visualisations. In: Rea, P.M. (ed.) Biomedical visualisation. Advances in experimental medicine and biology, vol. 1320. Springer, Cham (2020). https://doi.org/10.1007/978-3-030-47483-6_7
11. Syed, A., Siddiqui, M.S., Abdullah, H.B., Jan, S., Namoun, A., Alzahrani, A., Nadeem, A., Alkhodre, A.B.: In-depth review of augmented reality: tracking technologies, development tools, AR displays, collaborative AR, and security concerns. Sensors **23**(1), 146 (2023). https://doi.org/10.3390/s23010146
12. Viglialoro, R.M., Condino, S., Turini, G., Carbone, M., Ferrari, V., Gesi, M.: Augmented reality, mixed reality, and hybrid approach in healthcare simulation: a systematic review. Appl. Sci. **11**(5), 2338 (2021). https://doi.org/10.3390/app11052338

General Provisioning Strategy for Local Specialized Cloud Computing Environments

Piotr Orzechowski and Henryk Krawczyk

1 Introduction

In general, service providers establish a service level agreement (SLA [1]) covering the general terms and conditions in which they will work with customers. The SLA is not only a set of conditions for service providers, but it could be also a source of benefits for customers. The contract between the provider and customer describes different characteristics of the service, which makes the services comparable between different providers. The SLA should also contain methods of redressing service issues. Other topics mentioned in SLA documents include:

- client expectations according to his/her needs,
- detailed descriptions of every service offered, under all possible circumstances, with the turnaround times included,
- definition of quality measurement metrics and quality level assurance,
- compensation or payment if the provider cannot properly fulfill this SLA.

Cloud computing [2] is mainly built on top of virtualization, as cloud users typically rent virtual resources from cloud providers. A popular form of virtualization is the use of virtual units (virtual machines, containers) which are created to run on a host machine (typically a physical server). Thanks to this, cloud architectures integrate IT environments and share scalable resources across a network to deliver an online platform on which client applications can run. In general, cloud systems

P. Orzechowski (✉)
Centre of Informatics Tricity Academic Computer and Network, Gdansk University of Technology, Gabriela Narutowicza 11/12, 80-233 Gdansk, Poland
e-mail: porzechowski@task.gda.pl

H. Krawczyk
Faculty of Electronics, Telecommunications and Informatics, Gdansk University of Technology, Gabriela Narutowicza 11/12, 80-233 Gdansk, Poland
e-mail: hkrawk@pg.edu.pl

W. Zamojski et al. (eds.), *Dependable Computer Systems and Networks*, Lecture Notes in Networks and Systems 737, https://doi.org/10.1007/978-3-031-37720-4_18

are highly complex, as they deal with a range of distributed components, users, and deployment scenarios.

In the literature, different enhancement and approaches to cloud management are considered and various models are proposed [4]. In general, the service provider is responsible for managing the resources to fulfill the requests generated by users. Service providers employ suitable algorithms to manage the incoming client requests (services) and to manage their virtual resources efficiently. Management strategies make it possible for providers to maximize revenue by utilizing their available resources up to their limits. In practice, in terms of the performance of cloud computing resources, the choice of management strategy makes a pronounced difference.

Our consideration and experiments focus on the implementation of a provisioning algorithm for local cloud computing, with the assumption that these local providers possess a more limited number of services and resources available for clients. Typical local cloud architectures are implemented through commonly available open-source software (Unix, OpenStack, Kubernetes). The presented models have been simulated hosting the TASKcloud [5] service (based on OpenStack software) which is a cloud computing service developed and deployed in our institution. The paper focuses on some aspects related to the main management strategies regarding the allocation and provisioning of resources (virtual units). They can be defined in different ways under the accepted assumptions related to clients' requirements, cloud architecture, models of services and resources, and optimization criteria. Resource Allocation refers to the allocation (reservation) of a pool of resources represented by virtual machines or containers (virtual resources – VR) to satisfy the SLA previously accepted by both the user and the cloud provider, while Resource Provisioning is the effective provisioning of a portion of the reserved resources to execute the fixed set of services notified by the user. Figure 1 explains the proposed approach. When a cloud provider accepts a request from a customer, it has to create the appropriate number of VRs and allocate user services to run. A typical example of resource provisioning is the deployment of a new virtual machine by the consumer, which uses a subset of the physical resources to run the single service. Due to virtualization, we can allocate resources flexibly, based on the current client demands. Then we can estimate the optimal values of resources for a concrete client demand. In general, this will be a much lower value than the value estimated based on the SLA.

The paper considers local specialized clouds, such as the mentioned TASKcloud, and considers a resource provisioning strategy for a lower number of services and resources compared to the global clouds deployed by such Big Tech firms such as Google, Microsoft and IBM. In such a case, deterministic provisioning algorithms can be considered. Their behavior can be examined for different optimization criteria, i.e., load balancing, consolidation, and fault tolerance. The impact of many parameters related to heuristic algorithms can be eliminated this way. Moreover, a parallel version of the algorithm to increase its speed and to consider mixed optimization criteria has been prepared. The paper presents that mixed criteria could have a significant impact on the execution time compared to a single criterion. It makes it possible to estimate

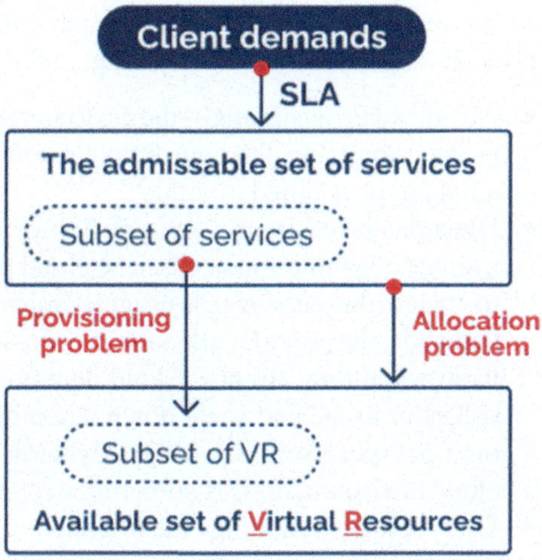

Fig. 1 Considered scope of cloud management. *Source* Authors

the assurance cost of t-faults tolerance, where t is the maximal number of faulty virtual units in the cloud.

The structure of the article is as follows. First, in Sect. 2, the state-of-the-art in the field of the practical implementation of management strategies in local cloud environments is discussed. In Sect. 3, the model of the assumed provisioning problem is described. Three criteria and the universal parallel algorithm to solve load balancing, consolidation, and fault tolerance problems are defined. The experiments are described, and the test results of the algorithm are discussed in detail in Sect. 4. Section 5 concludes the article.

2 Categories of Cloud Provisioning

Organizations can manually provide whatever resources and services they need, but public cloud providers offer tools to provision multiple resources and services, for instance: AWS CloudFormation, Microsoft Azure Resource Manager, and Google Cloud Deployment Manager. Such solutions concern global clouds and primarily concentrate on the administrative problems of cloud provisioning and offer many tools to support whole organizations rather than single customers. This paper focuses on local specialized clouds and considers the methods of allocating a cloud provider's resources and services to customers. In the current considered provisioning problems, the customer signs a formal contract of service with the cloud provider [7]. The provider prepares the agreed-upon resources or services for the customer and delivers them. It is a process that can be conducted using one of three delivery models. Each delivery model differs depending on the kinds of resources or services the organization purchases, how and when the cloud provider delivers those resources or services, and how the customer pays for them. The three models are advanced provisioning, dynamic provisioning, and user self-provisioning:

- Advanced Provisioning—the customer requests services from the provider and the provider prepares the appropriate resources in advance. The customer is charged a flat fee or is billed monthly,
- Dynamic provisioning—the customer can purchase cloud resources based on average consumption needs. The cloud provider deploys and adjusts the resources to match the customer's usage demands. Based on the customer's fluctuating demands, the provider allocates more resources when they are needed and removes them when they are not. Cloud deployments typically scale up to accommodate spikes in usage and scale down when demand decreases. The customer is billed on a pay-per-use basis. When dynamic provisioning is used to create a hybrid cloud environment, it is sometimes referred to as cloud bursting,
- User self-provisioning (also called cloud self-service)—the customer buys resources from the cloud provider through a web interface or portal and the cloud provider makes these resources available shortly after purchase. This usually involves creating a user account and paying for the resources with a credit card. Those resources are then quickly spun up and made available for use—within hours, if not minutes. Examples of this type of cloud provisioning include an employee purchasing cloud-based productivity applications via, e.g., the Microsoft 365 suite or G Suite.

A self-service provisioning model helps to streamline users' requests and manage cloud resources but requires strict rules to ensure they do not provision resources they should not. In this paper, the focus is on this type of provisioning model. The following metrics can be distinguished within the above provisioning approach:

- Scalability—there is no requirement for forecasting infrastructure needs; organizations can simply scale up and scale down their cloud resources based on short-term usage requirements,
- Provisioning speed—developers can quickly spin up a set of workloads on demand, removing the need for an IT administrator who provisions and manages the compute resources,
- Cost savings—many cloud providers allow customers to pay for only what they consume. However, the attractive economics presented by cloud services can present its own challenges, which organizations should address in their cloud management strategies.

It is a well-known fact that resource over-provisioning can cost users more than necessary and resource under-provisioning hurts application performance. In general, it is a complicated optimization problem, and there is a wide research avenue available for solving this. In the paper [8], some details about various optimization techniques for resource provisioning are presented. It has been shown that the cost-effectiveness of cloud computing strongly depends on how well the customer can optimize the cost of renting resources (Virtual Machines) from cloud providers. In the above paper, a framework is proposed. It is inspired by a cloud layer model, to enable the optimal provision of resources by combining the concepts of autonomic computing, linear regression and Bayesian learning. The efficacy of the proposed framework is evaluated using both the CloudSim toolkit and real-world workload traces from Google, followed by the traces from Clarknet. Such parameters as response time, SLA violations, virtual machine usage hours and cost were evaluated. In the paper [9], the authors propose a concrete solution for migrating physical servers to a cloud with the usage of the Azure cloud framework. The utilization of physical server resources on remote VM servers is considered. Such a migration process in the framework was implemented in two phases: first by integrating physical servers into virtual ones by creating virtual machines, and then by integrating the virtual servers into cloud service providers in a cost-effective manner. Two virtual machine instances were created using Microsoft Hyper-V on Windows Server 2016 R2. Applications that were installed on a workstation were migrated to the VM and the performance of this VM was monitored using a PowerShell script. Then Tableau was used to generate load and do analytical calculations to evaluate the physical server functionality.

The above papers concentrate on the IaaS level, considering VM-based environments, where a hypervisor will strictly allocate resources to the deployed VMs. The deployed VMs, however, can compete for the shared physical resources, but the hypervisor should detect and prevent this to not violate SLA requirements. In this paper, a more general approach is proposed to mitigate these constraints, where cloud services are assigned to virtual units. A dynamic provisioning approach is presented, and a deterministic algorithm is proposed. As was mentioned in Sect. 1, three different optimalization criteria are analyzed and some results are given. Moreover, the implementation of the algorithm is prepared in such a way that it can be used in the TASKcloud environment. Deployment of TASKcloud is fully automated using advanced Ansible playbooks and it can be adapted to change the scheduling mechanisms.

3 Model of Cloud Environments to Optimize Provisioning Strategies

As was shown in the previous section, there are many proposals on how to build a suitable provisioning strategy for cloud environments. However, heterogenous resources and the different methods of their cooperation, as well as the diversity, variability, and unpredictability of the required workload, and different needs of various cloud users make universal, simple, and effective methods most useful. They can be formulated based on general models, which can be used at different levels of cloud management strategies.

Consider a set P of permutations of services and virtual resources; their number can be $P = n * m$, where m is the number of all possible services and n is the number of all possible virtual resources. For the exact solution of the provisioning problem, all possible allocation modes must be reevaluated and the best mode chosen.

Due to the large number of exponential modes, the problem is an example of a set packing problem which is of NP-complete type. Moreover, in the proposed method, the provisioning of a set of services at one point in time is considered instead of queued services that are provisioned one by one. The assumed notation is presented below:

- $S = \{s_1, s_2, \ldots, s_m\}$—is the set of user demands representing services waiting to run in the computing cloud, where s_i, $i = 1, 2, \ldots, m$ is a user demand to run the $i - th$ service. It can be a single task or a scenario of tasks.
- $R = \{r_1, r_2, \ldots, r_n\}$—is the set of virtual resources available in the cloud, where r_j represents the $j - th$ resource which belongs to one of the cloud servers. Each virtual resource is supported by some physical resources described by computing capabilities, such as computational power, storage, and cloud services.
- $\psi(S, R)$—is the allocation matrix of required services S to the available cloud resources belonging to R, in brief:
- $\psi = [\psi_{ij}]$, where:

$$\psi_{ij} = \begin{cases} 1 - if\ service\ s_i\ is\ allocated\ to\ cloud\ resource\ r_j \\ 0 - otherwise \end{cases} \tag{1}$$

where $i \in \{1, 2, \ldots, m\} \wedge j \in \{1, 2, \ldots, n\}$

- δ—is the vector representing the current rate of use of all cloud resources R, before allocation of the new services from S, i.e., $\delta = [\delta_1, \delta_2, \ldots, \delta_n]$, $\delta_i \in \langle 0, 1.2\rangle$. In further considerations, a small 20% over-provisioning is allowed.

In practice, the rate of use, as a ratio of the amount of currently occupied resources to the amount of all resources available, can be calculated. In simple cases, the percentage of virtual machines currently active to all available space can be used here.

Let $\gamma = [\gamma_{ij}]$ be the vector determining the rate of the $j - th$ resource use when the $i - th$ service can be assigned to it; $\gamma_{ij} \in \langle 0, 1\rangle$. Note that γ can be calculated in the same way as δ. In practice, it means the percentage of engaged resources.

The load of the *j-th* resource after the allocation of services S on resources R according to ψ can be calculated in the following way:

$$\delta'_j = \delta_j + \sum_{i=1}^{m} \psi_{ij} \cdot \gamma_{ij} \tag{2}$$

Let T be the matrix of the given execution times for all services running on all resources, i.e.:

$$T = [t_{ij}] \tag{3}$$

where t_{ij} denotes the processing time of service s_i on resource r_j. It can be calculated empirically either by testing or by estimating the service properties and the characteristics of the resources.

Note that it is true where $\gamma_{ij} \leq 1 - \delta_j$, otherwise the processing time t_{ij} can be increased according to the following formula:

$$t'_{ij} = \left(\delta_j + \gamma_{ij}\right) \cdot t_{ij} \tag{4}$$

The execution time of all services S running on resources R for allocation $\psi(S, R)$ is denoted by $\tau(S, R)$. If the services are to be processed sequentially, then:

$$\tau(S, R) = \sum_{i=1}^{m} \sum_{j=1}^{n} t_{ij} \cdot \psi_{ij} \tag{5}$$

For the parallel execution of all services (which is possible when $n \geq m$):

$$\tau(S, R) = max_{i=1,\ldots,m} \sum_{j=1}^{n} t_{ij} \cdot \psi_{ij} \tag{6}$$

In general:

$$max_{i=1,\ldots,m} \sum_{j=1}^{n} t_{ij} \cdot \psi_{ij} \leq \tau(S, R) \leq \sum_{i=1}^{m} \sum_{j=1}^{n} t_{ij} \cdot \psi_{ij} \tag{7}$$

Let us consider provisioning problems for three different optimization criteria: load balancing, consolidation, and fault tolerance. Load balancing algorithms are used to distribute new demands of users (services) among the virtual resources to guarantee a well-balanced load across all cloud nodes. In contrast, consolidation is usually achieved by spreading the service workload over a smaller set of resources so the servers remaining unused can be powered down or put into standby mode. The first approach minimizes the total execution time of the set of services, in other words, to maximize the use of their resources at a lower overall client cost to increase their profit. The second approach copes better with highly fluctuating demands from clients. Moreover, having a set of free servers (nodes) that is not currently needed also makes it possible to design fault-tolerant systems. In that case, we plan the execution of each of the user tasks on more than one node. Below, we discuss the criteria in more detail and provide formal optimization criteria.

Load balancing algorithms are used to distribute new requests of users (services) in a cloud between the virtual units to guarantee an equal number of services allocated to each cloud server. However, each client demand can be expressed by service workloads to run on virtual units. Then, load balancing is a mechanism to balance the load by uniformly distributing the workload among the nodes [10]. Effective load balancing mechanisms will optimize the utilization of resources and improve the cloud's performance. There are various implementations of such mechanisms based on different load balancing algorithms [11]. In [12], capacity planning methods for cloud users and cloud service providers, and algorithms that combine the capabilities of different strategies which are more efficient, are considered. In consequence, load balancing algorithms seek to distribute service workloads across several virtual machines in a manner that minimizes the average time taken to complete the execution of those workloads, which typically results in server utilization being maximized and balanced.

The optimization problem for load balancing is defined as looking for (S, R), which minimizes $\tau(S, R)$, subject to:

$$\begin{cases} \delta_i \cong \delta_j \forall_{i \neq j} i, j\{1, 2 \ldots, \mathrm{n}\} \\ \sum_{i=1}^{m} \psi_{ij} 0 \\ \sum_{j=1}^{n} \psi_{ij} = 1 \\ \gamma_{ij} \leq 1 - \delta_j \end{cases} \tag{8}$$

Workload consolidation aims at maximizing the usage of servers by grouping services to run concurrently on fewer virtual units. Workload consolidation is one way to reduce resource wastage by clustering services on a subset of the pool of available machines. This technique is used to maintain control over the potentially high economic and environmental cost [13]. Many different approaches have been proposed for workload consolidation, but it is unclear which of the proposed approaches works best in each situation. In the paper [14], the authors showed that consolidation algorithms, whose goal is to maximize the number of empty physical machines, perform many virtual machine migrations, named eager migrations. These migration processes have a significant impact on the response times of the services deployed on those machines. The authors propose a new method and a heuristic to decide which virtual machines should be migrated. This solution takes into account the variability of the sizes of the virtual machines and prioritizes virtual machines with a steady capacity to be migrated first. In the paper [15], the authors proposed a solution to allocating a set of services based on a bin packing problem. The described framework is a semi-online workload management system which gathers incoming user requests to start a workload and packages them into sets. Then a whole group of services is taken into account during the allocation process. Such an allocation policy produces a saving of up to 40% of the resources compared to other consolidation algorithms.

For consolidation, the optimalization criterion of maximizing the number of empty nodes (value |I|) is proposed:

$$I = \left\{ j | \psi_{ij} = 0_i \right\}, \text{ |I| is cardinality of } I \tag{9}$$

subject to:

$$\gamma_{ij} \leq 1 - \delta_j \tag{10}$$

$$\forall j, j\{1, 2, \ldots, \mathrm{n}\}\backslash \mathrm{I} \sum_{i=1}^{m} \psi_{ij} 1 \tag{11}$$

$$\forall j, jI \sum_{i=1}^{m} \psi_{ij} = 0 \tag{12}$$

A fault-tolerant system works on one of two strategies. The replication strategy assumes that service replicas are running for each service in parallel and the result is obtained by majority voting. Alternatively, the redundant strategy assumes that the redundant servers or virtual units reside on an inactive mode unless and until any fault tolerance system demands their availability. Thus, if one part of the system fails, it has other instances that can be used in its place to keep it running. Extensive research efforts are consistently being made to implement fault tolerance in cloud infrastructures: the paper [16] gives a systematic and comprehensive elucidation of different fault types, their causes and various fault tolerance frameworks used in cloud implementations. Recently, cloud computing-based environments have presented new challenges to support fault-tolerance and opened up new paths to develop novel strategies, architectures, and standards. In the paper [17], the needs and solutions of fault-tolerance in cloud computing are discussed and future research directions specific to the development of cloud computing fault-tolerance are enumerated. In further considerations, an assumption has been made that t fault tolerance means that resources are allocated in a manner such that the impact of t failures (e.g., failures of virtual units) on the system performance is minimized or unimportant. The optimalization problem for t-faults tolerance, where $n \geq 2t + 1$ can be defined as minimizing $\tau(S, R)$, can be expressed as:

$$\sum_{j=1}^{n} \psi_{ij} \geq t + 1, \; \forall_i \in \{1, 2, \ldots, m\} \tag{13}$$

$$\delta'_j \leq 1.2 \tag{14}$$

As can be seen, each optimalization problem given above can be solved either separately or in different combinations, depending on the user needs. There are many available options based on genetic algorithms or artificial intelligence. They differ in some assumptions related to the features of the user requests and the services and capabilities of the cloud resources. In general, to minimize the total running time, the following properties are considered:

- For user needs—requirements contained in SLA agreements should be considered during management processes and their implementation requires consideration at three levels: global, local, and operating system level. In this report, we investigate the local level,

- For services—they are deterministic, their processing time preemptive without precedence constraints regarding the order of services, and each service cannot be further split into smaller subtasks,
- For resources—the processing capacity of the node remains unchanged but bounded, i.e., a limited number of services can be processed in sequential order of provisioning. The number of resources (nodes) can be invariant according to the user needs.

4 Experiments and Results

To evaluate the provisioning strategy, the following parallel provisioning algorithm is proposed:

Algorithm

```
Input Data: S,R,δ,T

do in parallel:
  create all allocations for ψ(S,R)
  select allocations satisfying criteria (8), (9), (13)
end
make selection of the best allocations
Output Data: ψ(S,R) and τ(S,R) according to the selected
criteria
```

Because the above algorithm is NP complete type, despite the proposed parallelization, only configurations with a maximum of 8 services and 4 resources (nodes) have been analyzed. It has been assumed that each service has its copy, which is a backup in case of a node failure. Services are named s_i and their copies are named s_i'. Let us assume the following input data for 8 services (Tables 1 and 2).

The solutions presented in Table 3 were obtained using the proposed algorithm.

Experiments were run for sets of 2, 4, 6, 8, 10 and 12 services and the results are presented in Figs. 2 and 3.

The randomly chosen values of Tables 1 and 2 were analyzed and the processing times of the sets of services were determined. It is shown that the most time-consuming criterion is fault tolerance, the second is consolidation, while the lowest processing time is consumed for the load balancing criterion. Moreover, the common consideration of the two criteria of load balancing and fault tolerance produced a slightly better result than the common consideration of consolidation and fault tolerance, which is also verified in practice. This provides a practical suggestion for the implementation of permission strategies.

Table 1 Values of matrix T

Services/Resources	r_1	r_2	r_3	r_4
s_1	4	5	3	4
s_1'	4	5	3	4
s_2	3	4	3	2
s_2'	3	4	3	2
s_3	5	4	6	3
s_3'	5	4	6	3
s_4	3	3	4	3
s_4'	3	3	4	3

Table 2 Values of matrix γ

Services/Resources	r_1	r_2	r_3	r_4
s_1	0.4	0.5	0.3	0.4
s_1'	0.4	0.5	0.3	0.4
s_2	0.3	0.4	0.3	0.2
s_2'	0.3	0.4	0.3	0.2
s_3	0.5	0.4	0.6	0.3
s_3'	0.5	0.4	0.6	0.3
s_4	0.3	0.3	0.4	0.3
s_4'	0.3	0.3	0.4	0.3

Table 3 Optimal allocation for the considered model and 8 services. Values of processing time are in brackets. (x^*) means acceptation of resource overload

Criteria/Resources	r_1	r_2	r_3	r_4
Load balancing (LB)	s_4, s_4'(6)	s_3'(4)	s_1, s_1'(6)	s_2, s_2', s_3(7)
Consolidation (CONS)			s_1, s_1', s_2, s_2' (12.6*)	s_3, s_3', s_4, s_4' (12.6*)
Fault tolerance (FT)	s_1', s_4' (7)	s_3' (4)	s_1, s_2' (6)	s_2, s_3, s_4 (8)
Load balancing and Fault tolerance (LBFT)	s_1', s_4'(7)	s_3' (4)	s_1, s_2' (6)	s_2, s_3, s_4 (8)
Consolidation and Fault tolerance (LBFT)	s_1, s_3, s_4 (12.6*)		s_1', s_2, s_4'(10)	s_2', s_3' (5)

5 Final Remarks

Three different cases of provisioning problems have been considered. The execution time of a set of independent services has been compared. A formal model that can be used at different levels of cloud resources—virtual units or physical units—has

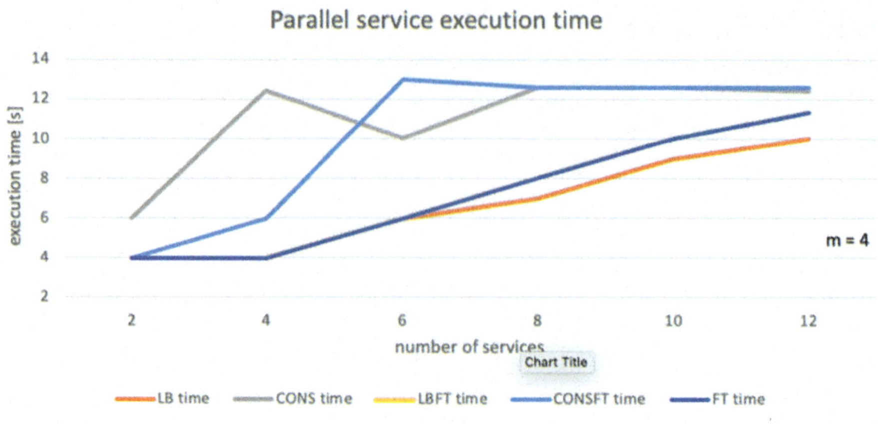

Fig. 2 Execution time of a set of services on different resources provisioned using different algorithms

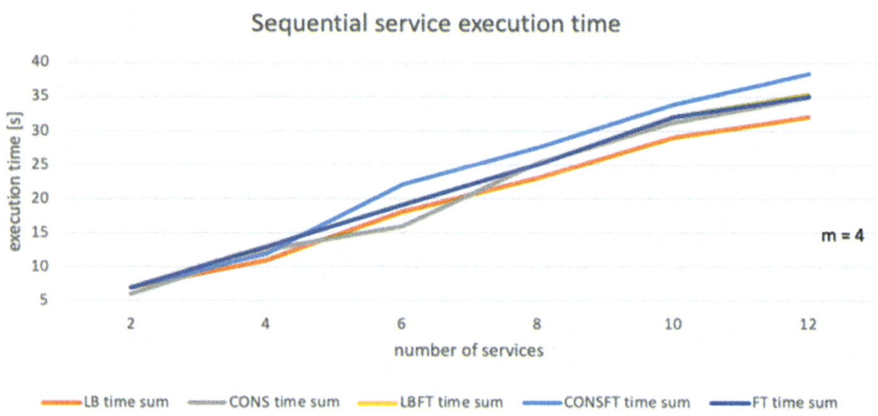

Fig. 3 Sum of execution times of services on different resources provisioned using different algorithms

been proposed. Three optimalization problems have been classified based on their mean processing time. A hybrid approach has been investigated, and load balancing with fault tolerance is shown to produce more promising results than consolidation with fault tolerance. The presented model makes it possible to analyze a series of service sets required to run in a cloud environment and achieve acceptable scalability. It makes it possible to determine the proper strategy of provisioning for changing user requirements or clients' demands in near real time. The sequential and parallel execution of services on one node can also be considered. To improve the provisioning speed of much bigger sets of services and resources (which will be interesting for global clouds), a heuristic algorithm should be considered. There is also the aspect of the influence of the services on each other, as can be seen in Eq. (4). This problem has been mentioned in [18, 19] and provides a possible path to further enriching the algorithm proposed in this paper. The resulting algorithm could minimize the interaction of services in different categories, which should positively impact cost savings for clients (services should execute with no delays). Such a solution will be considered in further research.

Provisioning problems have been tested in a TASKcloud test environment, which also confirms the presented simulation results. The next step is to implement the provisioning tool for this environment to utilize the natural possibilities of the platform.

Acknowledgements The Regional Operational Program of the Pomeranian Voivodeship for 2014–2020, Project Number RPPM.01.02.00-22-0001/17, "Establishment of the Competence Center STOS (Smart and Transdisciplinary knOwledge Services) in Gdansk in the field of R&D infrastructure."

References

1. Odun-Ayo, I., Udemezue, B., Kilanko, A.: Cloud service level agreements and resource management advances. Sci. Technol. Eng. Syst. J. **4**(5), 228–236 (2019)
2. Manvi, S., Shyam, G.K.: Cloud Computing Concepts and Technologies. 1st edn. CRC Press (2021)
3. Jennings, B., Stadler, R.: Resource management in clouds: survey and research challenges. J. Netw. Syst. Manag. **23**, 567–619 (2015)
4. Gonzalez, N., Carvalho, T., Miers, C.: Cloud resource management: towards efficient execution of large-scale scientific applications and workflows on complex infrastructures. J. Cloud Comput. **6** (2017)
5. Orzechowski, P.: Task cloud infrastructure in the centre of informatics—Tricity academic supercomputer & network. TASK Q. **22**, 313–319 (2018)
6. Beaumont, O., Eyraud-Dubois, L., Guermouche, A., Lambert, T.: Comparison of static and dynamic resource allocation strategies for matrix multiplication. In: 26th IEEE International Symposium on Computer Architecture and High Performance Computing (SBAC-PAD), Florianopolis (2015)
7. Sumalatha, K., Anbarasi, M.S.: Provisioning Cloud Resources: Optimization Techniques for Resource Provisioning in Cloud Environment. 1st edn. Lambert Academic Publishing (2019)
8. Panwar, R., Supriya, M.: Dynamic resource provisioning for service-based cloud applications: a Bayesian learning approach. J. Parall. Distrib. Comput. **168**, 90–107 (2022)
9. Perumal, K., Mohan, S., Frnda, J., et al.: Dynamic resource provisioning and secured file sharing using virtualization in cloud azure. J. Cloud Comput. **11** (2022)
10. Kashyap, D., Viradiya, J.: A survey of various load balancing algorithms in cloud computing. Int. J. Sci. Technol. Res. **3**(11), 115–119 (2014)
11. Mesbahi, M., Rahmani, A.: Load balancing in cloud computing: a state of the art survey. Int. J. Modern Educ. Comput. Sci. **8**, 64–78 (2016)
12. Sharma, M., Sharma, P.K., Sharma, S.S.: Efficient load balancing algorithm in VM cloud environment. Int. J. Comput. Sci. Technol. **3**(1), 439–441 (2012)
13. Ponto, R., Kecskeméti, G., Mann, Z.: Comparison of workload consolidation algorithms for cloud data centers. Concurr. Comput. Pract. Exp. 1–24 (2021)
14. Ferreto, T.C., Netto, M.A., Calheiros, R.N., Rose, C.A.D.: Server consolidation with migration control for virtualized data centers. Futur. Gener. Comput. Syst. **27**(8), 1027–1034 (2011)

15. Armant, V., De Cauwer, M., Brown, K.N., O'Sullivan, B.: Semi-online task assignment policies for workload consolidation in cloud computing systems. Futur. Gener. Comput. Syst. **82**, 89–103 (2018)
16. Hasan, M., Goraya, M.S.: Fault tolerance in cloud computing environment: a systematic survey. Comput. Ind. **99**, 157–172 (2018)
17. Rehman, A.U., Aguiar, R., Barraca, J.P.: Fault-tolerance in the scope of cloud computing. IEEE Access **10**, 63422–63441 (2022)
18. Orzechowski, P.: Complementary oriented allocation algorithm for cloud computing. TASK Q. **21**(4), 395–403 (2017)
19. Orzechowski, P., Proficz, J., Krawczyk, H., Szymański, J.: Categorization of cloud workload types with clustering. Proc. Int. Conf. Signal, Netw., Comput., Syst. **395**, 303–313 (2016)

Tabular Structures Detection on Scanned VAT Invoices

Paweł Pawłowski, Marek Bazan, Maciej Pawełczyk, and Maciej E. Marchwiany

1 Introduction

In this paper we present approaches to detection of the tables with items covered by VAT invoices. The used data set contains real and synthetic invoices represented as images. This problem is similar to table detection in scanned documents at ICDAR 2019 [1]. However, the IDCAR competition tasks did not contain real VAT invoices.

We describe three different approaches of tabular structures detection on VAT invoices:

1. CascadeTabNet [2],
2. Region Proposal Networks with Faster RCNN [3],
3. Our pipeline based on graph neural networks [4] followed by the postprocessing step performed by Faster R-CNN (ResNet101—backbone) on masks obtained from graph processing.

In this paper we present the performance of the above approaches on a non-public dataset of 400 real invoices from the JT Weston Company (Poland) and also on a synthetic dataset of 105 invoices that we publish in [5]. The performance is measured by the F_1 score and shown for different values of Intersection over Union (IoU) threshold values. The IoU is an evaluation metric for an object detector accuracy measure. IoU is calculated using an object detector output bounding box and ground truth bounding box by taking a ratio of the intersection area to the union area for these two boxes. The larger IoU is, the better detection models provide. The procedure for calculating the F_1 score is published in [5].

In the table detection task our goal was to have detections with the highest possible IoU. We are comparing selected models for different IoU thresholds, meaning that we treat detection as positive if the IoU is equal to or greater than the given threshold.

M. Bazan (✉)
Department of Computer Engineering, Wroclaw University of Science and Technology, ul. Janiszewskiego 11/17, 50-370 Wrocław, Poland
e-mail: marek.bazan@pwr.edu.pl

P. Pawłowski · M. Bazan · M. Pawełczyk · M. E. Marchwiany
JT Weston sp. z o.o., Atrium Plaza, al. Jana Pawła II 29, 00-867 Warszawa, Poland

W. Zamojski et al. (eds.), *Dependable Computer Systems and Networks*, Lecture Notes in Networks and Systems 737, https://doi.org/10.1007/978-3-031-37720-4_19

The performance on both synthetic and real invoices leads us to choose the model based on Region Proposal Networks as a baseline model.

To our knowledge, general table detection in scanned documents has not been a subject covered in the literature, with the exception of the detection of general tables in scanned documents. As shown in this paper the general-purpose methods, officially published in peer review journals, e.g. by Prasad et al. [2] with a publicly available source code, do not work sufficiently well in the application to VAT invoices.

The remaining part of the paper is organized as follows. In Sect. 2 we review of state-of-the-art of three types neural network approaches table detection, namely (a) Faster R-CNN architecture pretrained for object detection, (b) Cascade Mask R-CNN and (c) graph neural networks. Then in Sect. 3 we describe models that we configured (CascadeTabNet and Faster RCNN) and built (graph neural network) for table detecton in VAT invoice tasks. In Sect. 3.3 we present also results obtained for our dataset for synthetic, as well as real data. Finally, we conclude the paper by depicting the next step, which is processing and understanding the content of a detected table.

2 Related Work

The first approaches of the application of deep learning to table detection used convolutional neural networks was reported by Hao work [6]. The method proposed therein however is limited to PDF documents and tables with ruling. The first approach using deep-learning for table detection in images of scanned documents was proposed by Gilani et al. [7]. This method relies on a Faster R-CNN known from object detection tasks. The Region Proposal Network in the Faster R-CNN was ZFNet [8]. Independently, another method based on a Faster R-CNN was proposed by Schreiber et al. [9] where ZFNet has been use as the backbone of the region proposal network but also tested VGG-19 architecture [10]. Another method using Faster R-CNN networks with the VGG19 backbone was presented by Sun et al. in [11]. This method uses the same model for both tables course location the document and corner detection. The achieved results on the ICDAR 2017 task are better than the scores of the Faster R-CNN model with a number of pre- and post-processing steps that won the competition.

Even better results are achieved on the architecture similar to the Faster R-CNN proposed in [9] but using deformable convolutional networks instead of conventional ones in [12]. Instead of using VGG19 or ZFNet the authors used ResNet101 in its deformable version.

Recently, ResNetX [13] architecture was proposed also as a backbone for a Faster R-CNN network built on a Tablebank [14] dataset of 417K images containing tables in Word and Latex-generated images. Although the amount of data is large compared to other publically available datasets, the results on the test dataset extracted from it are on average at the level of $F_1 = 0.93$.

In paper, [15] famous YOLOv3 architecture was used for object detection on tasks from the ICDAR 2013 and 2017 competitions. Although the proposed architecture achieves results with comparable levels of F_1 and even slightly better than in [12], they are obtained using additional postprocessing steps.

A thorough comparative study of YOLO architecture vs yet another two object detection approaches, namely Retina Net and Mask RCNN is presented in [16]. The test tasks used there are also from ICDAR 2013, ICDAR 2017, ICDAR 2019 as well as on the Tablebank dataset. The results presented by the authors, however, are not as good as in other papers.

The Mask RCNN and its cascade extension are a crucial component of the architecture of CascadeTabNet and is the subject of the comparative study in the present paper. The architecture is described in details in Sect. 3.1. Cascade architectures with deformable convolutions are also used in the field of table detection in [17]. In [17] the results of CDecNet architecture on various publicly available datasets are presented for various intersection over union thresholds which is in accordance with our experience with cascade object datection type models, i.e. not such a performance for high intersection over union. It means that highly accurate detection is difficult with such networks.

Graph neural networks, to our best knowledge, were used for table detection only by Riba et al. in [4]. Our work was greatly inspired by that methodology. A graph construction corresponding to a document in the method presented in this paper is similar to that presented in [4], although, in the later paper, the authors only exploited neighboring nodes in X, Y belts in directions perpendicular to the analysed node. In the research presented in the current paper, we also tested a rectangular neighborhood but it turned out to give worse results. The Riba's architecture of the graph neural network is based on an introduced graph residual layer—inspired by the Residual layer for Convolutional Neural Networks introduced in [18]. Instead, we tested three architectures for graph neural networks: GCN (see [19]) GraphSage (see [20]) and GAT (see [21]). Additionally, the difference in our approach is also that we fix the misclassification of nodes using a Faster R-CNN classification on masks of nodes that follows graph processing.

In contrast to GAT [21], in the literature, GCN [19] architecture is also successfully applied in [22] and DGCNN in [23]. However, the application presented in those papers is table cells recognition, which is not the subject of this paper. In [22] Lohani et al. reported very high F_1 values for particular cell classification, even for the invoice column-based manner. However, the authors do not report what was the variability of patterns in the dataset. In our dataset, the tables contained rows that have multiple line descriptions of items and additionally descriptions may be centrally aligned. Such data do not occur in patterns in data generated in [23], therefore after observing that graph networks may misclassify the nodes and have to be postprocessed by Mask RCNN we decided to process tables themselves by deep learning architecture, the results of which will be the subject of a future publication.

3 Three Approaches to Table Detection in Scanned VAT Invoices

In this section we present three methods for table detection. We also compare the performance to those models.

3.1 CascadeTabNet

The only approach with a source code publicly available has been published by Prasad et al. in [2]. It is a three-stage cascade mask R-CNN network with a feature extractor based on a High-Resolution Net (HRNet) proposed in [24]. The Cascade Mask R-CNN network [25] is a neural network model extended by a separate branch for instance segmentation. In the CascadeTabNet the 'mask head' is connected to the last layer, which is a fully connected dense layer. CascadeTabNet has F_1 score 0.532 and 0.816 for real and synthetic data respectively whit IoU 0.5. For higher IoU performance of model drops dramatically. In Appendix C we present performance of the CascadeNet on a real invoice dataset and on a synthetic dataset.

3.2 Models for Object Detection Based on RPNs and Fast RCNN

Models based on object detection techniques can also be used for tables detection. We focused on a combination of a region proposal network and a fast region convolutional neural network. Such an approach was first introduced in [26] and was the main component of the winning solutions of ICDAR-2019 [9]. We tested the newer models for object detection that are available at Model Zoo [3].

The choice of the best model for table detection has been made in the following steps:

1. Generate a synthetic dataset based on real invoices. The examples of generated invoices with their templates. Are presented in Appendix A. A total number of 6090 augmentation images have been created based on 105 synthetic images during a single training epoch.
2. Build training dataset contains: all generated augmented synthetic data and 305 real invoice data per each epoch. The remaining 100 real invoices (unseen during a training) were put into the test dataset.
3. Train a number of different models selected from three meta architectures, i.e. SSD, Faster R-CNN and R-FCN. All models were trained with on-the-fly heavy augmentation with transformation shown in Table 4.
4. Compute the F_1 score with IOU equal to 0.5 on all trained models. Based on the F_1 score select the best model.
5. Test on test set invoices from the real invoice dataset.

3.2.1 Description of the Model Architecture for Object Detection

The schematic sketch of object detection architectures used in this paper can be found in [15]. The TensorFlow object detection Model Zoo contains three main groups of models which we investigated:

1. **Single Shot Detection Models** [27] rely on a single convolutional network for location detection of boxes anchors and classification of objects within them. Such architecture is a component of models which use two separate stages for box proposals regression and classification of the proposal box content such as FR-CNN and R-FCNN. Resnet-50 [18] and Inception v2 [28] are used for feature extraction [29].
2. **Faster R-CNN Models** [26] consist of two sub-networks: one for agnostic proposal boxes generation (called the Region Proposal Network) and the second for classification of these proposals (called Fast R-CNN). The two networks share the same convolutional feature extractor. The output of the extractor together with the anchors from a Region Proposal Network go to the Fast R-CNN sub-network. Inception v2 [28], Resnet 50, Resnet 101 [18] and Inception Resnet v2 are used for feature extraction [30].
3. **R-FCN Models** [31] are similar to the Faster R-CNN. This architecture differs from Faster R-CNN by the placement of the proposal region cropping mechanism in its pipeline. For Faster R-CNN cropping is done directly after the first stage. For R-FCN cropping is shifted just before the final dense layer. Resnet 101 [18] was used for feature extraction.

All feature extractors from Model Zoo were pre-trained on a Microsoft COCO dataset [32].

3.2.2 Results of Numerical Experiments for Faster RCNN

We trained models on the augmented synthetic dataset. The test set has synthetic invoices created from all templates. The results are presented in Table 1.

The best results were obtained for Faster R-CNN with the Inception Resnet v2 with the atrous algorithm for convolution calculation (see [33]) with a delation rate equal to 2.

For model testing we used 2 type of augmentation: (i) light where 4 random transformation was used and (ii) heavy with all 58 transformation (see Table 4). We used 3 testing scenarios:

1. real dataset with a light augmentation,
2. mixed real and synthetic dataset with a light augmentation,
3. mixed real and synthetic dataset with a heavy augmentation.

Table 1 Model's performance trained on a synthetic dataset

Model name	F_1(IoU = 0.5)	Model name	F_1(IoU = 0.5)
ssd mobilenet_v1	0.731	faster rcnn_inception_v2	0.772
ssd resnet_50_fpn	0.642	faster rcnn_resnet50	0.809
ssd mobilenet_v2	0.752	faster rcnn_resnet101	0.842
rfcn resnet101	0.724	**faster rcnn_inception_ resnet_v2_atrous**	**0.977**

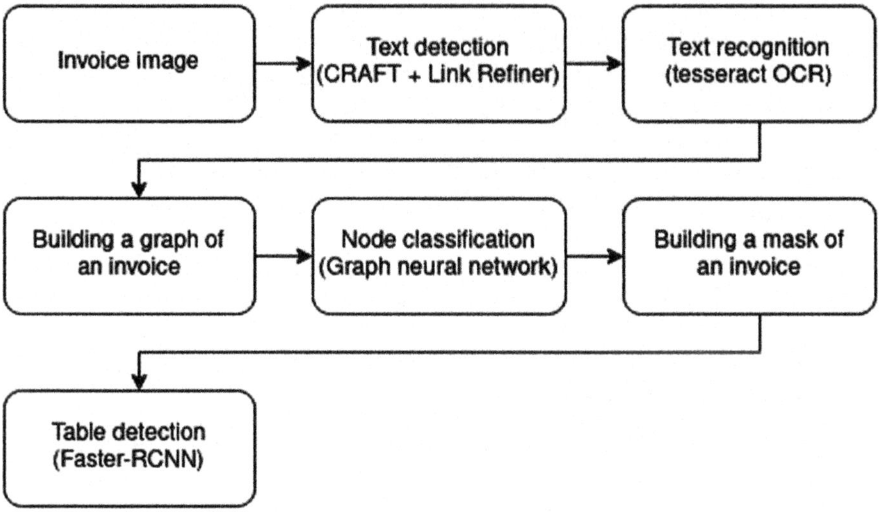

Fig. 1 An algorithm for table detection in an invoice image using a graph neural network

The best performance for F_1 score we obtained for the mixed real and synthetic dataset with a light augmentation. This data set has been used for training. Results for all augmentation as a function of IoU are presented in Appendix D.

3.3 *Graph Neural Networks*

In this paragraph, we present steps to build a document graph neural network for table extraction. We performed node classification for three classes: a header of a table, a table body and an outside of a table. The algorithm is presented with a control block diagram in Fig. 1.

3.3.1 Text Lines Detection

is performed by CRAFT and Link refiner models [34]. A review of other neural methods for the generation of text boxes may be found e.g. in [35]. Outputs of this stage are coordinates of text lines bounding boxes. This kind of detection enables us to represent the whole document as a graph, where each line of text is treated as one node.

3.3.2 Text Recognition

is performed by Tesseract [36]. We loop through all text lines detected by CRAFT and input them to tesseract.

3.3.3 Building Graph of an Invoice

Each bounding box (text line) is treated as one node. Each node is represented with a feature vector that describes its position, textual attributes, and distances to the nearest neighbours in the X and Y-axes. We considered two methods of connecting nodes and three methods of calculating the distance between connected nodes, as well as using no information about distances.

1. Nodes relations
    (a) Connecting node with all nodes presented in a rectangular area.
    (b) Connecting node only with the nearest neighbours from each side (left, right, up, down) (if the distance is not to big).
2. Distance
    (a) Euclidean distance between nodes
    (b) 1/(Euclidean distance between nodes)
    (c) [Distance in X-axis, Distance in Y-axis].

An example of a graph generated for a document is presented in Fig. 3.

3.3.4 Node Classification

Since using graph neural networks for the task of table detection is not common practice, there is a lack of proven solutions and architectures. However, the task can be solved as a node classification problem. Such an approach is presented it in this paper. In this section, we presented used architectures. All of them have been implemented in PyTorch Geometric [37].

In this paper we investigated two classification problems: (i) with two classes: table elements and text elements, and (ii) with three classes: text elements, header elements and table elements. We tested three graph convolution layer architectures:

1. Graph Convolutional Networks (GCN) [19]

$$\mathbf{h}_v^{(k)} = \sigma\left(W^{(k)} \sum_{u \in N(v)} \frac{\mathbf{h}_u^{(k-1)}}{|N(v)|}\right) \quad (1)$$

2. GraphSage [20]

$$\mathbf{h}_v^{(k)} = \sigma\left(W^{(k)} \cdot \left(\mathbf{h}_v^{(k-1)} \parallel \text{max\_pool}\left(\{\mathbf{h}_u^{(k-1)},\ \forall u \in N(v)\}\right)\right)\right) \quad (2)$$

   where $(\cdot \parallel \cdot)$ means a concatenation operator and max_pool is an element-wise max pooling operation over neighboring node features,
3. Graph Attention Networks (GATs) [21]

$$\mathbf{h}_v^{(k)} = \Big\Vert_{l=1}^{N} \sigma\left(\sum_{u \in N(v)} W^{(k)} \alpha_{vu}^{l} \mathbf{h}_u^{(k-1)}\right) \quad (3)$$

   where α_{vu}^{l} is an attention weight between nodes v and u with an l-th attention head. The $||_{l=1}^{N}$ is a concatenation over N attention heads. For the output layer instead of (3) node features are processed using Eq. (4)

$$\mathbf{h}_v^{(k)} = \sigma\left(\frac{1}{N}\sum_{l=1}^{N} \sum_{u \in N(v)} W^{(k)} \alpha_{vu}^{l} \mathbf{h}_u^{(k-1)}\right). \quad (4)$$

In Eqs. (1)–(4) $N(v)$ means neighboring nodes of the node v and $\mathbf{h}_v^{(k)}$ is a feature vector of node v at the k-th layer.

All the network architectures are trained with cross-entropy loss in the space of weights $W^{(k)}$ $k = 0, \ldots, M$ where M is a number of layers and attention coefficient α_{vu}^{l}. The ADAM optimizer was usage.

We tested several GNN architectures with respect to the number of layers, the number of hidden neurons on linear components and the number of attention heads. The best models with their performance for each architecture are summarised in Table 2. The highest accuracy models are built with GAT layers. The column headers describe the used architecture with the number of layers, type of layer, hidden size and the number of classification heads for GAT layers. In the rows we can see metrics calculated for each class. The architecture of our solution corresponds to that presented in [4]. However, we replaced residual layers used in model in [4] with one of the layers such as GCN, GraphSage or GATs.

Table 2 The results for three classes were calculated on a test dataset. The headers contain information about the network structure—the first number refers to the number of layers, then the type of used layers, the number of classification heads (in the case of GATs), and the hidden size. All architectures were optimized using the ADAM optimizer. The best achieved results for GAT architecture were obtained for the initial learning rate equal to $75 \cdot 10^{-5}$

	Two classes			Three classes		
	5GAT _head4_ hidden17	4Sage Conv_ hidden13	4GCN _hidden15	5GAT _head6_ hidden14	5Sage Conv_ hidden15	4GCN _hidden15
Accuracy	0.973	0.966	0.880	0.978	0.950	0.880
Precision of text	0.971	0.974	0.918	0.986	0.961	0.930
Precision of table	0.968	0.950	0.830	0.972	0.954	0.871
Recall of text	0.980	0.965	0.874	0.981	0.969	0.937
Recall of table	0.965	0.972	0.883	0.976	0.921	0.853
F1 of text	0.974	0.967	0.884	0.982	0.963	0.931
F1 of table	0.965	0.957	0.843	0.972	0.934	0.850

3.3.5 Building an Invoice Mask

Based on the outputs from the graph neural network we build a mask of the given invoice by marking each class bounding box with RGB colour.

3.3.6 Table Detection

Models used in pipleine works whit very high accuracy although some misclassified nodes can be found. That's why in training our Faster-RCNN R101-FPN model we expanded the training set with augmented masks created by switching classes in a small percentage of randomly selected nodes. After training the model we have tested its performance on the test set. The results are presented in Table 3.

4 Limitations

The performance of the proposed method on selected documents from the Marmot data set [38], which was unseen in training is worse on documents similar to the training data set. However, switching undetected tables with original tables in invoice documents gives correct detection. Moreover, switching invoice tables with tables in the Marmot documents also gives a correct detection. It means that at least one

Table 3 Results for Faster-RCNN as post-processing for a Graph Neural Network on the test dataset of real and synthetic invoices. The achieved F_1 scores for increasing IoU should be compared with those obtained for other methods and presented in Tables 5 and 6

	Real data			Synthetic data		
IoU	Prec.	Recall	F_1	Prec.	Recall	F_1
0.6	0.940	0.979	0.959	0.994	1.0	0.997
0.7	0.919	0.979	0.948	0.994	1.0	0.997
0.8	0.893	0.978	0.934	0.988	1.0	0.994
0.9	0.738	0.973	0.839	0.959	1.0	0.979

component of the invoice is crucial for table detection: (i) a table structure or (ii) the structure of a neighborhood of the table. If both elements are different from the invoices used for training, the pre-trained model cannot be used.

Another limitation of the proposed method is its throughput when considered separately from downstream document analysis tasks for which the method may be used. On average, one invoice is processed in about 19.5 s (where text detection takes 12.35 s). What gives about 180 invoices per hour on a single thread on a machine with an Intel Core i9 CPU. However, the text detection and recognition steps have to be done on any downstream tasks such as table structure and content recognition or more advanced invoice element classification. Moreover, this limitation can be easily overcome by parallel invoice processing. The parallelization has linear speed up with a number of physical threads exploited on CPU.

5 Conclusions and Future Work

The work presented in this document was targeted at table detection in VAT invoices. We showed that our model outperforms object detection concerning detection quality approaches in this task. Also, we presented possibilities of using our results in the general table detection problem.

We have presented a graph neural network-based approach to tabular structures detection on scanned VAT invoices. We investigated three common graph neural architectures as GCN, GraphSage and GAT. Based on the F_1 score for table location predictions with as high an Intersection Over Union threshold as possible, GAT architecture was chosen. The graph neural network approach was compared to the pre-trained CascadeTabNet and DeepDeSRT. These models are constructed from the Cascade Faster Mask R-CNN network with the publicly available source code and object detection models available in TensorFlow zoo, respectively. For training models we used the private dataset of real invoices and also the dataset of synthetic invoices that we published. Both datasets were in Polish.

For better model generalization for object detection models, two approaches for data augmentation were tested: (i) heavy augmentation with 58 transformations and (ii) light augmentation with 4 randomly selected transformations.

The method using GAT architecture outperforms methods based on CascadeTabNet and on F-RCNN networks. It allowed us to achieve a high F_1 score for high Intersection Over Union. Our model predicts tables' location in documents very precisely overlapped with true locations from a test dataset.

We discovered that graph neural network performance can be increased by an analysis of the vertically neighbouring nodes.

The presented method's novelty lies in incorporating the algorithm into the text processing pipeline, not into text preprocessing. It enables rapid processing of data from tables and smooth text processing.

Our future work on VAT invoice understanding is going to cover two fields: (i) column classification, rows segmentation and finally cell data extraction and (ii) classification of invoice data from outside the tables. The former problem has been discussed by many authors. In the context of graph neural networks interesting results were shown in [22, 23]. Classification of entities other than data in a table using machine learning methods was presented e.g. [39].

The approach presented in this paper will be the first step of a table daa ta extraction algorithm. In particular, a graph neural network classifier allows a table header distinction from the table body. The presented methodology also allows us to exclude tables from VAT invoice entities classification tasks.

6 Founding

This work was financed with European Union funds from the Smart Growth Operational Programme 2014–2020, Measure 1.1.: R&D projects of enterprises. Submeasure 1.1.1.: Industrial research and development carried out by enterprises.

Acknowledgements We want to thank Mr. Tomasz Gniazdowski from JT Weston company for performing execution time measurements for the proposed method.

Appendices

A. Example from a Real, Synthetic Dataset and Augmentation

See Fig. 2 and Table 4.

Table 4 Augmentation operations for invoice data. (a) A heavy augmentation covered all 58 transformations. (b) A light augmentation used 4 transformations randomly chosen for each invoice

Basic	Gamma	Mulitiply, add and flip
Fog	Gamma	Multiply075
Rain	GammaAdd25	Multiply125
Clouds	GammaAddM25	Add25
Snowy		AddM25
Blur		leftRightFlip
Snowflakes		upDownFlip

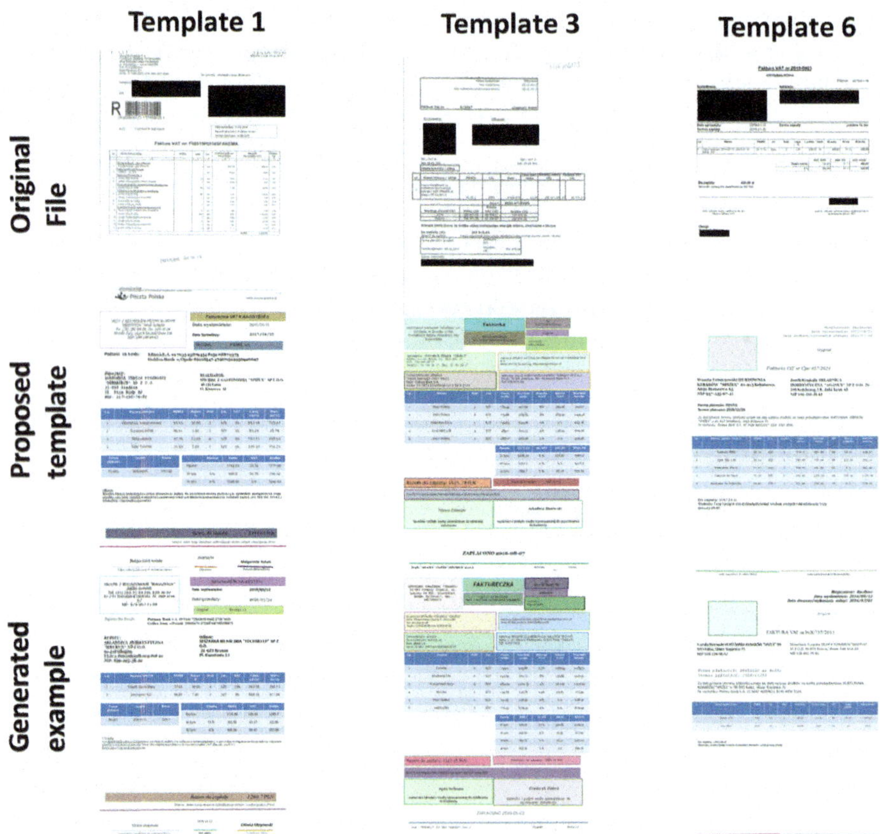

Fig. 2 Example from a synthetic dataset. The templates based on original invoices are in the top row, synthetic invoices are in the following rows

B. Example of a Graph Generated for a Single Document

See Fig. 3.

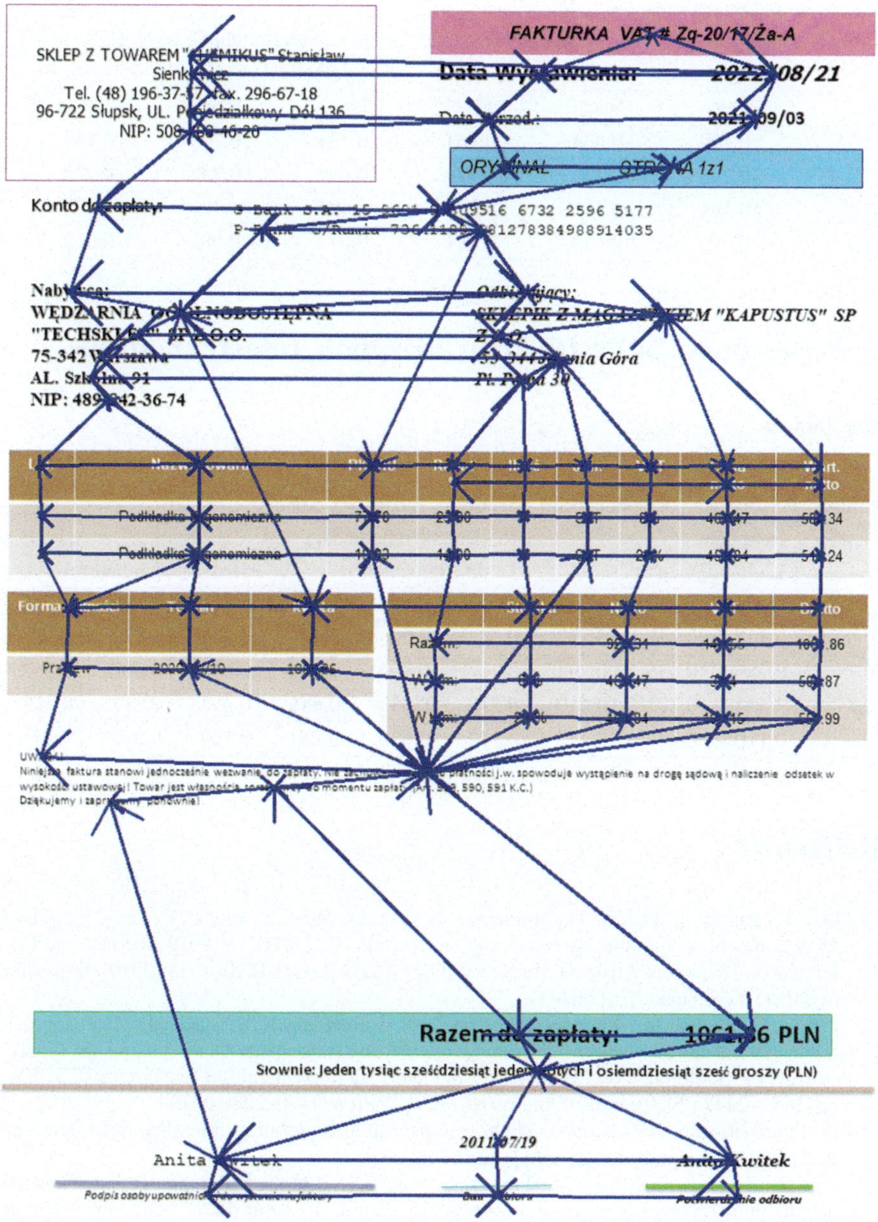

Fig. 3 Example of a graph generated for a single document

C. Performance of the CascadeTabNet on the Real and Synthetic Invoices

See Table 5.

Table 5 Performance of the CascadeTabNet model on the real and synthetic invoices

	Real data			Synthetic data		
IoU	Prec.	Recall	F_1	Prec.	Recall	F_1
0.6	0.409	0.528	0.461	0.704	0.793	0.746
0.7	0.280	0.433	0.34	0.677	0.787	0.728
0.8	0.194	0.346	0.249	0.493	0.729	0.588
0.9	0.065	0.150	0.091	0.017	0.085	0.028

D. Performance of Faster_rcnn_inception_resnet_v2_atrous

See Table 6.

Table 6 Performance of **faster_rcnn_inception_resnet_v2_atrous**

	Real data			Real and synthetic data					
	Light augmentation			Light augmentation			Heavy augmetation		
IoU	Prec.	Recall	F_1	Prec.	Recall	F_1	Prec.	Recall	F_1
0.6	0.827	0.847	0.837	0.823	0.906	0.863	0.801	0.879	0.838
0.7	0.756	0.835	0.794	0.702	0.892	0.786	0.691	0.862	0.767
0.8	0.504	0.771	0.610	0.518	0.859	0.646	0.493	0.817	0.615
0.9	0.110	0.424	0.175	0.177	0.676	0.281	0.162	0.595	0.255

References

1. Gao, L., Huang, Y., Déjean, H., Meunier, J.-L., Yan, Q., Fang, Y., Kleber, F., Lang, E.: ICDAR 2019 competition on table detection and recognition (CTDAR). In: 2019 International Conference on Document Analysis and Recognition (ICDAR), pp. 1510–1515 (2019). https://doi.org/10.1109/ICDAR.2019.00243
2. Prasad, D., Gadpal, A., Kapadni, K., Visave, M., Sultanpure, K.: Cascadetabnet: an approach for end to end table detection and structure recognition from image-based documents. In: 2020 IEEE/CVF Conference on Computer Vision and Pattern Recognition Workshops (CVPRW), pp. 2439–2447 (2020). https://doi.org/10.1109/CVPRW50498.2020.00294
3. Google, Tensorflow 1 detection model zoo. https://github.com/tensorflow/models/blob/master/research/object_detection/g3doc/tf1_detection_zoo.md
4. Riba, P., Dutta, A., Goldmann, L., Fornés, A., Ramos, O., Llad's, J.: Table detection in business document images by message passing networks. Pattern Recognit. **127** (2022). https://doi.org/10.1016/j.patcog.2022.108641

5. Weston, J.T.: Synthetic vat invoices data (2022). https://github.com/marekbazan/invoices
6. Leipeng, H., Liangcai, G., Xiaohan, Y., Zhi, T.: A table detection method for pdf documents based on convolutional neural networks. In: 2016 12th IAPR Workshop on Document Analysis Systems (DAS), pp. 287–292 (2016). https://doi.org/10.1109/DAS.2016.23
7. Gilani, A., Qasim, S., Malik, I., Shafait, F.: Table detection using deep learning. In: 2017 14th IAPR International Conference on Document Analysis and Recognition (ICDAR), pp. 771–776 (2017). https://doi.org/10.1109/ICDAR.2017.131
8. Zeiler, M.D., Fergus, R.: Visualizing and understanding convolutional networks. In: Fleet, D., Pajdla, T., Schiele, B., Tuytelaars, T. (eds.) Computer Vision. ECCV 2014, vol. 8689, pp. 818–833 (2014)
9. Schreiber, S., Agne, S., Wolf, I., Dengel, A., Ahmed, S.: Deepdesrt: deep learning for detection and structure recognition of tables in document images. In: 2017 14th IAPR International Conference on Document Analysis and Recognition (ICDAR), vol. 01, pp. 1162–1167 (2017). https://doi.org/10.1109/ICDAR.2017.192
10. Simonyan, K., Zisserman, A.: Very deep convolutional networks for large-scale image recognition. In: International Conference on Learning Representations, San Diego (May 2015)
11. Sun, N., Zhu, Y., Hu, X.: Faster R-CNN based table detection combining corner locating. In: 2019 International Conference on Document Analysis and Recognition (ICDAR), pp. 1314–1319 (2019). https://doi.org/10.1109/ICDAR.2019.00212
12. Siddiqui, S.A., Malik, M.I., Agne, S., Dengel, A., Ahmed, S.: Decnt: deep deformable CNN for table detection. IEEE Access **6**, 74151–74161 (2018). https://doi.org/10.1109/ACCESS.2018.2880211
13. Xie, S., Girshick, R., Dolláir, P., Tu, Z., He, K.: Aggregated residual transformations for deep neural networks. In: 2017 IEEE Conference on Computer Vision and Pattern Recognition (CVPR), pp. 5987–5995 (2017). https://doi.org/10.1109/CVPR.2017.634
14. Li, M., Cui, L., Huang, S., Wei, F., Zhou, M., Li, Z.: Tablebank: a benchmark dataset for table detection and recognition. In: Proceedings of the 12th Language Resources and Evaluation Conference, pp. 1918–1925 (2020)
15. Huang, J., Rathod, V., Sun, C., Zhu, M., Korattikara, A., Fathi, A., Fischer, I., Wojna, Z., Song, Y., Guadarrama, S., Murphy, K.: Speed/accuracy trade-offs for modern convolutional object detectors. In: 2017 IEEE Conference on Computer Vision and Pattern Recognition (CVPR), pp. 3296–3297 (2017). https://doi.org/10.1109/CVPR.2017.351
16. Casado-García, Á., Domínguez, C., Heras, J., Mata, E., Pascual, V.: The benefits of close-domain fine-tuning for table detection in document images. In: Bai, X., Karatzas, D., Lopresti, D. (eds.) Document Analysis Systems, pp. 199–215 (2020). https://doi.org/10.1007/978-3-030-57058-3_15
17. Agarwal, M., Mondal, A., Jawahar, C.V.: Cdec-net: composite deformable cascade network for table detection in document images. In: 2020 25th International Conference on Pattern Recognition (ICPR), pp. 9491–9498 (2021). https://doi.org/10.1109/ICPR48806.2021.9411922
18. He, K., Zhang, X., Ren, S., Sun, J.: Deep residual learning for image recognition. In: 2016 IEEE Conference on Computer Vision and Pattern Recognition (CVPR), pp. 770–778 (2016). https://doi.org/10.1109/CVPR.2016.90
19. Kipf, T.N., Welling, M.: Semi-supervised classification with graph convolutional networks. In: International Conference on Learning Representations (ICLR) (2017)
20. Hamilton, W.L., Ying, R., Leskovec, J.: Inductive representation learning on large graphs. In: Guyon, I., Luxburg, U.V., Bengio, S., Wallach, H., Fergus, R., Vishwanathan, S., Garnett, R. (eds.) Advances in Neural Information Processing Systems, vol. 30, pp. 1024–1034 (2017)
21. Veličković, P., Cucurull, G., Casanova, A., Romero, A., Liò, P., Bengio, Y.: Graph attention networks. In: International Conference on Learning Representations (2018)
22. Lohani, D., Beläd, A., Belaïd, Y.: An Invoice Reading System Using a Graph Convolutional Network, pp. 144–158 (2019). https://doi.org/10.1007/978-3-030-21074-8_12
23. Qasim, S.R., Mahmood, H., Shafait, F.: Rethinking table recognition using graph neural networks. In: 2019 International Conference on Document Analysis and Recognition (ICDAR), pp. 142–147 (2019). https://doi.org/10.1109/ICDAR.2019.00031

24. Wang, J., Sun, K., Cheng, T., Jiang, B., Deng, C., Zhao, Y., Liu, D., Mu, Y., Tan, M., Wang, X., Liu, W., Xiao, B.: Deep high-resolution representation learning for visual recognition. IEEE Trans. Pattern Anal. Mach. Intell. **43**(10), 3349–3364 (2020). https://doi.org/10.1109/TPAMI.2020.2983686
25. Cai, Z., Vasconcelos, N.: Cascade R-CNN: delving into high quality object detection. In: 2018 IEEE/CVF Conference on Computer Vision and Pattern Recognition, pp. 6154–6162 (2018). https://doi.org/10.1109/CVPR.2018.00644
26. Ren, S., He, K., Girshick, R., Sun, J.: Faster R-CNN: towards real-time object detection with region proposal networks. In: Cortes, C., Lawrence, N., Lee, D., Sugiyama, M., Garnett, R. (eds.) Advances in Neural Information Processing Systems, vol. 28. Curran Associates, Inc. (2015)
27. Liu, W., Anguelov, D., Erhan, D., Szegedy, C., Reed, S., Fu, C.-Y., Berg, A.C.: SSD: single shot multibox detector. Lecture Notes in Computer Science, pp. 21–37 (2016). https://doi.org/10.1007/978-3-319-46448-0_2
28. Ioffe, S., Szegedy, C.: Batch normalization: accelerating deep network training by reducing internal covariate shift. In: Proceedings of the 32nd International Conference on Machine Learning, vol. 37, pp. 448–456 (2015)
29. Sandler, M., Howard, A., Zhu, M., Zhmoginov, A., Chen, L.-C.: Mobilenetv2: inverted residuals and linear bottlenecks. In: 2019 IEEE/CVF Conference on Computer Vision and Pattern Recognition, pp. 4510–4520 (2019). https://doi.org/10.1109/CVPR.2018.00474
30. Szegedy, C., Ioffe, S., Vanhoucke, V., Alemi, A.: Inception-v4, inception-resnet and the impact of residual connections on learning. In: AAAI'17: Proceedings of the Thirty-First AAAI Conference on Artificial Intelligence, vol. 31, pp. 4278–4284 (2017)
31. Dai, J., Li, Y., He, K., Sun, J.: R-FCN: object detection via region-based fully convolutional networks. In: Lee, D., Sugiyama, M., Luxburg, U., Guyon, I., Garnett, R. (eds.) Advances in Neural Information Processing Systems, vol. 29, pp. 379–387 (2016)
32. Lin, T.-Y, Maire, M., Belongie, S., Bourdev, L., Girshick, R., Hays, J., Perona, P., Ramanan, D., Zitnick, C.L., Dolláir, P.: Microsoft coco: common objects in context. In: Fleet, D., Pajdla, T., Schiele, B., Tuytelaars, T. (eds.) Computer Vision–ECCV 2014, vol. 8693, pp. 740–755 (2014). https://doi.org/10.1007/978-3-319-10602-1_48
33. Chen, L.-C., Papandreou, G., Kokkinos, I., Murphy, K., Yuille, A.L.: Deeplab: semantic image segmentation with deep convolutional nets, atrous convolution, and fully connected crfs. IEEE Trans. Pattern Anal. Mach. Intell. **40**(4), 834–848 (2018). https://doi.org/10.1109/tpami.2017.2699184
34. Baek, Y., Lee, B., Han, D., Yun, S., Lee, H.: Character region awareness for text detection. In: 2019 IEEE/CVF Conference on Computer Vision and Pattern Recognition (CVPR), pp. 9357–9366 (2019). https://doi.org/10.1109/CVPR.2019.00959
35. Wang, P., Li, H., Shen, C.: Towards end-to-end text spotting in natural scenes. IEEE Trans. Pattern Anal. Mach. Intell. (2021). https://doi.org/10.1109/TPAMI.2021.3095916
36. Kay, A.: Tesseract: an open-source optical character recognition engine. Linux J. (159) (Jul 2007)
37. Fey, M., Lenssen, J.E.: Fast graph representation learning with PyTorch Geometric. In: ICLR Workshop on Representation Learning on Graphs and Manifolds (2019)
38. Marmot v1 dataset (2002). https://www.icst.pku.edu.cn/cpdp/sjzy/
39. Szegedi, G., Veres, D.B., Lendák, I., Horváth, T.: Context-based information classification on Hungarian invoices. In: ITAT (2020)

Automation of Deanonymization Queries for the Bitcoin Investigations

Przemysław Rodwald and Nicola Kołakowska

1 Introduction

Bitcoin, as a digital coin which is not issued by any central authority,[1] was first suggested in 2008 by Nakamoto [7], and became fully operational in January 2009. Since then, Bitcoin has been marketed as a global decentralised digital currency and has been seen as increasingly used as part of criminal activities. The private sector reports that the unlawful use of cryptocurrencies accounts for a small part of their overall use, accounting for the 0.15% cryptocurrency transaction volume in 2021 despite the raw value of illicit transaction volume reaching its highest level ever (0.62%) [2]. In contrast, research from academia estimated a much higher volume, reporting that about 23% of transactions are associated with criminal activities [4]. The significant difference in estimation might be partially due to the different approaches and methodologies in the transaction analysis. But research agrees that the proportion of cryptocurrency use associated with illicit activities compared to legitimate use decreased overtime while the absolute amount has continued to increase [3].

Recent years have seen cryptocurrency increasingly used as part of criminal activities and to launder criminal proceeds. Criminals have also become more sophisticated in their use of cryptocurrencies. In addition to using cryptocurrencies to obfuscate money flows as part of increasingly complex money laundering schemes, cryptocurrencies are increasingly used by criminals as a means of payment or as an investment fraud currency. The number of cases involving cryptocurrencies on the Polish domestic market increase as well. The author, as a court expert, was involved in dozens of cryptocurrency legal cases in 2022, compared with just a few in 2021. In cases where illicit transactions must have been analysed. Following transactions through the Bitcoin blockchain manually or using visualization tools is a vital skill, but an investigator needs to translate that work into real-world data [5]. That time consuming and monotonous work could be automated and improve the whole process of investigation, what is the main goal of the research presented in this paper.

[1] https://rodwald.pl/blog/182/BTC.py.

P. Rodwald (✉) · N. Kołakowska
Department of Computer Science, Polish Naval Academy, Gdynia, Poland
e-mail: p.rodwald@amw.gdynia.pl

W. Zamojski et al. (eds.), *Dependable Computer Systems and Networks*, Lecture Notes in Networks and Systems 737, https://doi.org/10.1007/978-3-031-37720-4_20

2 BTC Addresses Clustering

Cluster addresses is a key element of the deanonymization process. Knowing that a given address belongs to a specific entity, interested authorities (e.g. law enforcement) may request the identified entity to disclose the personal data of the owner of the investigated address. All methods that group addresses into clusters allow to identify a given entity when at least one address belonging to a given cluster is correctly assigned to a real entity (exchange, mining pool, gambling). In the blockchain itself, this information is not stored, sometimes except in mining transactions where mining cooperatives reveal their identities, mainly for marketing reasons. Internet resources come to our aid, where in many places (forums related to cryptocurrencies, websites of entities disclosing the addresses of their wallets for the purpose of making donations, etc.) you can find information that clearly links a given address to a particular entity.

The subject of address aggregation was discussed in the literature particularly intensively in 2013. At that time, Ron and Shamir [11] analyzed the transaction chain, showing statistical results for typical behavior of Bitcoin network users based on transaction graphs. In the same year, several papers [1, 6, 8, 9] show heuristics that allow grouping addresses in such a way that they belong to the same entity.

The first heuristic, called idioms-of-use or multi-input transactions, groups input addresses included in a single transaction. If two or more input addresses are part of a single transaction, they are assumed to be controlled by the same entity. This heuristic should be limited to only input addresses that require a single private key and is defined as follows:

Heuristic 1. If two or more input addresses requiring the use of a single private key are part of a single transaction, they are controlled by the same entity.

The practical use of this heuristic is relatively easy. If among the input addresses of a certain transaction there is an address previously identified by the analyst and assigned to a specific entity, then new input addresses can also be assigned to a given entity.

The second heuristic groups one of the output addresses, the so-called rest address, with the input addresses included in a single transaction. If a transaction contains two output addresses, it is assumed that one of them is controlled by the same entity as the input addresses. This heuristic is called shadow addresses or change closures in the literature.

Heuristic 2. If the transaction consists of two output addresses, one of them is controlled by the same entity as the input addresses.

This heuristic is additionally based on the assumption that users rarely transfer funds to two different recipients in one single transaction. The main difficulty here is to correctly identify the shadow address of the two output addresses. Several scenarios explaining how to solve this problem have been presented in [10].

These theoretical foundations behind clustering addresses are used in the tools supporting the deanonymization process, both commercial (e.g.: Chainalysis, CipherTrace, Qlue, MerkleScience, Scorechain, Coinfirm, Cheksy, Iknaio) and free (e.g.: walletexplorer.com, sydeus.rodwald.pl, graphsense.info, ikna.io. These tools, thanks to the use of above heuristics, link addresses with real-world entities.

3 Project Design Rationale

The process of designing the solution was guided by the following design assumptions:

- open source nature of the project—a free development environment and a final result as a free open source license,
- "lightness" of the solution—not storing transaction data (currently, a full bitcoin node with transaction data is about 450GB large) but using data provided by block explorers,
- resistance to limitations imposed by data sources—the script should stop its operation for a period forced by query limits, and then automatically resume its operation,
- variety of data sources—the script should obtain data from many free data sources, both regarding deanonymization data and sanction data,
- flexibility—implementing the script in such a structure that in the future it can be easily expanded with other potential data sources.

4 Sources of Transaction Data

The first issue when designing the solution was to choose the source of transaction data, i.e. data in the bitcoin blockchain. It was decided to duplicate transaction data, and blockchain.info and blockcypher.com were selected as two independent data sources.

Blockchain.info is one of the most popular block explorers. It provides a free API without requiring an access key. In order to download transaction data for individual addresses, attention must be paid to the limit of downloaded transactions, which may not exceed 50 transactions per query. In order to download and analyze all transactions for the analyzed address, subsequent queries should download portions up to a maximum of 50 transactions.

Blockcypher.com is the second block explorer that was decided to use in this solution. It provides well documented API.[2] Depending on the purchased package, it offers different query limits, while the free package has the following limitations:

[2] https://www.blockcypher.com/dev/#blockchain-api.

2000 queries per day, 200 queries per hour and 3 queries per second. Also here, attention must be payed to the transaction limit for a particular address (20 transactions per query).

5 Sources of Deanonymization Data

One of the most important elements implying the effectiveness of the proposed solution was the selection of data sources providing data on the entities to which the addresses belong. It was decided to approach this issue as broadly as possible and acquire data from four data sources: walletexplorer.com, sydeus.rodwald.pl, graphsense.info and ikna.io.

WalletExplorer.com is one of the more valuable free projects for cryptocurrency investigators. It provides information about identified entities, currently for nearly 37 million[3] BTC addresses. The author of the site, AleÅ¡ Janda, currently works for the commercial solution chainanalysis.com. The system provides an API,[4] where an e-mail address should be provided as the API key. There is no need to register a new API key. The limitation is two queries per second for the required parameter "caller", which is basically any string of characters. This approach makes it easy to use a mechanism to bypass the restrictions—different values of the "caller" parameter can be given in queries.

Sydeus.rodwald.pl is a dedicated system, designed, implemented and used by the author to carry out expert opinions for the needs of law enforcement. Currently, there are over 142 million identified BTC addresses in the system's database, and for each new block added to the bitcoin blockchain, the system runs cron with scripts of the discussed heuristics. For efficiency reasons, access to the website (and API) has been limited only to logged-in users, who are law enforcement authorities that obtained access data as part of the expert reports performed by the author.

GraphSense.info is an open source cryptoasset analytics research platform, developed under the leadership of the Austrian Institute of Technology (AIT), with funding from the Austrian Research Promotion AgencyâŁ™s KIRAS program KRYPTOMONITOR. GraphSense is provided as an open source software, under the terms of the MIT license. The AIT offers free access to a hosted version of the software at demo.graphsense.info. The system provides well-documented API.[5] The current number of identified BTC addresses exceeds 143 million, and the query limits are 1000 queries per hour.

Ikna.io is a website provided by the Iknaio company founded in 2021, providing operational services around the open source analytics platform GraphSense. The

[3] The estimation was made using a dedicated script parsing data directly from walletexplorer.com website.

[4] Detailes about API are available on request send to the walletexplorer autor, https://www.walletexplorer.com/info.

[5] https://api.graphsense.info.

system, like its prototype, also provides access via API.[6] Currently, there are nearly 260 million tagged BTC addresses in the system database, and the query limits depend on the selected package, starting from 1000 queries per month for the free package. Paid packages, in addition to the public tags, additionally provide proprietary tags.

6 Sources of Sanctions Data

The last category of acquired data is data on addresses subject to sanctions. It was decided to obtain this data from two sources: chainalysis.com and bitcoinabuse.com. Sanctioned entities refer to entities listed on economic embargo lists, such as by the US, EU, or UN, with which anyone subject to those jurisdictions is prohibited from dealing. This includes the Specially Designated Nationals list of the US Department of the TreasuryâŁ™s Office of Foreign Assets Control. *Chainalysis.com* offers a free tool for checking the presence of addresses on sanctions lists called "Free Crypto Sanctions Screening Tools",[7] which in addition to the web interface also gives you the ability to call API queries. *BitcoinAbuse.com* is a public database of bitcoin addresses used by hackers and criminals. The endpoints have a rate limit of 30 requests per minute or one request every two seconds on average.

7 Implemented Solution

It was decided to implement the solution using a script written in Python programming language. During the implementation of the project, a number of problems were encountered and solved, including those related to the limits of queries to individual data sources.

To run the script, the Python environment must be installed. The script is run with three parameters as an input data: the BTC address the investigator want to analyze, the data sources services (more precisely, the first letters of the names of the services) and the direction of the search (IN or OUT, IN for incoming transactions, OUT for outgoing transactions—default value). Currently, the script supports queries to the four deanonymization services discussed above (their calls are as follows: walletexplorer.com [W]—default value, sydeus.rodwald.pl [S], graphsense.info [G], ikna.io [I]) and two sanctions services (chainalysis.com [C], bitcoinabuse.com [B]). A sample execution command looks as follows:

```
python BTC.py 115DL5MannhGS3rsmYYxCCZcHHekw8WDSP SG OUT
```

---

[6] https://api.ikna.io.

[7] https://www.chainalysis.com/free-cryptocurrency-sanctions-screening-tools.

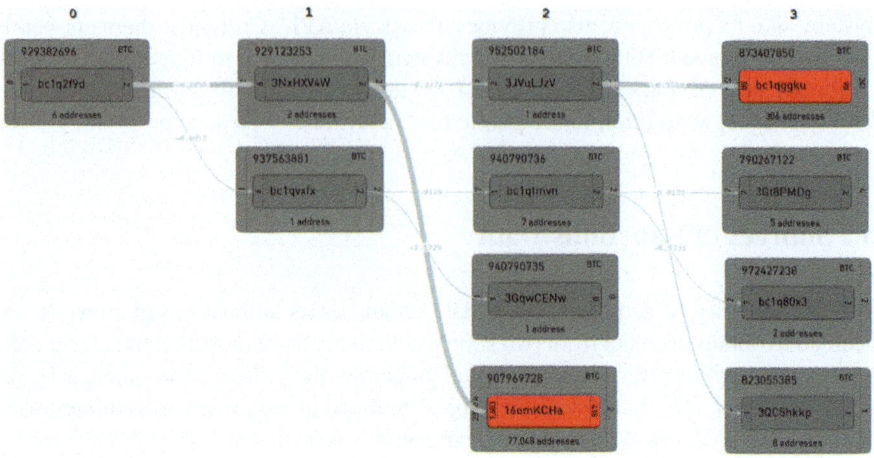

Fig. 1 Visualization of parameters *epoch* = 3 and *max* = 50 for the bitcoin transaction graph in GraphSense

Two additional parameters have also been used in the script: *epoch*—specifies the maximum number of consecutive epochs for which the script will be executed, *max*—specifies the maximum number of input/output addresses for which next epoch will be analysed. The visualization of both parameters is shown in Fig. 1 for the parameters *epoch* = 3 and *max* = 50. The analyzed bc1q2f9d... address is first checked in deanonymization and sanction data sources. This address transfers funds to two addresses 3NxHXV4W... and bc1qvxfx... which in the next epoch (1) are analyzed in the data sources. From address 3NxHXV4W... funds are transferred, among others, to address 16omKCHa... in the next epoch (2), which due to the number of output addresses (418) exceeding the set *max* parameter will not be further analyzed in the next epoch (3) (marked in red in the figure, similarly to address bc1qggku... for which also the number of output addresses (90) exceeds the *max* value). The run of the whole algorithm is terminated at the third epoch due to the set parameter *epoch* = 3. The total number of 11 addresses has been analysed in this scenario.

A strongly recommended pre-action before usage the script is to request services blochchair.com, chainalysis.com, graphsense.info, ikna.io for an individual access API key. Most of the listed systems have dedicated forms for this purpose: [B],[8] [C],[9] [I].[10] For graphsense [G],[11] an e-mail from the government domain must be sent to obtain the key. Lack of API keys will result in the execution of the script for the default configuration: transaction data is downloaded from the blockchain.info, deanonymization data from [W], sanctions data is not requested by default. After

---

[8] https://www.bitcoinabuse.com/register.

[9] https://www.chainalysis.com/free-cryptocurrency-sanctions-screening-tools.

[10] https://www.ikna.io/order/free.

[11] email to contact@graphsense.info.

obtaining the API keys, they must be inserted in the appropriate, clearly described, places inside the script:

```
apikey_blockcypher = "de183fd2..."
apikey_sydeus      = "6071f9bb..."
apikey_chainalysis = {'X-API-Key':'b88847f0...','Accept':'
application/json'}
apikey_graphsense  = {'Authorization':'b/EVrO1e...'}
...
```

The script itself, except for requesting web services for deanonymization and sanction data, tries to aggregate obtained data. Data aggregation is desirable because, due to the use of different data sources, the results obtained from them do not have to indicate the same source of origin. In addition, the data obtained are not always identical despite pointing to the same entity (for example, YoBit.net and yobit). Finally, an analysed BTC address could not be identified by all asked websites but by some of them. In the result file a sign (+ or –) informs about the result of data aggregation. For this purpose a python class SequenceMatcher is used. A sample content of the *result.txt* file with positive aggregation for all addresses looks as follows:

```
Address|Sydeus|Graphsense|Similarity|PreviousAddress|Epoch

115DL5MannhGS3rsmYYxCCZcHHekw8WDSP|10xBitco.in|10xBitco.in|
+ |0

14vWwBC2VFGF7nFLDWxSQWoaPiDuPeHfBk|FaucetBOX.com|faucetbox|
+ |115DL5...|1
18fDVfCh7jQPKPe7sXrijcRwTKSyB2JBiT|10xBitco.in|10xbitco.in|
+ |115DL5...|1
184GPXhgWr1e3VxPM57NZ1mnoogtpwEQPk|10xBitco.in|10xbitco.in|
+ |115DL5...|1
1KremojSu99KDq4QwCjBcnJb6GwdXUivS8|10xBitco.in|10xbitco.in|
+ |115DL5...|1
1Q8gdk8j2gTxVcRpJeDV4SmcJVXSH9K9UL|10xBitco.in|10xbitco.in|
+ |115DL5...|1
1AW2EpbxFVLbugMNYJzpyWeBXcXsUk7k3M|10xBitco.in|10xbitco.in|
+ |115DL5...|1
38ZHSPP5J2mojAmQaafeZctC8BMrPNZ5L1|Cubits.com|cubits.com|
+ |115DL5...|1
3Mw3Bei7T18s6MDB19FCXs3nPgXmACLfXg|Cubits.com|cubits.com|
+ |115DL5...|1
1Dyyp7oxDnyPJV7hiUkpqh61YLyydZYwrQ|Coinbase.com||
+ |115DL5...|1
```

8 Summary

The growing popularity of cryptocurrencies, also for committing crimes involving extortion or their use in money laundering, results in a noticeable increase in the number of criminal cryptocurrency cases. The proposed solution is intended to assist the law enforcement to identify entities to which BTC addresses may belong. It should be noted that this activity is just the first step in a three-stage attempt to identify the real owner of the address. In the second step, the authority should ask the identified entity to provide information about the data stored in the system associated with a given address (name, address, ID card scan, login IP addresses, etc.). In the third step, the obligated institution should make the data in its possession available to the law enforcement, what is not always easy to execute.

The author, as part of his research and court expert activities, plans further development of the proposed solution, in particular the implementation of scripts for other cryptocurrencies and extension to other data sources.

References

1. Androulaki, E., Karame, G.O., Roeschlin, M., Scherer, T., Capkun, S.: Evaluating user privacy in bitcoin. In: International Conference on Financial Cryptography and Data Security, pp. 34–51. Springer (2013)
2. Chainalysis: the 2022 crypto crime report (Feb. 2022). https://go.chainalysis.com/2022-Crypto-Crime-Report.html
3. Europol: Cryptocurrencies-Tracing the Evolution of Criminal Finances. Publications Office of the European Union (2021)
4. Foley, S., Karlsen, J.R., Putniņš, T.J.: Sex, drugs, and bitcoin: how much illegal activity is financed through cryptocurrencies? Rev. Financ. Stud. **32**(5), 1798–1853 (2019)
5. Furneaux, N.: Investigating Cryptocurrencies: Understanding, Extracting, and Analyzing Blockchain Evidence. Wiley (2018)
6. Meiklejohn, S., Pomarole, M., Jordan, G., Levchenko, K., McCoy, D., Voelker, G.M., Savage, S.: A fistful of bitcoins: characterizing payments among men with no names. In: Proceedings of the 2013 Conference on Internet Measurement Conference, pp. 127–140 (2013)
7. Nakamoto, S.: A peer-to-peer electronic cash system (2008). https://bitcoin.org/bitcoin.pdf
8. Ortega, M.S.: The bitcoin transaction graph-anonymity. Ph.D. thesis, Master's thesis, Universitat Oberta de Catalunya (2013)
9. Reid, F., Harrigan, M.: An analysis of anonymity in the bitcoin system. In: Security and Privacy in Social Networks, pp. 197–223. Springer (2013)
10. Rodwald, P.: Kryptowaluty z perspektywy informatyki śledczej. Polish Naval Academy (2021). ISBN: 978-83-959756-7-7
11. Ron, D., Shamir, A.: Quantitative analysis of the full bitcoin transaction graph. In: International Conference on Financial Cryptography and Data Security, pp. 6–24. Springer (2013)

Structural Models for Fault Detection of Moore Finite State Machines

Valery Salauyou

1 Introduction

Currently, unmanned aerial vehicles (drones) are widely used in many areas of human activity, including in military conflicts. One of the ways to combat drones is to influence the drone with an electromagnetic pulse in order to disable its control system [1]. A finite state machine (FSM) is a mathematical model of sequential circuits as well as control devices.

Figure 1 shows the traditional structural model of an FSM, where X is the input signals (an input vector); Y is the output signals (an output vector); R is the state register in which the code of the present (current) state is stored; Φ is the combinational circuit (logic) determining the code of the next state; Ψ is the combinational circuit determining the values of the output signals. Note that the output of the register R is connected by the feedback to the input of the combinational circuit Φ. For Moore FSMs, the values of the output signals are formed on the basis of the present state code, and for Mealy FSMs, the values of the output signals are formed on the basis of the present state code and the values of the input signals. So for Mealy FSMs, the inputs X are connected to inputs of the combinational circuit Ψ. The clock signal clk for the register R is generated using the generator Oscillator.

As a result of exposure to an electromagnetic pulse, failures may occur:

- at the inputs: an invalid input vector;
- in the R register: an invalid present state code;
- in the feedback circuit: an invalid present state code;
- in the logic Φ: an invalid next state code;
- in logic Ψ: an invalid output vector;
- in the clock circuit: no clock signal;

V. Salauyou (✉)
Bialystok University of Technology, Wiejska 45A, 15-351 Bialystok, Poland
e-mail: v.salauyou@pb.edu.pl

W. Zamojski et al. (eds.), *Dependable Computer Systems and Networks*, Lecture Notes in Networks and Systems 737, https://doi.org/10.1007/978-3-031-37720-4_21

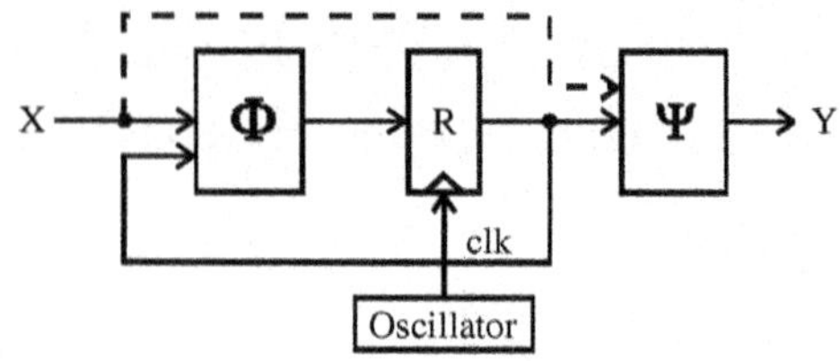

Fig. 1 The traditional structural model of FSMs

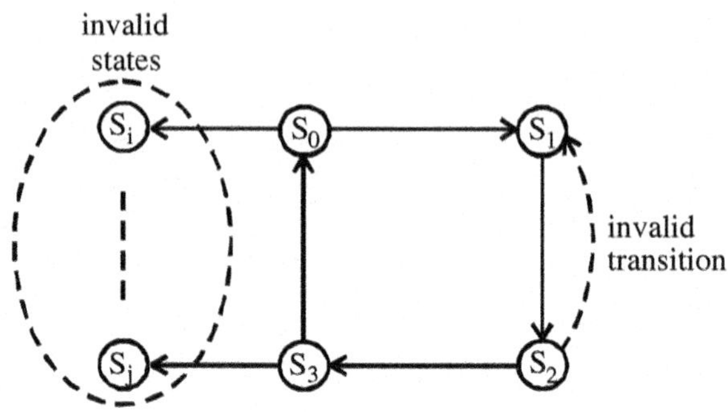

Fig. 2 Possible faults of the FSM

- in the generator Oscillator: failure of the clock generator.

In addition, as a result of exposure to an electromagnetic pulse on the FSM, the transitions of the FSM to invalid states are possible, as well as invalid transitions to valid states (Fig. 2).

There are quite a lot of methods for building fault-tolerant FSMs. Most of them are devoted to countering single event upsets (SEUs) caused by radiation and cosmic rays, which change the contents of flip-flops or memory cells, while the primary inputs always considered as correct. However, the electromagnetic pulse is characterized by the following features of impact on drones:

- significant duration of the exposure time, compared to the space particle;
- impact on all circuit components at once;
- generates not short-lived SEUs, but long-lasting multiple faults;
- affects mainly on wires (input, output, feedback);
- rarely changes the contents of registers or memory cells.

Note that modern methods of designing digital devices [2] differ significantly from the approaches used a few decades ago. Recently, hardware description languages (HDLs), such as VHDL, Verilog, SystemVerilog, etc., have received significant development. As a result, the synthesis of the device is reduced to a correct description of the design behavior in some HDL [3]. At the same time, the traditional stages of the FSM synthesis are eliminated: encoding of states; forming of logical (Boolean) equations for the combinational circuit Φ and Ψ; minimization, factorization and decomposition of the logic. All these actions are performed automatically by the synthesis tools. In addition, the synthesis tools can optimize the logic that is needed to handle errors. Therefore, new approaches to the design of fault-tolerant FSM are needed.

The research area of this paper is the design of fault-tolerant FSMs. But before correcting the FSM failure by any method, the failure must be detected. Since the model of the Moore FSM is most often used in engineering design practice, the purpose of this paper is to develop structural models for detecting faults of the Moore FSMs.

Designing fault-tolerant computational and control systems, which are FSMs, has always been topical since the advent of computers. The problem of the FSM protection from cosmic rays was intensively studied in the late 1960s. The traditional solution to the problem is multiple duplication of the FSM architecture, while the method of triple modular redundancy (TRM) for protection against SEU has become the most widespread [4]. In general, the fault tolerance of a digital system can be provided by the architecture redundancy, the runtime increase, and the data redundancy [5].

The use of field programmable gate arrays (FPGAs) to design FSMs offers several advantages over application-specific integrated circuits (ASICs): low cost, short time to market, and possibility of reprogramming. However, FPGAs are more susceptible to SEUs caused by space particles than ASICs. Therefore, Xilinx has released the special FPGAs of the Virtex family, which support the TMR method at the hardware level [6].

Separate works are devoted to methods of encoding states in the synthesis of fault-tolerant FSMs. In [7], four methods of the state encoding for fault-tolerant FSMs are considered: binary, one-hot, Hamming with distance 2 (H2) and Hamming with distance 3 (H3); this compares a fault tolerance and resource utilization. In [8], the methods of state encoding (binary, one-hot and H3) to eliminate SEU in the state register are investigated; it is recommended to manually set the logic for recovery from an invalid state.

Recently, the interest in the design of fault-tolerant FSMs has not weakened. In [9], the proposed method improves the fault tolerance of FSMs by selectively applying the TMR method according to the importance of the state. In [10], the quasi-delay-sensitive architecture of an FSM is proposed and compared to TMR.

Consideration of the above-mentioned works shows that most synthesis methods of fault-tolerant FSMs are devoted to improvement of the TMR method for correcting SEU in the state register. It is assumed that the primary inputs, the logic of generating the next state code, logic of forming outputs, as well as additional logic for detecting and correcting errors do not have failures.

At the same time, structural models of FSMs different from traditional ones are very effective for improving performance, reducing implementation cost and power consumption in implementing FSMs on FPGAs [11]. This paper proposes structural models of the Moore FSM for detecting multiple faults and preventing their negative impact on the controlled object.

The paper is organized as follows. Section 2 describes an example of the Moore FSM, which is used to explain the proposed approach. Section 3 presents the proposed structural models of FSMs. The experimental results and their analysis are presented

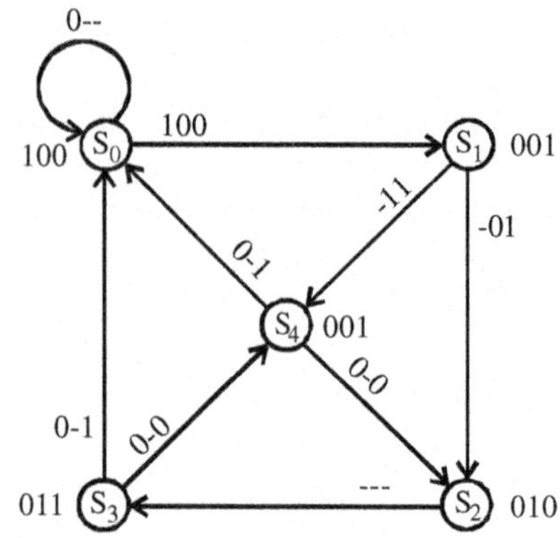

Fig. 3 The graph of the Moore FSM

in Sect. 4. Section 5 provides recommendations for the practical use of the proposed models when considering specific examples. The Conclusion summarizes the results and indicates the direction of further research.

2 The Demonstration Example

As an example, consider the Moore FSM, the graph of which is shown in Fig. 3. Our FSM has 5 states, 3 inputs and 3 outputs. The vertices of the graph correspond to the states $S_0,\ldots, S_4$, and the arcs of the graph correspond to the transitions of the FSM. The input vector that initiates the transition is written near each arc of the graph. Near each vertex of the graph is written the output vector, which is formed in this state. Here the hyphen ("-") can take any bit value: 0 or 1.

The valid transitions between states and the valid output vectors in each state are determined directly from the FSM graph. The valid input vectors in each state are defined by the developer based on the behavior of the device. Let the valid input vectors in each state for our example be defined as follows: S_0 and S_2—any input vectors; S_1—001, 101, 011, and 111; S_3 and S_4—000, 010, 001, and 011.

3 Proposed Structural Models of FSMs

Structural models of FSMs for fault detection and prevention of negative effects of faults on the controlled object are shown in Fig. 4.

The structure in Fig. 4a allows to detect invalid FSM transitions. For this purpose the combinational circuit VT (valid transitions) is added to the traditional FSM structure (Fig. 1). The input of the combinational circuit VT receives the code *state* of the present state, which is stored in the R register, and the code *next* of the next state, which is generated by the logic Φ. The output of the combinational circuit VT

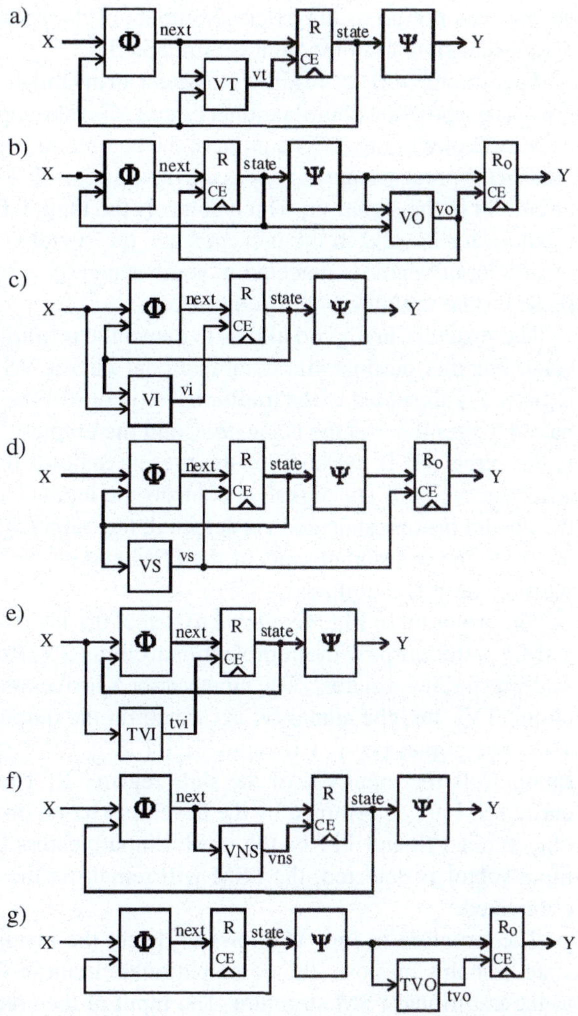

Fig. 4 Structural models of FSMs for detecting: **a**—invalid transitions; **b**—invalid output vectors in each state; **c**—invalid input vectors in each state; **d**—invalid code of the present state; **e**—invalid input vectors for the whole FSM; **f**—invalid code of the next state; **g**—invalid output vectors for the whole FSM

is the signal vt, which controls the input CE (clock enable) of the state register R. If the transition is valid, the signal vt = 1, otherwise vt = 0. If an invalid transition is detected (vt = 0) the FSM remains in the present state until the fault is corrected.

The structure in Fig. 4b allows to detect the invalid output vector in each state. For this purpose the combinational circuit VO (valid outputs) is added to the structure in Fig. 1, as well as the output register R_O. The input of the combinational circuit VO receives the code *state* of the present state and the values of outputs generated by logic Ψ. The output of the combinational circuit VO is the signal vo, which controls the input CE of the registers R and R_O. If the output vector is valid, the signal vo = 1, otherwise vo = 0. If an invalid output vector is detected (vo = 0) the invalid output vector does not go to the external outputs of the FSM and the FSM will remain in the present state until the fault is eliminated.

The structure in Fig. 4c allows to detect an invalid input vector for a particular state. For this purpose the combinational circuit VI (valid inputs) is added to the traditional FSM structure in Fig. 1. The input of the combinational scheme VI receives the code *state* of the present state as well as the input vector X. The output of the combinational circuit VI is the signal vi, which controls the input CE of the state register R. If the input vector is valid in the present state, the signal vi = 1, otherwise vi = 0. If the invalid input vector is detected in some state (vo = 0), the FSM will remain in the present state until the fault is eliminated.

The structure in Fig. 4d allows to detect the invalid code *state* of the FSM present state. For this purpose the combinational circuit VS (valid states) and the output register R_O are added to the traditional structure of the FSM. The input of the combinational circuit VS is the code *state*, and the output of the combinational circuit VS is the signal vs. If the code of the present state of the FSM is valid, then vs = 1, otherwise vs = 0. The signal vs controls the inputs CE of the registers R and R_O. If the invalid present state code is detected, the output signals generated by logic Ψ do not go to the external outputs of the FSM and the FSM remains in the present state until the fault is eliminated.

The structure in Fig. 4e allows to detect the invalid input vector X for the whole FSM. For this purpose the combinational circuit TVI (total valid inputs) is added to the traditional FSM structure. The input vector X arrives on the input of the combinational circuit TVI, and the signal tvi is formed on the output of the combinational circuit TVI. The signal tvi = 1 if input vector X is valid, otherwise tvi = 0. The signal tvi controls the input CE of the state register R. The invalid input vectors for the entire FSM are determined by the developer based on the behavior of the controlled object. Let 110 and 111 be the invalid input vectors for our example. If the invalid input vector is detected, the FSM will remain in the present state until the fault is eliminated.

The structure in Fig. 4f allows to detect the invalid code *next* of the FSM next state. For this purpose, the combinational circuit VNS (valid next states) is added to the traditional FSM structure. The input of the combinational circuit VNS is the next state code *next* generated by the logic Φ. The output of combinational circuit VNS is the signal vns. The signal vns = 1 if the next states code is valid, and vns = 0 otherwise. The signal vns controls the input CE of the state register R. For our example, the valid codes are the codes of the states $S_0,\ldots, S_4$. If an invalid next state code is detected, the FSM will remain in the present state until the fault is eliminated.

The structure in Fig. 4g allows to detect the invalid output vector for the whole FSM. For this purpose the combinational circuit TVO (total void outputs) is added to the traditional FSM structure. The inputs of the combinational circuit TVO are the signals generated by the logic Ψ, and the output of the combinational circuit TVO is the signal tvo. The signal tvo = 1 if the generated output vector is valid, and tvo = 0 otherwise. The signal tvo controls the input CE of the output register R_O. For our example the valid output vectors are the output vectors generated in the states of

Table 1 Possible causes of the faults detected by FSM structural models

Causes of failure	VT	VO	VI	VS	TVI	VNS	TVO
Φ	*	*	*	*		*	*
X	*	*	*	*	*	*	*
Feedback	*	*	*	*		*	*
Ψ		*					*
R	*	*	*	*		*	*

the FSM: 100, 001, 010, 011, and 001. If the invalid output vector is detected, the register R_O will retain its previous value until the fault is corrected.

Table 1 summarizes the possible causes of faults that are detected by the structural models in Fig. 4, where Φ is a failure in the logic Φ; X is the invalid input vector; *feedback* is a failure in the feedback circuit; Ψ is a failure in the logic Ψ; R is a failure in the state register R.

The diagnostic signals vt, vo, vi, vs, tvi, vns, and tvo, which are generated in the FSM structural models in Fig. 4, can be output to the external FSM outputs to detect the fault location. The structures in Fig. 4 can be arbitrarily combined together for the most efficient detection of the FSM faults, if desired. Note, however, that the more diagnostic signals in the FSM structure, the easier it is to locate the fault.

Note that all the FSM structural models considered are not targeted for implementation on a particular electronic component: each structural model can be implemented on both ASIC and FPGA.

4 Experimental Results

All considered structural models for our example of the Moore FSM were described in Verilog language in the style with three processes [3]. As a result, the following FSM designs were created for our example:

- based_FSM—the traditional (basic) structural model of the Moore FSM (Fig. 1);
- VT_FSM—the FSM with check of the valid transitions between states (Fig. 4a);
- VO_FSM—the FSM with check of the valid output vectors in each state (Fig. 4b);
- VI_FSM—the FSM with check of the valid input vectors in each state (Fig. 4c);
- VS_FSM—the FSM with check of the valid code of the present state (Fig. 4d);
- TVI_FSM—the FSM with check of the valid input vectors for the whole FSM (Fig. 4e);
- VNS_FSM—the FSM with check of the valid code of the next state (Fig. 4f);

Table 2 Results of experimental studies of the structural FSM models

Design	L	L/L_b	F	F/F_b	Notes
based_FSM	11	1.00	355	1.00	
VT_FSM	15	1.36	277	0.78	
VO_FSM	12	1.09	439	1.24	The delay per one clock cycle
VI_FSM	14	1.27	323	0.91	
VS_FSM	12	1.09	447	1.26	The delay per one clock cycle
TVI_FSM	12	1.09	349	0.98	
VNS_FSM	12	1.09	389	1.10	
TVO_FSM	12	1.09	438	1.23	The delay per one clock cycle

- TVO_FSM—the FSM with check of the valid output vectors for the whole FSM (Fig. 4g).

To avoid the exclusion of the additional logic as a result of the automatic optimization by synthesis tools, the output signals of the additional combinational circuits are declared with the attribute "keep".

In order to evaluate resources and performance, the designs has been synthesized using Quartus Prime tool (version 22.4) from Intel on the Cyclone 10 LP FPGA. The experimental results are given in Table 2, where L is the number of used logic elements (LUTs) or area; F is the maximum frequency of functioning (in megahertz) or speed; L_b and F_b are similar parameters for the basic structural model (the design based_FSM); L/L_b and F/F_b are relations of corresponding parameters.

Table 2 shows that the structures VT (36% increase in area) and VI (27% increase in area) have the largest area overhead, while the other structures slightly increase the area of the FSM (9% increase in area).

In terms of performance, the most costly structure is VT, which reduces performance of the FSM by 22%, followed by structure VI (reduction of performance by 9%). The structure TVI slightly reduces performance (only by 2%), and the structure VNS even increases performance by 10%. The structures VO, VS, and TVO significantly increase performance of the FSM (by 23–26%). This is explained by the introduction of the output register R_O in the basic FSM structure. However, the addition of the register R_O leads to delay of the output signals for one clock cycle.

5 Recommendations for the Practical Use

The selection of the FSM structural models and their possible merging is performed by the designer of the control device. The following factors must be taken into account when selecting FSM structural models:

- faults that are detected by the model;

- the ability to locate the fault;
- admissible increase in area;
- admissibility of delay per one clock cycle of the FSM operation;
- admissibility of the FSM performance reduction.

Table 3 summarizes the properties of the proposed structural FSM models, which can be taken into account when selecting the most appropriate the FSM structural model.

For example, let it be necessary to select the structural model of the Moore FSM that allows detecting the largest number of faults and does not allow the delay per one clock cycle of the FSM operation, while it is desirable to minimize the increase in the area and decrease the FSM performance. Since the FSM does not allow the delay per one clock cycle of the output signals, according to Table 3, the structures VT, VI, TVI, and VNS should be chosen. We exclude the structure TVI because it (Table 1) detects only one fault (the invalid input vector X). The remaining structures detect the same number of possible causes of faults (Table 1). From the structures VT, VI, and VNS only structure VNS (Table 3) slightly increases the area and does not reduce the performance, compared to the basic structure, so for the implementation we choose the structure VNS. As a result, we solved the task without increasing the area and without reducing the performance of the FSM.

When choosing the FSM structure for implementation, the number and complexity of detected faults should also be taken into account. Suppose, for example, the same task is set, but the delay per one clock cycle of the output signals is allowed. The structures VO and TVO reveal the greatest number of causes of the faults (Table 1). The structures VO and TWO slightly increase the area. However, since the structure TVO detects invalid output vectors for the entire FSM (there are not very many such vectors), and the structure VO detects invalid output vectors for each state, and detecting such faults is much more difficult. Therefore, we choose the structure VO for implementation, since it detects a greater number of more complex faults, while the FSM area does not increase, and the FSM performance even increases by 28%.

If it is necessary to specify the location and cause of the fault most accurately, then all structures in Fig. 4 should be combined together for implementation, and the diagnostic signals vt, vo, vi, vs, tvi, vns, and tvo should be output to the external pins of the FSM.

Table 3 Properties of the proposed FSM structural models

Property	VT	VO	VI	VS	TVI	VNS	TVO
Does not delay per one clock cycle	*		*		*	*	
Slightly increases the FSM area		*		*	*	*	*
Increases the FSM performance		*		*		*	*

6 Conclusions

This article presents structural models of the Moore FSMs that allow to detect the following faults:

- the invalid transitions between the states;
- the invalid output vectors generated in each state;
- the invalid input vectors in each state;
- the invalid present state code;
- the invalid next state code;
- the invalid input vectors for the entire FSM;
- the invalid output vectors for the entire FSM.

In addition, the considered structural models allow to detect multiple failures in the logic Φ determining the code of the next state, in the logic Ψ determining the values of the output signals, in the state register R, and in the feedback circuit.

Future research will focus on developing structural models that allow to correct of the FSM failures.

Acknowledgements The present study was supported by a grant WZ/WI-III/5/2023 from Bialystok University of Technology and founded from the resources for research by Ministry of Science and Higher Education.

References

1. Park, S., Kim, H.T., Lee, S., Joo, H., Kim, H.: Survey on anti-drone systems: components, designs, and challenges. IEEE Access **9**, 42635–42659 (2021)
2. Solov'ev, V.V.: ASMD–FSMD technique in designing signal processing devices on field programmable gate arrays. J. Commun. Technol. Electron. **66**(12), 1336–1345 (2021)
3. Salauyou, V., Zabrocki, Ł.: Coding techniques in Verilog for finite state machine designs in FPGA. In: IFIP International Conference on Computer Information Systems and Industrial Management, pp. 493–505. Springer, Cham (2019)
4. Lyons, R.E., Vanderkulk, W.: The use of triple-modular redundancy to improve computer reliability. IBM J. Res. Dev. **6**(2), 200–209 (1962)
5. Aviziens, A.: Fault-tolerant systems. IEEE Trans. Comput. **100**(12), 1304–1312 (1976)
6. Carmichael, C.: Triple module redundancy design techniques for Virtex FPGAs. Xilinx Application Note XAPP197, 1 (2001)
7. Burke, G. R., Taft, S.: Fault tolerant state machines. nasa.gov (2004)
8. Berg, M.: A simplified approach to fault tolerant state machine design for single event upsets. In: Mentor Graphics Users' Group User2User Conference (2004)

9. Choi, S., Park, J., Yoo, H.: Area-efficient fault tolerant design for finite state machines. In: 2020 International Conference on Electronics, Information, and Communication, pp. 1–2. IEEE (2020)
10. Verducci, O., Oliveira, D. L., Batista, G.: Fault-tolerant finite state machine quasi delay insensitive in commercial FPGA devices. In: 2022 IEEE 13th Latin America Symposium on Circuits and System, pp. 1–4. IEEE (2022)
11. Klimowicz, A.S., Solov'ev, V.V. Structural models of finite-state machines for their implementation on programmable logic devices and systems on chip. J. Comput. Syst. Sci. Int. **54**(2), 230–242 (2015)

Application of Generative Models to Augment IMU Signals in Gait Biometrics

A. Sawicki and K. Saeed

1 Introduction

In recent years, a dynamic development of human movement analysis applications using IMU sensors can be observed. The large increase in the number of works in this area is mainly related to the high availability of this type of sensors. Accelerometers and gyroscopes manufactured using microelectromechanical system (MEMS) technology are commonly embedded in mobile devices such as smartphones and smartwatches [1]. This offers new opportunities for the implementation of smart systems. Currently, classical machine training methods are pervasive and represent the major trend in existing applications. At the same time the use of deep training methods, which shows promise in the field of image processing, is rather limited. Of course, there are pioneers in the scientific field [2–4], but the research area is not extensively explored.

In this paper, we would like to present the author's results of research on data augmentation methods and their impact on biometrics person verification system. In our experiments, we proposed the architecture of a Density Network-type generative model [5–7] and benchmarked the achieved results against competing methods—TimeVAE [8] and Riemannian Hamiltonian VAE [9]. Validation of the solution was performed for a CNN classifier with an attention mechanism [4]. The research was carried out using the author's data corpus, which is available on request. The data

A. Sawicki (✉) · K. Saeed
Faculty of Computer Science, Bialystok University of Technology, Bialystok, Poland
e-mail: a.sawicki@pb.edu.pl

K. Saeed
e-mail: k.saeed@pb.edu.pl

K. Saeed
Department of Computer Science and Electronics, Universidad de la Costa, Barranquilla, Colombia

W. Zamojski et al. (eds.), *Dependable Computer Systems and Networks*, Lecture Notes in Networks and Systems 737, https://doi.org/10.1007/978-3-031-37720-4_22

corpus is characterized by essential features such as: 100 participants; data collected during two motion tracking sessions (over two days).

The carried out experiments helped answer the following questions, whether data augmentation involving generative models can have a beneficial effect on gait based biometrics systems? Can data generation using density network models be more effective than using variational autoencoder-based models?

2 Literature Review

The following section describes in detail signal generation methods based on distribution models such as LSTM-MDN, and briefly describes VAE based architectures [8, 9].

Among the publications dealing with the modelling of distribution parameters, the scientific articles [5–7] are particularly worth mentioning. In [5], the LSTM-MND network was applied to generate handwriting signatures. The first component of the network was used as a feature extractor, while the second one was responsible of data augmentation. Proposed network architecture was able to produce parameters describing movement of the pen on the screen (bivariate normal distribution) as well as touch of the digital-pen on the screen (Bernoulli distribution). In the presented approach, k different distributions were modelled. The probability of using a particular one was described by the parameter $\boldsymbol{\pi}$, which was also modelled by the network.

The presented solution has several disadvantages. Firstly, a significant training set is required due to the recursive LSTM module. Number of modelled k distributions is determined by the choice of the researcher. This can lead to selecting an overly extensive model. Furthermore, the presented network is not fully scalable to higher dimensions of data. The presented approach shows the modelling of dependent signals in 2D space. For higher dimensionality, the cost function becomes significantly more complicated and requires the determination of an inverse covariance matrix.

The LSTM-MDN approach was also presented in [6], where it was used to generate specifically IMU signals. In presented paper [6], the possibility of generating an accelerometer signal for Human Activity Recognition (*HAR*) data is presented. The deterministic nature of the LSTM layers was cited as the only reason for using the Density Network layer. It should be noted that the authors described the generated signal as difficult to evaluate, and compare with real signals. Therefore it is difficult to evaluate the similarity between the generated output and acquired signals.

On the other hand, publication [7] presents the architecture of an MLP network implementing Distributrional Regression. The posted example describes a network architecture capable of reconstructing a sinusoidal signal by modelling the variance and mean value, at any time t/x. The model is trained on N synchronized noisy sinusoidal signals. In contrast to previously discussed solutions, the network was only capable of evaluating fixed-length samples. Furthermore, the network was only used to model the parameters of one-dimensional signals. Finally proposed solution focus only on the theoretically aspects completely ignoring the issues of synthetic generation of real signals.

Among the competing VAE solutions, scientific articles [8, 9] are particularly worth mentioning. In general VAE models consist of two components—encoder and decoder. The first is used for feature extraction, while the second is used to reconstruct the input data. In the training process, the original signal is applied to the model input. The output is expected to reconstruct the input data with as small difference as possible. The so-called bottleneck in the middle of the model makes the model perform a compression function by remembering the most important features of the signal. In the case of variational autoencoders, the bottleneck is used to model normal distributions. Synthetic data generation, is performed by applying random parameters to the decoder module. In the present work, two VAE models are used, i.e. TimeVAE [8] dedicated to working with time series and Riemannian Hamiltonian VAE [9] as a state of the art solution in the field of multi-dimensional data augmentation.

Moving to the details, what differentiates the TimeVAE model from the baseline VAE is highlighting worthy. This type of architecture has dedicated "Trend" and "Seasonality" Blocks layers in the encoder layer. This stands in opposition to the classical approach where there are fully-connected layers. According to the authors, this approach makes the model perform more effectively for input data in the form of time series. In the case of RHVAE, the modification consisted of two changes the first of which concerned the so-called latent space. In the case of the mentioned architecture, it is modelled as a Riemannian manifold and uses Riemannian metric learning. The second important element was the use of the so-called Riemannian Hamiltonian Monte Carlo Sampler, which is expected to ensure better quality of the generated data especially for low-volume datasets.

The proposed augmentation method is a combination of the methods presented in [5, 7]. Our pilot experiments with MDN showed a very large value of the parameter π for one of the k models. On the other hand, other $k-1$ models had a π factor very close to zero. This indicated that the model was overfitting the input data. We therefore decided to carry out further experiments on a network modelling a single normal distribution. In contrast to [7], it was decided that one architecture should model the signal in more than two dimensions. In the research carried out the data represents 6-dimensional space (as a result of using the measurement readings of the triaxial accelerometer and triaxial gyroscope).

3 Methodology

3.1 Dataset Description

The study uses a proprietary human gait database dedicated to the analysis of biometrics systems. The corpus is characterized by the number of 100 participants and the availability of two motion tracking sessions within two separate days. During a single motion tracking session, the participant performed 20 gait trials under laboratory conditions at a distance of approximately 3 m.

Data acquisition was carried out using Notiom's professional motion capture Perception Neuron system. This device consisted of 17 sensors capable of sampling signals such as acceleration, angular velocity, magnetic field strength and orientation at 120 Hz frequency rate. Due to the use of so-called 9DOF sensors (3 axis accelerometer, gyroscope and magnetometer), it can be assumed that orientation was characterized by high precision. The entire body suit was equipped with sets of motion sensors mounted on various parts of the body.

It was decided to use measurement data from just one sensor located on the right thigh, as it reflects well the measurement values of the sensor built into the mobile phone (located in the trouser pocket). This approach is quite close to the real-life scenario.

The measurement data collected during data acquisition was subject to preprocessing

1. segmented according to the algorithm developed in our earlier work [10];
2. transformed from the sensor reference system to the global system [11];
3. interpolated to a fixed length of 128 samples.

Figure 1 presents example data for the selected participant. The individual subplots present all available gait cycles for the selected sensor axis.

3.2 Author's Data Augmentation Method

This paper presents the concept of signal augmentation through the use of generative models. In the first stage ('*TRAINING PHASE*'), the models were trained using samples of historical data. In the second ('*SAMPLING PHASE*'), the models were used to generate new samples, which eventually augmented the training set of classifiers (Fig. 2).

For the experiment described and the author's database (100 subjects), 100 instances of the model were created. Each was able to generate measurement signals of a 3-axis gyroscope and a 3-axis accelerometer specific to a particular participant.

The "training phase" involved training 100 instances of the model. Each had 12 outputs modeling parameters such as mean value as well as standard deviation. In the "sampling phase" 960 or 1820 synthetic samples were generated for each participant.

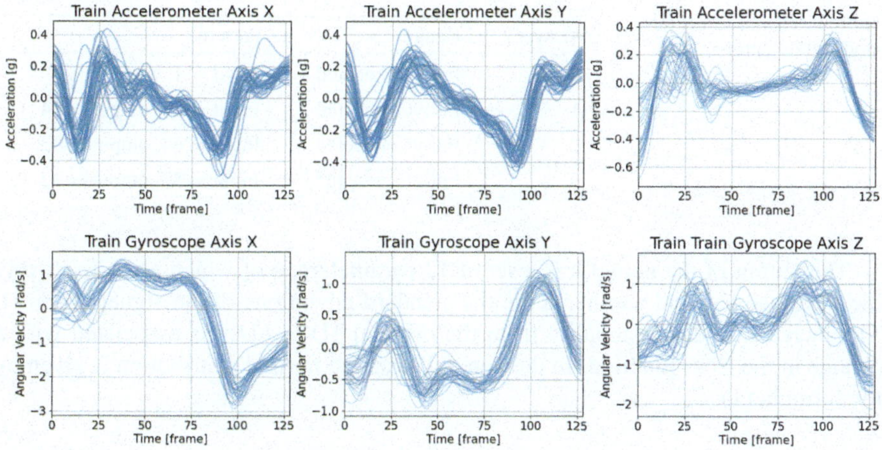

Fig. 1 Examples of all gait cycles available to the selected participant

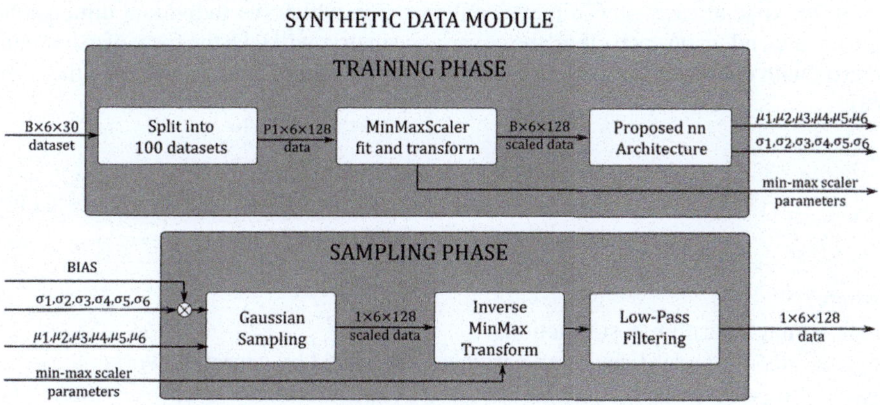

Fig. 2 Data augmentation mechanism diagram

In real condition each participant performed about 30 complete gait cycles. It can be summarised that there were 32 or 64 new samples per original gait cycle.

Futhermore, in the "sampling phase", it is possible to manually increase the *BIAS* parameter. This indicator is responsible for amplifying the standard deviation.

Table 1 summarises the architecture of the neural network. It should be noted that it consists of typical fully connected layers. However, the simultaneous modelling of 12 outputs, as well as a dedicated cost function, distinguishes such a network from typical models. The Adam Algorithm was used as the network optimizer, with a learning rate parameter equal to 0.001. The training process took 200 epochs and was the only stopping criterion. It was decided not to use the cost function of the validation set, as the validation and test sets have significant distribution shift.

Table 1 Proposed neural model architecture

#Layer	Range	Pattern
0	FullyConnected	Input = 1, output = 20
1	FullyConnected	Input = 20, output = 64
2	FullyConnected	Input = 64, output = 32
3	FullyConnected	Input = 32, output = 12

The output of the network is a set of 12 parameters, 6 of which are respectively the mean values, and another 6 are the standard deviations of the signal at time t. The increase in dimensionality from the original 1D to 6D was associated with a change in the Loss function. In the general case, the normal distribution is given by the formula (1):

$$f_{\mu,\Sigma} = \frac{1}{(2\pi)^{n/2}|\Sigma|^{1/2}} \cdot exp\left(-\frac{1}{2}(X-\mu)^T\Sigma^{-1}(X-\mu)\right) \tag{1}$$

In the special case—a 6-dimensional space, in which the individual dimensions are not dependent on each other, the covariance matrix will take the form of a diagonal matrix with coefficients $\sigma_1^2, \ldots, \sigma_6^2$. The logarithmic cost function, will take the form (2):

$$loss = -\log\left(8\pi^3|\sigma_1||\sigma_2||\sigma_3||\sigma_4||\sigma_5||\sigma_6|\right)^{-1} + \frac{1}{2}\left[(y_1-\mu_1)^2 + (y_2-\mu_2)^2 + (y_3-\mu_3)^2 + (y_4-\mu_4)^2 + (y_5-\mu_5)^2 + (y_6-\mu_6)^2\right] \tag{2}$$

where:
$loss$—final logarithm cost function;
$\sigma_1, \ldots, \sigma_6$,—Standard deviation parameters modeled by neural network;
$\mu_1, \ldots, \mu_6$—Average parameters modeled by the neural network;
$y_1, \ldots, y_6$—Accelerometer and Gyroscope measurements reading/neural network input.

Figure 3a presents the original data, together with the model-determined signal means and standard deviations. Figure 3b shows an example of the 100 generated data

3.3 Gait Cycle Classification

Research on the persons verification has been carried out using a deep neural network with an attentional mechanism [4]. This classifier provides a state-of-the-art architecture in the analysis of signals acquired by the accelerometer and gyroscope. The data collected during the first day was used as the training set. As a data sample, we refer to a single block of data with a dimension of 6×128. This is the segmented gait

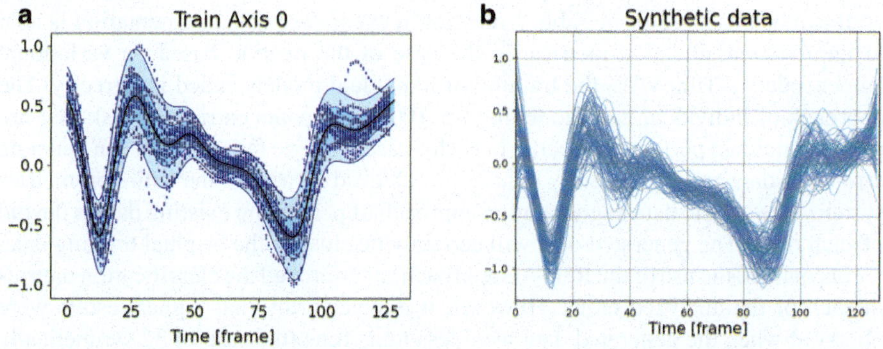

Fig. 3 Model-determined distribution parameters (**a**) generated synthetic data (**b**)

cycle that was transferred to the global system [11] and interpolated to a fixed length. Identically preprocessing was applied to second day samples which were used as the test set. It should be noted that special care was taken to ensure that no data leakage occurred at any stage. In addition, due to the non-deterministic nature of the setting of the initial weights, each training session of the network was repeated 10 times.

4 Results

The study used a tenfold simple validation. Data collected during the first day (3376 gait cycles) was used as a training set, while the second day (3321 gait cycles) data was used as a test set. Validation of the augmentation methods was carried out using the F1-score metric (weighted type). The results achieved are related to VAE literature methods [8, 9]. Table 2 shows the results according to the augmentation method (individual rows). Two columns consider two cases of the number of synthetically generated data.

Table 2 F1-score metrics grouped by augmentation methods

Augmentation method	F1-score (32 samples per g.c.)	F1-score (64 samples per g.c.)
Proposed architecture (bias = 1.5)	0.759 ± 0.004	**0.775 ± 0.014**
Proposed architecture (bias = 1.0)	0.754 ± 0.007	0.75 ± 0.028
RHVAE [9]	0.753 ± 0.016	0.749 ± 0.013
timeVAE[8]	0.188 ± 0.021	0.186 ± 0.01
Baseline (No aug.)	0.73 ± 0.019	

It can be observed from Table 1 that not in every case data augmentation is able to improve classification metrics. In the case of the method based on variational autoencoders—TimeVAE, the training of individual models ended incorrectly. The approach of individual gait modelling for 100 participants carries the risk that not every model will perform correctly. In such a case, a single failed model can generate erroneous data and significantly reduce the achieved metrics of the entire biometrics system. The classic data augmentation approach of perturbing existing data is devoid of such risks. The generated data will certainly be close to the original training data.

In contrast, the use of the RHVAE approach [9] improved the classification metrics in each of the analyzed cases. However, it is noteworthy that higher results were observed when the generated data were less numerous (the case of 32 synthetically generated data per original gait cycle, against the case of 64 synthetically generated data). This shows that the generated data has too much variability. The more data generated, the greater the "noise," making the training process more difficult.

Autoencoders lossy compress the original signal by training its features and are able to reflect the main trends of the signal (typically without local fast variations). In the case of a very small number of training data, this approach gives positive results. At the same time, in the presented case of biometrics data, there is a certain distribution shift between the training and test sets. An augmentation model based on variational autoencoders is not very robust to such a situation. Models of this type are only capable of generating data that is close to the training data. Autoencoder-based augmentation does not guarantee the integration of input data features that are essential for biometrics systems. These may be omitted in the compression stage. The aim of autoencoders is the reconstruction or extraction of the signal features which are necessary for the reconstruction.

The highest result was achieved for the proposed network architecture, with a standard deviation bias gain parameter equal to 1.5. This means that increasing the standard deviation by 50%, allowed to improve the classification rates. It is worth asking what is the classification result for other values of the parameter. Will the data generation with a basic value of the std parameter benefit the biometrics system? Table 3 summarizes the impact of the bias parameter on F1-score metrics.

The results presented in Table 3 show several important dependencies. First of all, even the basic generation of signals with the original value of the standard deviation has a beneficial effect on metrics. The second noticeable trend indicates that a

Table 3 Impact of the bias parameter on F1-score result

Bias parameter	F1-score (32 synthetic samples per g.t.)	F1-score (64 synthetic samples per g.t.)
1.0	0.754 ± 0.007	0.75 ± 0.028
1.5	0.759 ± 0.004	$\mathbf{0.775 \pm 0.014}$
2.0	0.74 ± 0.013	$\mathbf{0.73 \pm 0.033}$
4.0	0.738 ± 0.01	0.731 ± 0.016
Baseline (No aug.)	0.73 ± 0.019	

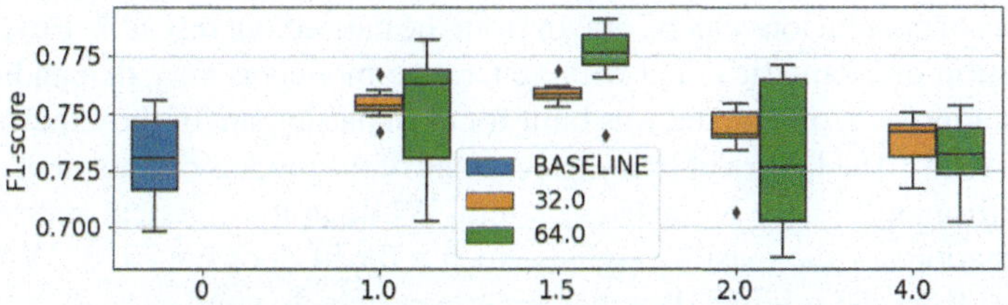

Fig. 4 Biometrics system verification metrics for the proposed network architecture and various BIAS amplification parameters

continuous increase in the bias parameter will not affect people's verification rates positively. It is profitable to increase the index to about 1.5 of the original value. The final point to add is that not in each of the studied cases the increase in the number of generated samples has positively affected the biometrics system.

The results of the experiment are also presented graphically in Fig. 4. The vertical axis shows the classification measures in the form of the F1-score metric. The horizontal axis shows the effect of the bias parameter. In the first position, the baseline case without data augmentation is illustrated. The box plots shows two variants of data generation—the case of 32/64 synthetic samples per original gait cycle.

The box plots presented in Fig. 4 show several important dependencies. First of all, for a higher number of generated samples, the results are characterized by a much higher variance. For example, for a BIAS equal to 2 and 64 generated samples, the achieved classification result could be even worse than in the case of no data augmentation. On the other hand, the median results indicate that it is particularly worthwhile to perform data augmentation for a bias parameter equal to 1.5 or 1.0.

5 Conclusions

This paper presents the impact of data augmentation on the performance of biometrics gait system. The carried out work demonstrates the concept of using generative models in the form of data distribution modelling neural networks as data augmentation mechanism. The aim of the research conducted is to verify whether the synthetic generation of new samples, that extends the classifier's training set would have a beneficial effect on the metrics of the biometrics system.

The performance of the proposed data augmentation mechanism is compared with results based on variational autoencoders RHVAE [9] and timeVAE [8], validation of the experiments is carried out using the author's corpus of human gait cycles (100 subjects) collected over two days. A CNN classifier with an attention mechanism [4] was applied in the person classification process. These architectures are dedicated to work with IMU measurement data. Validation is performed with 10 times repeated simple validation. The study is able to increase the performance of the biometrics system from a baseline 0.73 ± 0.019 F1-score to 0.775 ± 0.014 F1-score.

A series of conclusions can be drawn from the carried out research. First of all, the augmentation of biometrics data with generative models is very demanding. In the presented approach, one model was built for each participant of the experiment (100 models in total). The failure of training even one of the models can result in a decrease in the performance of the biometrics system. While in the classical approach based on perturbations of the existing signals, such a threat does not exist. New samples are created from the original data that are comparatively similar.

On the other hand, the limited amount of data for training biometricss results that more complex models are not able to generalize to the original data. Table 1 shows that the use of TimeVAE models in the augmentation process achieved low classification rates. While the model is able to reconstruct the original training data, the generation of new signals does not perform well.

For the proposed model and the original parameter σ (bias equal to 1), the results obtained are very close to the results of the RHVAE (Table 2). For a smaller count of generated samples it is 0.754 ± 0.007 (Proposed) and 0.753 ± 0.016 (RHVAE). For a larger number of synthetized gait cycles it is 0.75 ± 0.028 (Proposed) and 0.749 ± 0.013 (RHVAE). In the presented case, it can be concluded that the obtained results are equivalent. This relationship changes when the parameter σ for the proposed network architecture is artificially amplificated (Table 3). Modification of these parameters can model situations in which the subjects' gait velocities have changed. In this case, an increase in the amplitude of the accelerometer is associated with an increase in walking speed. This leads us to believe that it is the modification of the BIAS parameter that allows an advantage over competing methods (Table 3).

Acknowledgements This work was supported by grant 2021/41/N/ST6/02505 from Białystok University of Technology and funded with resources for research by National Science Centre, Poland. For the purpose of Open Access, the author has applied a CC-BY public copyright license to any Author Accepted Manuscript (AAM) version arising from this submission.

References

1. Sprager, S., Juric, M.B.: Inertial sensor-based gait recognition: a review. Sensors (Basel) **15**(9), 22089–220127. https://doi.org/10.3390/s150922089 (2015)
2. Zou, Q., Wang, Y., Wang, Q., et al.: Deep training-based gait recognition using smartphones in the wild. arXiv:1811.00338 (2020)
3. Delgado-Escano, R., Castro, F.M., Cozar, J.R., et al.: An end-to-end multi-task and fusion CNN for inertial-based gait recognition. IEEE Access **7** (2019)
4. Huang, H., Zhou, P., Li, Y., et al.: A Lightweight attention-based CNN model for efficient gait recognition with wearable IMU sensors. Sensors **21**, 2866 (2021). https://doi.org/10.3390/s21082866(2021)
5. Graves, A.: Generating sequences with recurrent neural networks. arXiv:1308.0850 (2013)
6. Alzantot, M., Chakraborty, S., Srivastava, M.:SenseGen: A deep training architecture for synthetic sensor data generation. In: 2017 IEEE International Conference on Pervasive Computing and Communications Workshops, Kona, HI, USA (2017)
7. Dürr, O., Sick, B., Murina, E.: Probabilistic deep training with python, keras and tensorflow probability, October. ISBN 9781617296079 (2020)

8. Desai, A., Freeman, C., Wang, Z., et al.: TimeVAE: a variational auto-encoder for multivariate time series generation. arXiv:2111.08095 (2021)
9. Chadebec, C., Thibeau-Sutre, E., Burgos, N., et al.: Data augmentation in high dimensional low sample size setting using a geometry-based variational autoencoder. arXiv:2105.00026 (2021)
10. Sawicki, A., Saeed., K.: Application of LSTM networks for human gait-based identification,theory and engineering of dependable computer systems and networks. In: Proceedings of the Sixteenth International Conference on Dependability of Computer Systems DepCoS-RELCOMEX, Advances in Intelligent Systems and Computing, vol. 1389 (2021)
11. Sawicki, A., Saeed, K.:Comparison of orientation invariant inertial gait matching algorithms on different substrate types. In: Zamojski, W., Mazurkiewicz, J., Sugier, J., Walkowiak, T., Kacprzyk, J. (eds.) New Advances in Dependability of Networks and Systems. DepCoS-RELCOMEX (2022)

Ant Colony Optimization Algorithm for Finding the Maximum Number of d-Size Cliques in a Graph with Not All m Edges between Its d Parts

Krzysztof Schiff

1 Introduction

Generally complete d-partite graph ($d \geq 2$) is a graph whose all vertices can be divided into d disjoint subsets $V_1,\ldots, V_d$, called sections, without no common vertex, in such a way that there is no edge connecting vertices belonging to the same subsets V_i, $V(G) = \bigcup_{i\in 1\ldots k} V_i(G)$, such that $\forall_{u,v\in Vi}\{u, v\} \notin E(G)$ and $V_i \cap V_j = \varnothing$. The complete d-partite m-vertex graph $G = (V, E)$ can be divided into m d-dimensional cliques C_i ($i = 1,\ldots, m$), where m is the number of vertices in each set V_i. For complete graph there is a graph density $q = 1$, for incomplete graph there is a graph density $0 <= q <= 1$. The graph is incomplete when there is no all edges between any pair of vertices from different sets V_i and V_j. The edge between any pair of vertices from different sets V_i and V_j exists with probability q. The incomplete d-partite m-vertex graph can be divided into some number of d-dimensional cliques. Number of d-dimensional cliques in incomplete graph is less than m if m is a number of vertices in each section of graph. It is difficult to find maximum number of d-dimensional cliques into incomplete d-partitioning d-vertex graph and there is no exact polynomial-time algorithm for this problem. Generally finding the maximum clique problem is NP-hard and finding their maximum number is also NP-hard [1, 2].

In d-partite graph, a certain number of maximum cliques can be found very easily, but it is difficult find their maximum number since event the maximum cliques problem is very hard to solve, so we should use some optimization algorithm, and the ant colony algorithm is a perfect choice for such a combinatorial optimization problem [3]. There are no exact polynomial time algorithms for this problem, but there are many heuristic algorithms [4–6]. For the problem of finding the maximum number of 3-vertex cliques in 3-partite graph an ant algorithm has been elaborated

K. Schiff (✉)
Department of Automatic Control and Computer Science, Cracow University of Technology, 31-155 Cracow, Poland
e-mail: kschiff@pk.edu.pl

W. Zamojski et al. (eds.), *Dependable Computer Systems and Networks*, Lecture Notes in Networks and Systems 737, https://doi.org/10.1007/978-3-031-37720-4_23

[7], but not for finding the maximum number of d-vertex cliques in d-partite graph, so this paper presents for the first time ant algorithms for that problem.

This work is a continuation of the research o, the results of which are presented in [7], or rather the presentation of a completely new ant algorithm for determining the maximum number of cliques in a d-partite graph. The article [7] presents the results of determining the maximum number of 3-vertex cliques in graph with 3 sections using the ant algorithm. The problem of finding all d-cliques in d-partite graphs can be encounter in the textile industry [8, 9] and in data mining, particularly for clustering of categorical attributes over d-domains [10], identification of coherence in protein–protein interaction network [11], where protein interaction network is represented by a d-partite graph that is mined for d-cliques and in the analysis of heterogeneous functional genomics data [12].

This work is organized as follows. The Sect. 2 describes the problem and the Sect. 3 briefly the ant colony algorithm: the transition probability and the pheromone global updating rule. The experimental results for different dynamic transition probabilities and their discussion are presented in Sect. 4, and conclusions are given in Sect. 5.

2 Graphical Representation of the Problem

A graphical representation of the problem is shown in Fig. 1. Here we have a graph with 4 section, so 4-partite graph. In each section we have 5 vertices, in Fig. 1 these are the color vertices e.g. blue, red, yellow, black and green. For example: two red edges are missing, which has been marked with a red dashed line, and the other missing edges have not included at all for the sake of clarity. Thus, in Fig. 1, three cliques can be designated: the blue, the green and the yellow, but the maximum number of cliques can be 4 if the red and yellow cliques are correctly designated. The black clique cannot be determined because there is not enough black edges. So you can see that there is a problem of determining the maximum number of cliques, and as the number of vertices or the number of sections increases, the problem gets worse and harder to solve.

3 Ant Colony Optimization

3.1 Brief Description of Ants Behavior

Ants always take the shortest route to their food source. This is done in such a way that they cooperate with each other by leaving a so-called chemical trace: the pheromone. The amount of pheromone depends on the frequency of walking on a given path that leads to food. The ants will leave on the paths they follow an amount of pheromone

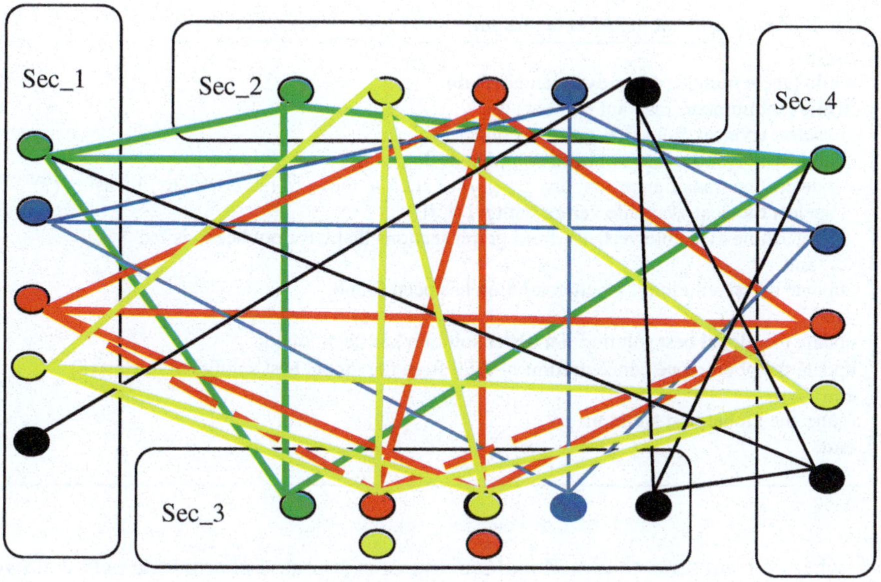

Fig. 1 There are 4 sections and the blue, red, green and yellow clique (solid line) or only the blue and green clique are detected. No all edges are shown, $q < 1$

depending on how plentiful the food source is and thus attract others ants to follow the same path.

In the ant algorithm, when applied to data represented by graphs that are sets of vertices and edges, ants move from vertex to vertex along the edges. The selection of the next vertex by the ant in order to visit it is carried out using the probabilistic rule defined by (Eq. 1):

$$P_k(i, j) = \frac{(\tau_{ij})^\alpha (\eta_{ij})^\beta}{\sum_{j \in allowed} (\tau_{ij})^\alpha (\eta_{ij})^\beta} \tag{1}$$

where τ_{ij} represent the quantity of pheromone trail on the edge that links the vertex i and j. Further, η_{ij} is a desire function, which represents the inclination of the ant to choose the edge (i, j). This is a heuristic information and it could include different properties of data structures. Coefficients α and β helps to set relative significance of pheromone and desire function.

When all ants have completed their solution during one cycle of algorithm, the quantity of pheromone lying on edges is updated according to the formula (Eq. 2):

$$\tau_{ij} = (1 - \rho)\tau_{ij} + \sum_{k=1}^{m} \Delta\tau_k(i, j) \tag{2}$$

Table 1 Pseudo-code of the ACO algorithm

```
begin
while (cycle number <max cycle number) do
 while (ant number<max ant number) do
    select a vertex from first graph section
    while (a solution has not been completed) do
    determine transition probability p_k (i,j) of ant k to move from vertex i to j by using (7)
    select the next available vertex with p_k (i,j)
    determine available vertices  from graph sections that have not been visited
     end
  update a best solution if a better solution has been found;
  end
update the global best solution if a better solution has been found;
update the pheromone concentration τ_ij ∀ i,j from the global best solution by using (8)
end
return the global best solution
end.
```

- where ρ represents the fractional amount of evaporated pheromone between two cycles of the ACO algorithm, $\Delta\tau_k$ *(i, j)* represents the amount of pheromone deposited by ant k on edge *(i, j)* and this is given by formula (Eq. 3):

$$\Delta\tau_k(\mathrm{i},\mathrm{j}) = \frac{Q}{L_k}, \text{ if ant k travels througth path (i, j)} \tag{3}$$

- here Q is a constant and L_k represents the total travel length of ant k.

The pseudo-code of the general ant algorithm is given in Table 1.

3.2 The Dynamic Desire Function

The aim of the desire function is to help ants in a better selection of vertices from graph sections to the created clique. The higher the value of desire function, the more ant is inclined to select such an object.

The desire function can be expressed by different mathematical formulas. The desire function can be static or dynamic. The static desired function is computed at the beginning of algorithm. The dynamic desire function is computed at each step of ant and its value is constantly changes each time a vertex is include in the clique created.

Let looks at Fig. 2. We have a red clique of 3-vertecs. Now we can add one off the available vertices from the graph Sect. 4 and up to graph section n. In order to probably select the "one best" vertex to be include into the blue clique from all those available vertices ant should have additional information. It should be only a quantity

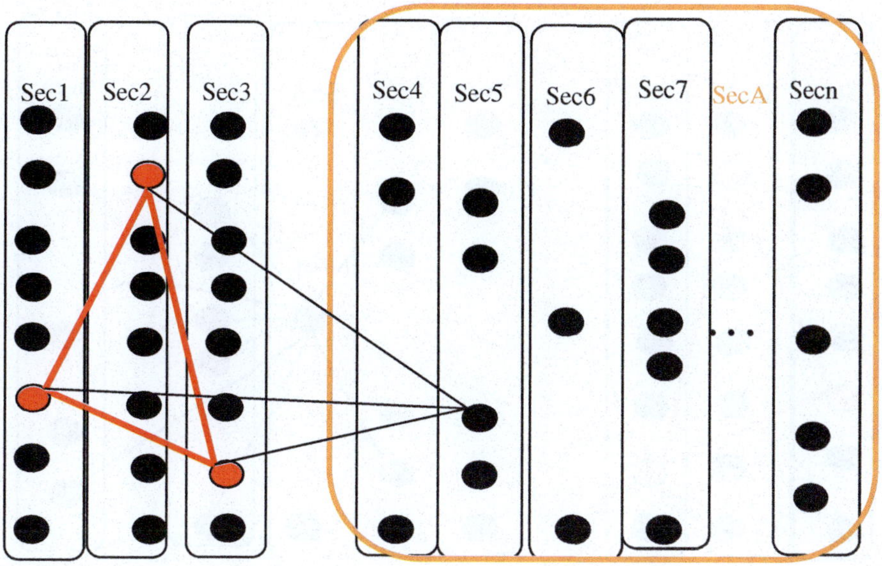

Fig. 2 Illustration how the desire function is determined

of pheromone or another type of information, so-called desire function. The "one best" means the most suited to build clique in order to receive the maximum number of such cliques.

These additional information is:

- the degree of vertex *j*, *d*[*j*] (see Fig. 3), it is the degree of each vertex j connected by an edge to vertex *i*, in which the ant is located and decides which of the vertices j should choose according to the rule (7)
- the maximum degree of a vertex from all vertices *j*
- the number of 3-vertex clique lk3w[i][j] in vertex *j* (see Fig. 4)
- the maximum number of 3-vertex clique lk3w + max.

Different combinations of those additional information create different desire functions $n(i, j)$ (see Table 2) and thus we can distinguish different ant colony optimization algorithms, as it is shown in Table 2.

4 Experiments

The conducted research consists of 6 main parts. Each time in each part of experiment, 8 algorithms were compared in terms of the maximum number of d-vertex cliques included in the best solution obtained. These comparisons were made with different parameters of the ant algorithm, as well as with different graph density, different numbers of graph parts (sections) and different number of vertices in each graph part

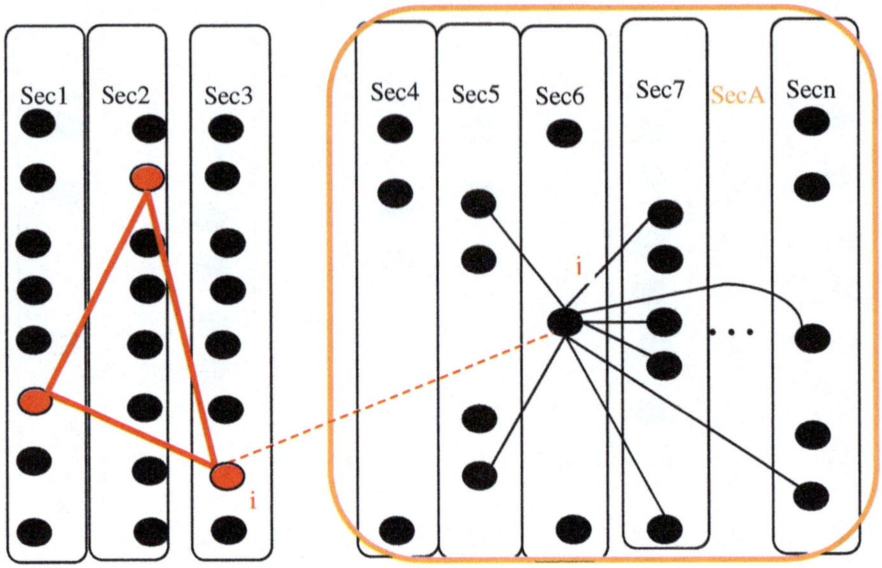

Fig. 3 The vertex e.g. j and its degree d[j] (edge connection)

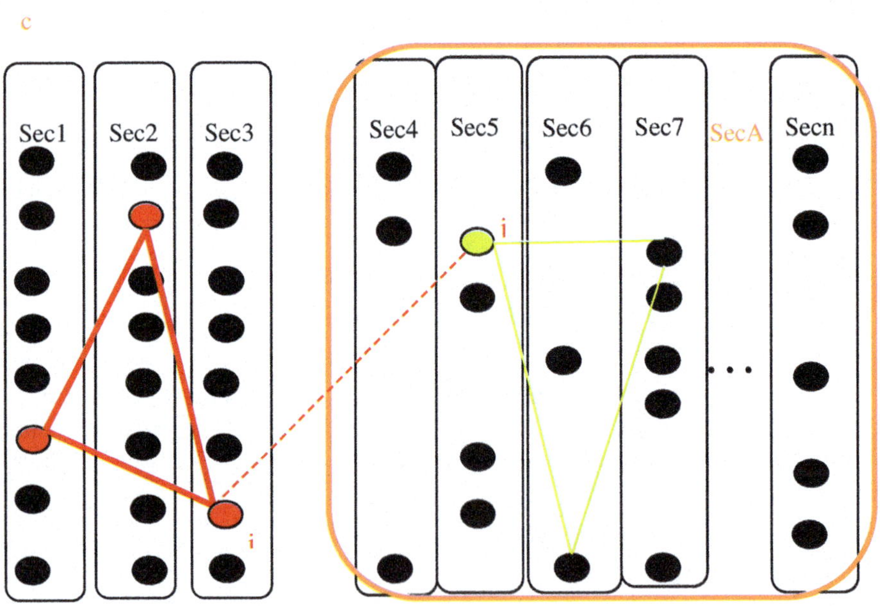

Fig. 4 The vertex j and e.g. its clique lk3w[i][j]

Table 2 Desire functions $\eta(i, j)$ of different ant algorithms

Algorithms	Formulas of desire functions
ALG_1	Without a desire function
ALG_2	$\frac{(d[j]+1)}{d\_max}$
ALG_3	$\frac{(lk3w[i][j]+1)}{lk3w\_max}$

(section). The results are presented in Tables 3, 4, 5, 6, 7 and 8. These are average values from 10 measurements. Those algorithms are ALG_1, ALG_2, ALG_3.

In the first part of the experiments the number of parts (sections) of the graph changes from 5 to 15, with the number of vertices in each part of the graph being 25, the density of the graph being 0.65 the number of ants being 10, the number of cycles being 30 and the evaporation rate being 0.995. The average maximum number of d-vertex cliques obtained as a result of the operation of the algorithms is presented in

Table 3 The average maximum number of d-vertex cliques obtained

d parts of the graph	5	8	10	12	15
ALG_1	57.5	40.1	19.8	5.5	0.6
ALG_2	57.6	40.4	20.3	5.5	0.7
ALG_3	57.8	41.8	21.8	6.6	1.0

Table 4 The average maximum number of d-vertex cliques obtained

Number of vertices in each graph part	20	40	60	80	100
ALG_1	2.4	9.1	19.8	32.8	49.1
ALG_2	2.5	9.6	20.3	33.3	49.4
ALG_3	2.9	9.8	21.8	36.0	53.0

Table 5 The average maximum number of d-vertex cliques obtained

Graph density	0.5	0.6	0.65	0.7	0.8	0.9
ALG_1	2.0	15.8	32.8	52.3	73.3	79.0
ALG_2	2.0	16.4	33.3	53.2	73.4	79.0
ALG_3	2.7	18.5	36.0	54.3	74.0	79.0

Table 6 The average maximum number of d-vertex cliques obtained

Number of cycles	10	20	30	40	50
ALG_1	32.4	33.1	32.8	33.4	33.7
ALG_2	32.5	32.7	33.3	34.6	34.3
ALG_3	34.2	36.3	36.0	36.5	36.0

Table 7 The average maximum number of d-vertex cliques – number of ant changes

Number of ants	10	20	30	40	50
ALG_1	32.8	34.1	34.9	34.5	35.0
ALG_2	33.3	34.3	34.9	34.7	35.1
ALG_3	36.0	36.5	36.7	37.6	37.3

Table 8 The average maximum number of d-vertex cliques – the evaporation coefficient changes

Evaporation 1-ρ	0.991	0.993	0.995	0.997	0.999
ALG_1	33.2	33.6	32.8	34.1	33.7
ALG_2	33.5	33.6	33.3	34.1	34.0
ALG_3	35.9	35.7	36.0	35.5	35.9

Table 3. The best algorithm is the ALG_3 algorithm. It can be seen that the average number of d-vertex cliques decreases while the number of graph parts increases and this is because it is much more difficult to obtain d-vertex cliques in d-part graph with increasing d.

In the second part of conducted experiments the number of vertices in each part of graph (section) is changing from 20 to 100 by 20 while the number of graph parts (sections) is equals 30, the number of ants is equals 10, the number of cycles is equals 30, the evaporation rate is equals 0.995 and the graph density is equals 0.65. The average maximum number of d-vertex cliques obtained as a result of the operation of the algorithms is presented in Table 4. The best performing algorithms is the algorithm ALG_3. As the number of vertices in each of the d parts of the graph increases, more cliques with d vertices can be determined.

In the third part of the experiments, the density of the graph changes from 0.5 to 0.9, where the number of graph parts (sections) is 30, the number of vertices in each graph parts is 25, the number of ants is 10, the number of cycles is 30 and the evaporation rate is 0.995. The average maximum number of d-vertex cliques obtained as a result of the operation of the algorithms is presented in Table 5. The best performing algorithms is the algorithm ALG_3. The existence of edges in a graph affects the existence of cliques in that graph. The more sparse the graph, the fewer cliques can be determined, and also larger cliques are more difficult to determine. In the experiment, as the density of a graph with d parts decreases, the number of cliques with k vertices decreases.

In the fourth part of experiments the number of cycle is changing from 10 to 50 by 10 while the number of graph parts (section) is equals 30, the number of vertices in each graph part is 25, the number of ants is 10, the evaporation rate is 0.995 and the graph density is 0.65. The average maximum number of d-vertex cliques obtained as a result of the operation of the algorithms is presented in Table 6. The best performing algorithms is the algorithm ALG_3. The number of cycles does not significantly affect the maximum number of d-vertex cliques that can be obtained.

In the fifth part of the experiments, the number of ants changes from 10 to 50 by 10, while the number of parts of the graph (sections) is 30, the number of vertices in each part of the graph is 25, the number of cycles is 30, the evaporation rate is 0.995 and the density of the graph is 0.65. The average maximum number of d-vertex cliques obtained as a result of the operation of the algorithms is presented in Table 7. The best performing algorithms is the algorithm ALG_3. The number of ants does not significantly affect the maximum number of d-vertex cliques that can be obtained.

In the sixth part of experiments the evaporation rate changes from 0.991 to 0.999 by 0.002 while the number of parts of the graph is 30, the number of vertices in each part of the graph is 25, the number of ants is 10, the number of cycles is 30 and the density of the graph is 0.65. The average maximum number of d-vertex cliques obtained as a result of the operation of the algorithms is presented in Table 8. The best performing algorithms is the algorithm ALG_3. The evaporation rate does not affect the maximum number of d-vertex cliques that can be obtained.

5 Conclusions

This article presents new ant algorithms with new dynamic desire functions. In this paper several desire functions and their influence on performance on ants colony optimization algorithms were investigated and it has been shown that desire function used in the algorithm ALG3 outperform the others and let us to receive the best result, i.e. the largest number of d-vertex cliques obtained. So we have seen that various desire functions have an effect on the maximum number of cliques with d vertices obtained, we have also seen that graph parameters such as the number of parts of the graph, the number of vertices in each part of the graph, and the density of the graph affect the maximum number of cliques obtained, and that parameters of the ant algorithm, such as: the number of cycles and the number of ants slightly affect the maximum number of cliques obtained, and the evaporation rate in the tested range showed no effect on the maximum number of cliques obtained. Of course, it is about the maximum number of cliques with d vertices in a graph with d parts.

References

1. Garey, M.R., Johnson, D.S.: Computer and Intractability. A Guide to the Theory of NP-Completeness. Freeman, New York (1979)
2. Karp, R.M.: Reducibility among Combinatorial Problems. In: Miller, R.E., Thatcher, J.W. (eds.) Complexity of Computer Computation, pp. 85–103. Plenum Press, New York (1972)
3. Dorigo, M., Caro, G.D., Gambardella, L.M.: Ant algorithms for discrete optimization. Artif. Life **5**(2), 137–172 (1999)
4. Biro, P., McDemid, E.: Three-sided stable matching with cyclic preferences. Algorithmica **58**(1), 5–18 (2010)

5. Chen, J.: Iterative expansion and color coding: an improved algorithm for 3D-matching. ACM Trans. Algorithms 6.1_6.22 (2012)
6. Eriksson, K., et al.: Three-dimensional stable matching with cyclic preferences. Math. Soc. **52**(1), 77–87 (2006)
7. Schiff, K.: An ant algorithm for the maximum number of 3-cliques in 3-partite graphs. Contr. Cybern. 347–358 (2021)
8. Grunert, T., Irnich, S., Zimmermann, H.J., Schneider, M., Wulfhorst, B.: Finding all k-cliques in k-partite graph, an application in textile engineering. Comput. Oper. Res. **29**(2), 13–31 (2002)
9. Mirghorbani, M., Krokhmal, P.: On finding k-cliques in k-partite graphs. Optimiz. Lett. **7**, 1155–1165 (2013)
10. Peters, M.: CLICK: clustering categorical data using k-partite maximal cliques. In: IEEE International Conference on Data Engineering (2005)
11. Liu, Q., Chen, Y.: High functional coherence in k-partite protein cliques of protein interaction networks. In: Proceedings of the IEEE International Conference on Bioinformatics and Biomedicine, pp. 111–117 (2009)
12. Phillips, C.A., Wang, K., Bubier, J., Baker, E.J., Chesler, E.J., Langston, M.A.: Scalable multipartite subgraph enumeration for integrative analysis of heterogeneous experimental functional genomics data. In Proceedings of the 6th ACM Conference on Bioinformatics, Computational Biology and Health Informatics, Atlanta, Georgia, 9–12 September, pp. 626–633 (2015)

Partitioning of an M-Part Weighted Graph with N Vertices in Each Part into N Cliques with M Vertices and the Total Minimum Sum of Their Edges Weights Using Ant Algorithms

Krzysztof Schiff

1 Introduction

The r-parts graph ($r \geq 2$) is a graph whose set of vertices can be divided into r disjoint subset $V_1,\ldots,V_r$, in such a way that there is no edge connecting the vertices belonging to the same subsets V_i, $V(G) = \bigcup_{i \in 1 \ldots k} V_i(G)$, such that $\forall_{u,v \in V_i}\{u, v\} \notin E(G)$. The graph $G = (V, E, W)$ can be divided into n maximal cliques C_i ($i = 1,.., n$), where n is the number of vertices in V_i. Each clique C_i represents one object. If there are m cameras, then each of these cliques C_i should have m vertices, $v_i \in V$ and $i = 1,..m$. Each edge $e_{ij} \in E$ has its own weight $w_{ij} \in W$ and each edge e_{ij} connects two vertices v_i and v_j, $i = 1,\ldots, n$, and $j = 1,\ldots, n$. Each clique C_i has $n*(n-1)$ edges and for each clique the sum of the weights of all edges can be computed. The problem under study, called the general maximum minimum clique problem (GMMCP), is to divide graph $G = (V, E, W)$ in such a way as to obtain n cliques of dimension m and so that their total sum of weights is as small as possible [1]. The problem can also be formulated as a multi-dimensional assignment problem [2–4]. Others algorithms have been elaborated for this problem: ant colony optimisation algorithms [5–7] and algorithms, that are based on network flow model [8–14], on binary integer programming [15–18], minimum cost disjoint path [19, 20], neural network [21, 22], particle swarm optimisation [23] and even genetic algorithms [24]. This article present new ant algorithms with new desired functions.

K. Schiff (✉)
Department of Automatic Control and Computer Science, Cracow University of Technology, 31-155 Cracow, Poland
e-mail: kschiff@pk.edu.pl

W. Zamojski et al. (eds.), *Dependable Computer Systems and Networks*, Lecture Notes in Networks and Systems 737, https://doi.org/10.1007/978-3-031-37720-4_24

2 Ant Colony Optimisation Algorithm

2.1 *Pseudo-Code of the Ant Algorithm*

The ants are looking for food. They always use some path to get to the food. The ants cooperate with each other using a pheromone, placing the pheromone on the path and thus helping the next ants to take that path. Suppose the *k*-th ant is starting its journey from vertex *i*. Ant *k* undertakes a random walk from vertex *i* to the next vertex *j* that is probabilistically chosen based on the probability formula in Eq. (1).

$$P_k(i,j) = \frac{\left(\tau_{i,j}\right)^{\alpha}\left(\eta_{i,j}\right)^{\beta}}{\sum_{j\in allowed}\left(\tau_{i,j}\right)^{\alpha}\left(\eta_{i,j}\right)^{\beta}} \tag{1}$$

where $\tau(i,j)$ represents the fractional quantity of the pheromone trail on the edge that links vertex *i* to *j*. Further, $\eta(i,j)$ is a desire function, which represents the inclination of the ant to choose the edge (*i*, *j*). Coefficients α and β help to set the relative significance of the pheromone and the desire function.

When all the ants have completed their journey, the pheromone concentration is updated on the edges according to the formula in Eq. (2).

$$\tau(i,j) = (1-\rho)\tau(i,j) + \sum_{k=1}^{m}\Delta\tau_k(i,j) \tag{2}$$

At the beginning, the first vertex is selected, from which the ant starts to form a clique. At each step, the ant adds another randomly selected vertex from the next part of the graph, as shown in Fig. 1. This next part of the graph is also chosen randomly. So when an ant adds m vertices from all *m* parts of the graph, one vertex after another to the created set, it will get a clique. Each ant creates *n* such cliques. The pseudo-code of the general ant algorithm is given below.

Pseudo-code of the ACO algorithm.

```
begin
while (cycle number <max cycle number) do
    while (ant number<max ant number) do
        select the first vertex to start the journey of the kth ant
        while (a solution has not been completed) do
        determine transition probability Pk (i, j) of ant k to move from vertex i to j by using (1)
        apply roulette wheel selection criteria on Pk (i, j) ∀ i,j to select the next available vertex
        determine the available vertices: vertices from cameras that have not been visited
        end
    update athe best solution if a better solution has been found;
    end
update the global best solution if a better solution has been found;
update the pheromone concentration τ(i,j) ∀ i,j from the global best solution by using (2)
end
return the global best solution
end.
```

2.2 Dynamic Desire Function

The desire function is designed to help each ant select the next vertex to be included in the created clique and, as a result, achieve all cliques with the lowest possible total sum of weights. Various so-called heuristic information contained in the desire function causes the total sum of weights of all cliques to vary depending on this information. The desire function allows each ant to choose the next vertex for the created clique that best corresponds to the solution sought. The greater the value of the desire function, the greater the likelihood that each ant will be willing to select a vertex.

We start with a blue clique of 3 vertices in Fig. 1 (solid blue lines). Now we can add one of the 3 available vertices from part_2, one of the 3 vertices from part_4 and one of the 3 vertices from part_6. If we add a vertex from part_6 we really add 3 edges to the created blue clique, which connect this vertex with the other vertices from the blue clique. In order to possibly select the "one best" vertex to be included in the blue clique from all 9 of those vertices from the 3 graph parts: part_2, part_ 4 and part_6, the ant can use additional available information, meaning heuristic information—the desire function. The "one best" means the most suitable to build the blue clique in order to get the total sum of all clique weights as small as possible.

These additional information used in desire functions are: the number of edges of potentially selected vertex j (3), the weights of all edges of a potentially selected vertex j– weight of c_j (4), the weight of the edge connecting the potentially selected vertex j with the vertex i previously added to the created clique—weight of $e(i, j)$.

$$\textit{number of edges adjacent a potential vertex}\, j = \sum_{i \in Clique} e(i, j) \tag{3}$$

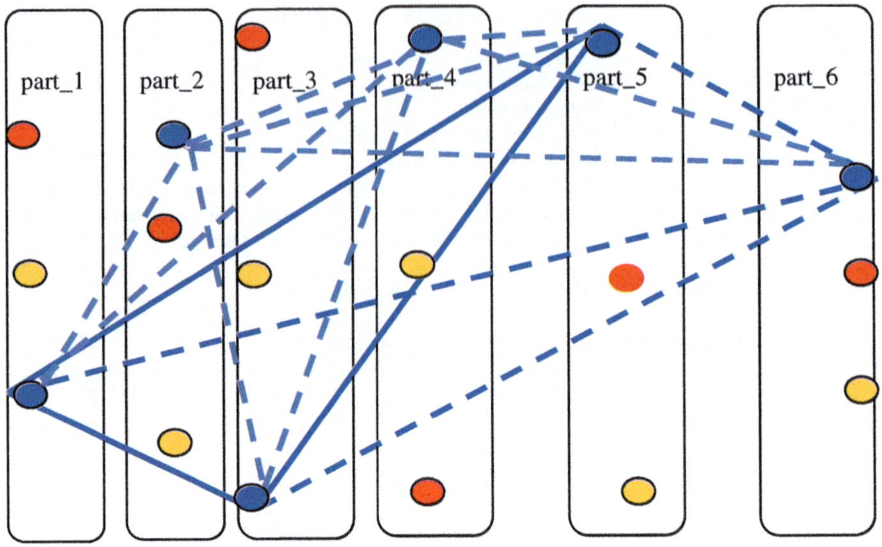

Fig. 1 Illustration of available vertices

$$weight\ of\ c_j = \sum_{i \in Clique} weight\ of\ e(i,j), j - is\ a\ potential\ vertex \tag{4}$$

Different combinations of those additional information create different deThe results are presented insire function $n(i, j)$, as shown in Tables 1 and thus we can distinguish different ant algorithms.

Table 1 Desire function $\eta(i, j)$ of different ant algorithms

Algorithms	Formulas of desire functions
ALG_1	$\frac{1}{(weight\ of\ e(i,j))^3}$
ALG_2	$\frac{1}{(weight\ of\ c_j)^3}$
ALG_3	$\frac{1}{weight\ of\ e(i,j)*(weight\ of\ c_j-weight\ of\ e(i,j))^2}$
ALG_4	$\frac{1}{(weight\ of\ e(i,j))^2*(weight\ of\ c_j-weight\ of\ e(i,j))^2}$
ALG_5	$\frac{1}{(weight\ of\ e(i,j))*(weight\ of\ c_j-weight\ of\ e(i,j))}$
ALG_6	$\frac{1}{(weight\ of\ e(i,j))^2*weight\ of\ c_j*(weight\ of\ c_j-weight\ of\ e(i,j))}$
ALG_7	$\frac{1}{(weight\ of\ e(i,j))*weight\ of\ c_j*(weightofc_j-weight\ of\ e(i,j))}$
ALG_8	$\frac{number\ of\ to\ be\ included\ edges\ from\ potential\ vertex\ j}{(weight\ of\ e(i,j))^3*(weight\ of\ c_j)^3}$

3 Experiments

The experiments consists of 5 parts and in each part of the experiments, 8 algorithms were compared in terms of the total sum of weights of all edges included into the best solution. Comparisons were made with different parameters of the ant algorithm. The results are presented in Tables 2, 3, 4, 5, and 6. They are the averages from 10 measurements.

In the first part of the conducted experiments the number of graph parts was changing from 10 to 30 while the number of vertices in each part of the graph was 25, the number of ants was 10, the number of cycles was 30 and the evaporation rate was 0.995. The sum of the weights of all cliques is shown in Table 3. The best algorithms are algorithm ALG_3 and ALG_4 with the lowest total minimum sum of weights of all cliques. We can also see that algorithm ALG_2 performed better than algorithm ALG_1 up to 25 graph parts, but algorithm ALG_1 performed better than algorithm ALG_2 for 30 graph parts. This means that algorithm ALG_1 performed better than algorithm ALG_2 in cases where the size of cliques was greater than 30

Table 2 Total sum of weights of all cliques–number of graph parts changes

A number of graph Algorithm\parts	10	15	20	25	30
ALG_1	45 412,3	114 495,7	214 631,3	347 313,8	510 340,0
ALG_2	43 290,6	111 817,5	211 276,5	341 614,7	523 179,7
ALG_3	42 437,9	108 849,2	206 684,2	336 193,1	496 921,0
ALG_4	42 533,1	109 048,9	206 837,7	336 332,5	496 914,1
ALG_5	44 034,2	112 103,2	211 091,0	342 225,1	504 924,9
ALG_6	44 315,3	112 145,6	211 577,6	342 468,4	504 836,6
ALG_7	45 472,0	114 738,5	215 286,4	347 716,7	511 392,7
ALG_8	44 448,4	112 465,7	211 992,0	342 474,1	505 035,2

Table 3 Total sum of weights of all cliques –number of vertices in graph parts changes

Number of vertices in graph parts	10	15	20	25	30
ALG_1	204 484,3	306 748,9	408 755,1	510 340,0	612 545,2
ALG_2	208 254,3	313 469,4	418 008,5	523 179,7	628 303,3
ALG_3	200 876,8	299 771,5	398 383,9	496 921,0	594 860,6
ALG_4	200 843,2	299 510,6	398 392,5	496 914,1	595 560,8
ALG_5	202 948,2	303 581,3	403 810,9	504 924,9	604 856,4
ALG_6	203 177,3	303 625,7	403 209,9	504 836,6	605 258,0
ALG_7	204 858,5	307 754,6	409 554,3	511 392,7	613 552,3
ALG_8	202 983,3	303 879,5	404 274,4	505 035,2	604 750,4

Table 4 Total sum of weights of all cliques–number of cycles changes

Number of cycles	10	20	30	40	50
ALG_1	511 208,0	510 754,6	510 340,0	510 657,1	510 147,6
ALG_2	525 267,9	524 186,8	523 179,7	523 254,6	523 318,6
ALG_3	497 984,1	497 246,9	496 921,0	497 132,9	496 158,4
ALG_4	498 223,8	497 557,7	496 914,1	497 023,2	496 647,8
ALG_5	505 674,1	504 774,5	504 924,9	504 638,8	503 809,7
ALG_6	505 359,3	505 673,5	504 836,6	504 436,9	504 164,2
ALG_7	511 818,3	512 170,6	511 392,7	511 622,0	511 272,2
ALG_8	505 221,6	505 669,6	505 035,2	505 426,3	504 598,1

Table 5 Total sum of weights of all cliques–number of ant changes

Number of ants	10	20	30	40	50
ALG_1	510 340,0	510 318,9	509 816,0	509 332,2	509 710,8
ALG_2	523 179,7	523 017,9	522 254,1	522 632,1	521 699,7
ALG_3	496 921,0	490 080,5	496 608,4	495 503,8	495 766,7
ALG_4	496 914,1	496 078,4	496 362,3	495 746,8	496 280,5
ALG_5	504 924,9	503 866,6	504 105,0	502 868,5	503 752,5
ALG_6	504 836,6	503 497,5	504 208,2	503 119,7	502 686,8
ALG_7	511 392,7	510 621,9	510 902,6	510 121,6	510 507,1
ALG_8	505 035,2	503 779,5	503 572,2	504 009,8	503 365,8

Table 6 The total sum of weights of all cliques–the evaporation coefficient changes

Evaporation 1-ρ	0.991	0.993	0.995	0.997	0.999
ALG_1	511 103,9	510 399,1	510 340,0	510 913,8	510 523,3
ALG_2	523 285,6	523 593,7	523 179,7	523 494,4	522 925,4
ALG_3	497 053,2	496 729,4	496 921,0	496 102,8	496 870,2
ALG_4	496 429,8	497 086,4	496 914,1	496 932,3	497 377,0
ALG_5	504 160,5	504 443,1	504 924,9	504 367,1	504 412,6
ALG_6	504 914,6	504 368,9	504 836,6	504 734,7	504 601,6
ALG_7	511 640,3	512 058,7	511 392,7	511 346,9	511 437,1
ALG_8	505 126,5	504 906,1	505 035,2	504 807,2	504 220,9

vertices, which prompted further research on the dynamic desire function and the results of this further research are presented in this paper.

In the second part of the conducted experiments the number of vertices in each part of the graph was changing from 10 to 30 while the number of graph parts was 30, the number of ants was 10, the number of cycles was 30 and the evaporation rate was 0.995. The total sum of the weights of all cliques is shown in Table 4. The best performing algorithms were algorithm ALG_3 and ALG_4. We can also see that algorithm ALG_1 performed better than algorithm ALG_2 and that the number of vertices in each part of the graph did not influence the comparison of this performance, which means the number of obtained cliques did not influence this comparison.

In the third part of the conducted experiments the number of cycles was changing from 10 to 50 while the number of parts of the graph was 30, the number of vertices in each part of the graph was 25, the number of ants was 10 and the evaporation rate was 0.995. The total sum of the weights of all cliques is shown in Table 5. The best performing algorithms were algorithms ALG_3 and ALG_4.

In the fourth part of the conducted experiments the number of ants was changing from 10 to 50 in increments of 10 while the number of graph parts was 30, the number of vertices in each part of the graph was 25, the number of cycles was 30 and the evaporation rate was 0.995. The total sum of the weights of all cliques is shown in Table 6. The best performing algorithms were algorithm ALG_3 and ALG_4.

In the fifth part of the conducted experiments the evaporation rate was changing from 0.991 to 0.999 while the number of graph parts was 30, the number of objects was 25, the number of ants was 10 and the number of cycles was 30. The total sum of the weights of all cliques is shown in Table 6. The best performing algorithms were algorithms ALG_3 and ALG_4.

4 Conclusions

This article presents new ant algorithms with new dynamic desire functions. Firstly algorithm ALG_2 was shown to perform better than algorithm ALG_1 only for numbers of graph parts lower than 30. Secondly, in this paper, several desire functions and their influence on the performance of ants colony optimisation algorithms were investigated, and it shown that the desire functions used in algorithms ALG3 and ALG4 produced the best results, which means the lowest total sum of all cliques weights.

References

1. Dehgan, A., Assari, S.M., Shah, M.: GMMCP Tracker: Globally optimal generalized maximum multi clique problem for multiple object tracking. Conf. Comput. Vis. Pattern Recognit. **1**, 4091–4099 (2015)
2. Deb, S., Yeddanapudi, M., PattiPati, K., Bar-Shalom, T.: A generalized S-D assignment algorithm for multi-sensor-multi-target state estimation. IEEE Trans. Aerosp. Electron. Syst. **33**, 523–538 (1999)
3. Feremans, C., Labbe, M., Laportee, G.: Generalized network design problem. Eur. J. Oper. Res. **148**, 1–13 (2003)

4. Koster, A.M.C.A., Hoesel, S.P.M., Kolen, A.W.J.: The partial constraint satisfaction problem: Facets and lifting theorems. Oper. Res. Lett. **23**, 89–97 (1998)
5. Bozdogan, A.O., Yilmaz, A.E., Efe, M.: Performance Analysis of swarm optimization approaches for the generalized assignment problem in multi-target tracking applications. Turkish J. Elec. Eng. Comp. Sci **18**, 1059–1076 (2010)
6. Bozdogan, A.O., Efe M.M.: Ant colony optimization heuristic for the multidimensional assignment problem in target tracking. IEEE. Nat. Radar. Conf., (2008)
7. Schiff, K.: Ant colony optimization for object identification in multi-cameras video tracking systems in New advances in dependability of networks and systems. In: Seventeenth International conference on dependability of computer systems: DepCoS-RELCOMEX. Wrocław, Poland (2022)
8. Liang, Y., Lu, X., He, Z., Zheng, Y.: Multiple object tracking by reliable tracklets. Signal, Image Video Process., 823–831 (2019)
9. Berclaz, J., Fleuret, F., Turetken, E., Fua, P.: Multiple object tracking using k-shortest paths optimization. IEEE Trans. Pattern Anal. Mach. Intel. **33**, 1806–1819 (2011)
10. Zhang, L., Li, Y., Nevatia, R.: Global data association for multi-object tracking using network flows. In: Proceedings of the 2008 IEEE Conf on Comp Vision and PR (pp 1–8). (2019)
11. Pirsiavash, H., Ramanan, D., Fowlkes, C.C.: Globally-optimal greedy algorithms for tracking a variable number of objects. In: Proceedings of the 2011 IEEE conference on computer vision and pattern recognition (pp 1201–1208). (2019)
12. Gaidon, A., Wang, Q., Cabon, Y., Vig, E.: Virtual worlds as proxy for multi-object tracking analysis. In: Proc IEEE computing society, conf on comp. and PR (pp 4340–4349). (2016)
13. Lenz, P., Geiger, A., Urtasun, R.: Follow Me: efficient online min-cost flow tracking with bounded memory and computation. Conf. Comp. Vision., 4364–4372 (2015)
14. Chari, V., Lacoste-Julien, S., Laptev, I., Sivic, J.: On pairwise costs for network flow multi-object tracking. In: Proc IEEE comput soc conf comp vision and PR (pp. 5537–5545). (2015)
15. Ristani, E., Tomasi, C.: Tracking multiple people online and in real time. In: Asian conference on computer vision (pp. 444–459). Springer, Cham, Switzerland (2014)
16. Kumar, R., Charpiat, G., Thonnat, M.: Multiple object tracking by efficient graph partitioning. In: Asian conference on comp vision (pp. 445–460). Springer, Switzerland (2014)
17. Tang, S., Andres, B., Andriluka, M., Schiele, B.: Subgraph decomposition for multi-target tracking. Conf. Comput. Vis. Pattern Recognit., 5033–5041 (2015)
18. Morefield, C.L.: Application of 0–1 integer programming to multi-target tracking problems. IEEE Trans. Autom. Control. **22**, 302–312 (1971)
19. Poore, A.: Multidimensional assignment and multi-target tracking. Partitioning data sets. In: Cox, I.J., Hansen, P., Julesz, B.B. (eds.) American mathematical society, pp 169–196. (1995)
20. Poore, A.B.: Multidimensional assignment formulation of data association problems arising from multi-target tracking and multi-sensor data fusion. Comput. Optim. Appl. **3**, 27–57 (1994)
21. Yoon, K., Kim, D.Y., Yoon, Y.C., Jeon, M.: Data association for multi-object tracking via deep neural networks. Sensors **19**(3), 1–17 (2019)
22. Lee, B., Erdenee, E., Jin, S., Rhee, P.K.: Efficient object detection using convolutional neural network-based hierarchical feature modeling. Signal. Image. Video **10**(8), 1503–1510 (2016)
23. Kennedy J., Eberhart, R.: Particle swarm optimization. In: Proc IEEE international conference on neural networks, vol. 4, pp. 1942–1948. (1995)
24. Chen, G., Hong, L.: A genetic based multi-dimensional data association algorithm for multi sensor multi target tracking. Math. Comput. Model. **26**, 57–69 (1997)

A Study of Architecture Optimization Techniques for Convolutional Neural Networks

Artur Sobolewski and Kamil Szyc

1 Introduction

Computer vision is becoming increasingly important in a variety of industries, including entertainment, manufacturing, and healthcare. There is a growing need to deploy vision algorithms on edge devices such as mobile phones or embedded computing platforms. To achieve this, we often need to adapt computer vision models to work with the limited infrastructure. However, the multitude of proposed techniques in literature makes it challenging to do this deliberately and effectively.

In this paper, we show how to effectively adapt Convolutional Neural Networks (CNNs) [1] to edge devices. The literature typically proposes a completely closed architecture [2, 3] using many different approaches to achieve this objective. The available methods focus on modifying the model to reduce the number of parameters and required computations (FLOPS). The additional challenge is to balance those changes made with the drop in accuracy achieved. The techniques of changing CNN architecture and model optimization are used to achieve this goal. Each of them can be applied using a variety of approaches. Many papers use multiple strategies without carefully analyzing which approach impacts the results most.

Our main contribution is the ablation study of how different approaches designed for mobile and embedded vision applications affect the final results regarding the reduced number of model parameters, FLOPS, and unwanted accuracy drop. Our study consists of two popular CNN architectures (VGG [4] and ResNet [5]) and successive addition of examined approaches monitoring the assumptions tested in this ablation study. We tested practiced approaches from the literature, such as n $\times$ 1 and 1 $\times$ n kernel convolutions [6], depthwise separable convolutions [7], SENet blocks [8], and model pruning with post-training quantization [9].

The paper is organized as follows. Section 2 presents related work within a description of the studied approaches. Section 3 presents details of our implementation

A. Sobolewski · K. Szyc (✉)
Wrocław University of Science and Technology, Wrocław, Poland
e-mail: kamil.szyc@pwr.edu.pl

W. Zamojski et al. (eds.), *Dependable Computer Systems and Networks*, Lecture Notes in Networks and Systems 737, https://doi.org/10.1007/978-3-031-37720-4_25

methodologies of studied techniques and the results obtained in these experiments. Section 4 summarizes the implemented methods with their suitability conclusion related to our research results.

2 Related Works

There are many suggestions on how to adapt computer vision models to work with the limited infrastructure, most by post-training techniques [9] and changing the model architecture [10]. We see the lack of ablation study on how different approaches affect the end outcomes regarding the reduced number of model parameters, FLOPS, and undesirable accuracy decline, or they just try to optimize a particular model using pre-selected mechanisms [11].

In this work, we focused on CNNs architecture [12]. Most of the proposed solutions for using these networks in edge devices focus on architectural changes. The pioneering work on the topic of this paper is SqueezeNet [13]. The fundamental element of the architecture of this model is the Fire module. Its goal is to reduce the number of filters—and thus computations—by using low-cost 1×1 kernel size convolution [14]. It reduces the dimensions of the feature map before more expensive operations such as 3×3 kernel size convolution.

The standard of modern CNNs uses Global Average Pooling [14] with only one final fully connected layer, replacing multiple of them. This simple change significantly improves model optimization and makes the model less prone to overfitting. In the original paper, the authors proposed an extreme version with no fully connected layer at all—by keeping the number of channels in the last convolutional operation equal to the number of classes.

Some papers suggest using low-level tricks for further optimization. For instance, the Inception v3 [6] model proposed uses a factorization of the standard convolution, replacing a single $n \times n$ filter with two successive $n \times 1$ and $1 \times n$ filters. This modification reduces the number of parameters for a single symmetric filter by even 33% (for $n = 3$) while maintaining the same feature map size and similar efficiency.

The significant step in reducing the number of operations was the proposal of depthwise separable convolution [7] (we sometimes refer to it as DWconv). This solution is based on the assembly of successive depthwise and pointwise convolutions. The former is a convolution performed independently on each input channel for feature extraction, while the latter is designed to project the result channels after depthwise convolution into a new space that integrates features from separate channels. Low-cost feature filtering and control over the number of channels in the feature map are the main incentives for this approach.

The authors of the ShuffleNet [2] further developed the depthwise separable convolution. Pointwise 1×1 convolution works on all channels of the feature map, which is the main reason for the large resource requirements. They used pointwise convolutions performed on only a subset (group) of channels. Depthwise 3×3 convolutions deal with the same group, but previously the feature map channels generated by the grouped pointwise convolutions were shuffled by specific operations. With this mechanism, the output channels are associated with all input channels, instead of being limited to information from the corresponding group. Later we will refer to it as the shuffle mechanism.

The authors of MobileNetV3 [3] found that the output of the pointwise convolution of the last bottleneck block(used by many architecture such as ResNet [5]) could be fed directly into Global Average Pooling. It eliminates the expensive computation needed to expand the number of channels for output generation. The first layer is also one of the slowest layers because the filters work on the input image with the highest resolution due to massive computations. They experimentally proved that keeping it small allows to maintain the performance of the model. The MobileNetV2 [15] and MobileNetV3 models use inverted residual bottlenecks, which maintain narrow connections between bottlenecks and expand channels of a feature map to the inner filtering with a 3×3 kernel. This modification brings most of the computation resource savings.

In large CNNs, there is often redundancy in the feature maps obtained at different stages of image processing, i.e., it is possible to obtain one channel by performing a low-cost operation on another. Taking advantage of this fact, the authors of the GhostNet [16] decided to make the network more efficient by using simple affine operations to generate new channels. Thus, using a simple convolution (usually 1×1), they obtain internal feature maps with a small number of channels, from which new "ghost" features emerge, usually using depthwise separable convolution. The output of the module is a tensor as a concatenation of the "ghost" and previous untransformed features. We used complete components proposed in GhostNet, called ghost bottlenecks or separated ghost modules.

Because applying the aforementioned simplifications can lead to performance losses, special mechanisms can be used to overcome these effects (even though increased parameters), such as the Squeeze-and-Excitation (SE) [8] block. We denoted it as SE connections. The response of this network fragment represents weights that indicate the relevance of a given channel of the feature map. This vector is expected to provide the network with a more efficient extraction of the representation of the input data in feature space.

Once the model is trained, techniques of model optimization can be applied. The most popular are pruning and further quantization [9]. Pruning involves eliminating model components that have little effect on predictions. Quantization involves reducing the precision of the parameters, which are typically represented as 32-bit floating-point integers.

3 Research on Compression Methods

This section describes how we performed experiments with architecture optimization techniques. We present our strategies for implementing the compression methods described in Sect. 2.

First, we focused on changes to the model architecture—the results we presented in Tables 1 and 2. We evaluate each experiment by the number of model parameters (absolute value in millions and as a percentage of the baseline), FLOPS (absolute value in millions and as a percentage of the baseline), model size after saving, and the final accuracy achieved. Next, we tested the pruning and quantization approaches.

We evaluated the impact of selected methods on the performance of well-known CNN architectures trained on popular image classification benchmarks—the ResNet-101 model trained on CIFAR-10 and the VGG-19 model trained on CIFAR-100. The selected architectures are well-studied by literature. Moreover, they do not contain modules specific for edge devices, which newer models, such as EfficientNet, increasingly eagerly use. Both datasets have 60 000, 32px by 32px resolution, color RGB images equally distributed in 10 or 100 classes. The ratio of train to test subsets is equal to 5:1.

The main assumption for our research was to maintain the original high-level structure of the models as possible. By this, we mean not to change the most significant network properties such as the depth (closely related to the number of layers), width (number of channels of feature maps), and resolutions (width and height of each feature map). Taking this into account, the usage of a particular compression method depends on individual CNN architectures.

The important characteristic of a ResNet is a high-level structure consisting of bottleneck blocks with residual skip connections, which are assembled into larger pieces called layers due to the shape of the feature maps that each convolution works on. Therefore, any compression method that affects the architecture does not interfere with it. The situation is similar to the VGG modification. It does not use residual connections and bottlenecks or other block-like elements, so we did not change it.

3.1 ResNet-101 on CIFAR-10

For ResNet trained on CIFAR-10, we prepared 14 experiments with different combinations of optimization approaches. We present the obtained results in Table 1. The percentages of number of parameters and FLOPS refer to the baseline model (Experiment No. 1). Below is a detailed description of each experiment with a comment on the results.

Fire Module The 3×3 convolution filter that occurs in the ResNet base model has been replaced by a corresponding filter size convolution of the 3×3 expand part of the Fire module. To adapt these blocks to the characteristics of the architecture, we assumed the preservation of the size of the feature maps that occur on the connections of the baseline bottleneck blocks. The squeezing part reduced the number of dimensions by 4 times, while the number of filters that went into the two convolutions of the expanded part was half of the target output of the block. The way that the Fire modules worked proved to be unsuitable for use in the ResNet architecture. For this reason, we abandoned further use of this module.

Table 1 Optimization techniques for ResNet-101 trained on CIFAR-10

No.	1	2	3	4	5	6	7	8	9	10	11	12	13	14
Approach														
Fire modules	–	✓	–	–	–	–	–	–	–	–	–	–	–	–
No FC classifier	–	–	–	–	–	–	✓	✓	✓	–	✓	✓	✓	✓
N × 1 to 1 × N Conv.	–	–	✓	–	–	–	–	–	–	–	–	–	–	–
DWconv	–	–	–	✓	✓	✓	✓	✓	✓	✓	✓	✓	✓	✓
Shuffle mech.	–	–	–	–	✓	–	–	–	–	–	–	–	✓	✓
Inv. bottlenecks	–	–	–	–	–	✓	✓	✓	✓	✓	✓	✓	✓	✓
Optimization before GAP	–	–	–	–	–	–	–	✓	✓	–	✓	✓	✓	✓
Less channels on the early stage	–	–	–	–	–	–	–	–	✓	–	✓	✓	✓	✓
Ghost bottleneck	–	–	–	–	–	–	–	–	–	✓	✓	✓	✓	✓
SE connections	–	–	–	–	–	–	–	–	–	–	–	✓	–	✓
Parameters [M]	42.513	59.061	35.397	21.240	7.536	2.255	1.187	0.873	0.869	0.450	0.400	1.424	0.151	1.174
Parameters [%]	100.00	138.92	83.26	49.96	17.73	5.30	2.79	2.05	2.04	1.06	0.94	3.35	0.35	2.76
FLOPS [M]	2520.22	3480.01	2141.58	1279.53	420.75	79.21	62.39	57.37	53.00	32.83	29.76	31.48	12.80	14.52
FLOPS [%]	100.00	138.08	84.98	50.77	16.69	3.14	2.48	2.28	2.10	1.30	1.18	1.25	0.51	0.58
Size [MB]	162.77	225.41	135.64	81.62	29.36	8.96	4.87	3.65	3.64	1.96	1.76	7.70	0.82	4.75
Accuracy [%]	94.33	92.78	94.24	93.03	92.87	93.09	93.07	93.01	93.00	92.70	92.43	92.55	90.82	90.74

N × 1 to 1 × N Convolution The first effective model compression comes with applying a 3 × 1 to 1 × 3 combination of convolutions instead of all previous 3 × 3. The number of parameters has been reduced by almost 17% and FLOPS by ≈15% with only a 0.1 pp. decrease in accuracy.

Depthwise Separable Convolution Singularly adapting the previous 3 × 3 convolutions to work on separate channels halved the size and required computations. Since the baseline ResNet bottlenecks already require a pointwise convolution to project feature maps onto fewer channels that directly follows the internal 3 × 3, there is no need to apply an additional pointwise convolution. We focused on adapting the depthwise separable convolution because the 3 × 1 to 1 × 3 convolution was underperforming for the compression goal.

Shuffle mechanism The aforementioned pointwise convolutions present in our baseline ResNet bottlenecks or in adjusted ghost bottlenecks we modified by applying the parameter *group* equal 4. We chose this value to be close to the observed (groups = 3) in the ShuffleNet paper [2] and to match the actual network width. After all the tuned convolutions, we also added the shuffle operation.

In three cases of application of this solution we achieved almost 3 times smaller models. The decrease in efficiency in these cases is not proportional. We believe that pointwise group convolution with channel shuffling (experiment No. 13—Table 1) is not an appropriate operation for computing feature maps from which ghost features are generated. The decrease in effectiveness can also be explained as a consequence of the strong—280-fold parameter reduction (151k parameters remain).

Inverted Bottlenecks Inspired by MobileNetV3 [3], we determined the number of channels in links of bottlenecks while maintaining their inner convolution shapes as 10/64/10, 22/128/128, 42/256/42, and 86/512/86 for ResNet four layers. In the case of ResNet architecture, this solution improved the accuracy even with other solutions, such as depthwise separable convolution, so we decided to apply it in other model structure experiments. To stay as close as possible to the original model, after the last bottleneck, we added a 1 × 1 convolution with Batch Normalization and activation to expand the channels to 2048.

Optimization at early and late model stages Reducing the number of kernels in the first layer or discarding unnecessary layers to save computation is worth considering if the model is already simplified. Completely discarding fully connected layers decreased accuracy by 0.02 pp. We also adjusted the last bottleneck first convolution to produce channels equal to the number of classes, leaving only 16 kernels in the first convolution. For example, all of these changes resulted in a decrease of only 0.09 pp, reducing the already smaller model (No. 9 compared to No. 6 in Table 1) by 1,386M parameters and 26.21M FLOPS.

Ghost Bottlenecks In the next stage of experiments, we applied the full ghost bottleneck proposed by the authors of GhostNet. This modification involved the use of two ghost modules (the first expanding the number of channels, the second reducing) introduced in Sect. 2, with depthwise separable convolution between when downsampling is needed, without interfering blocks skip connections. With this solution, we achieved an even smaller model in comparison with using inverted residual blocks (No. 6 form Table 1)—1.06% of baseline model parameters and 1.3% of FLOPS.

With respect to the baseline model, we obtained a loss in effectiveness of only 1.63 pp.

Squeeze-and-excitation In our opinion, adding the SE connection could bring benefits in terms of accuracy with a strong model reduction (in our case configuration No. 12 and No. 14 from Table 1). We placed these blocks just before the second ghost module in all ghost bottlenecks. These consisted of two linear layers with ReLu in between and the number of neurons depending on the feature maps (the first had 1/4 of the number of channels, the second equal to this number). In both cases, this operation increased the number of parameters by about 1M, with a small impact on FLOPS (about 2 pp). The benefits of this modification were not proportional to the increase in the number of parameters.

3.2 *VGG-19 on CIFAR-100*

For VGG trained on CIFAR-100, we prepared 12 experiments using different optimization approaches. The experiments cannot be identical to the ResNet due to differences in the baseline literature. We present the obtained results in Table 2. As before, the percentages of number of parameters and FLOPS are relative to the base model (Experiment No. 1), and each experiment is described in detail below.

Global Average Pooling VGG is a type of architecture that uses FC layers for a classification process based on late extracted features. Newer models replace these layers with GAP, as ResNet does. We can see that this simple improvement reduced the number of parameters by almost twice, with a marginal decrease in accuracy (0.18 pp). This reduction (almost half of the previous number of parameters) involves the removing of a large number of connections between 3 large fully connected layers, leaving only one linear with a number of neurons equal to the number of classes. This modification was applied to all other experiments with VGG, as it successfully improved the performance of the model.

Replacement of convolutions For ResNet, depthwise separable convolution made the VGG model much smaller (almost 10 times smaller in number of parameters and even larger in FLOPS). Again, we found that 3×1 to 1×3 convolutions underperformed the depthwise separable convolution in terms model compression. However, with about a 6-fold better reduction effect, we received worse accuracy than with 3×1 to 1×3 convolutions (4.67 pp).

Fire Modules VGG turned out to be more suitable for the use of Fire modules. To make the change meaningful, we replaced all the original convolutions with Fire modules, with the first squeeze part reducing the channels by a factor of 8 relatives to the baseline. Half of the block result was also generated by cheap 1×1 convolutions. Thus, using Fire modules save more computational resources than depthwise separable convolutions with higher accuracy. Because of this, we used Fire modules in the rest of the experiments with VGG.

No FC classifier We can observe that after removing the remaining FC layer of the classifier, we obtained an accuracy loss of 1.97 pp with minimal reduction (in

Table 2 Optimization techniques for VGG-19 trained on CIFAR-100

No.	1	2	3	4	5	6	7	8	9	10	11	12
Approach												
GAP	–	✓	✓	✓	✓	✓	✓	✓	✓	✓	✓	✓
N × 1 to 1 × N Conv.	–	–	✓	–	–	–	–	–	–	–	–	–
DWconv	–	–	–	✓	–	✓	✓	✓	✓	✓	✓	✓
Fire modules	–	–	–	–	✓	✓	✓	✓	✓	✓	✓	✓
No FC classifier	–	–	–	–	–	–	✓	✓	✓	✓	✓	✓
Shuffle mech.	–	–	–	–	–	–	–	✓	–	✓	✓	✓
Ghost modules	–	–	–	–	–	–	–	–	✓	✓	✓	–
SE connections	–	–	–	–	–	–	–	–	–	–	✓	✓
Parameters [M]	39.328	20.087	13.150	2.332	1.848	0.520	0.520	0.198	0.443	0.388	1.588	1.398
Parameters [%]	100.00	51.08	33.44	5.93	4.70	1.32	1.32	0.50	1.13	0.99	4.04	3.55
FLOPS [M]	418.02	398.79	256.08	46.8	40.27	11.04	11.19	4.99	9.62	8.73	10.23	6.49
FLOPS [%]	100.00	95.40	61.26	11.20	9.63	2.64	2.68	1.19	2.30	2.09	2.45	1.55
Size [MB]	150,10	76.70	50.24	8.97	7.21	2.14	2.14	0.90	1.85	1.64	6.23	5.50
Accuracy [%]	71.91	71.73	70.93	66.26	67.77	65.19	63.22	59.75	60.56	61.27	61.83	60.08

the presented measure, only 0.15 M less FLOPS). The possible reason could be an insufficient model capacity for the more challenging CIFAR-100 dataset.

Shuffle mechanism All of the pointwise convolutions occurring in the Fire modules we adjusted to work on separated groups as we did with ResNet. After the squeeze convolution, we added the shuffle operation, and after the concatenation in the output (this shuffle module works on a doubled number of groups to ensure full information flow between groups). Unlike the ResNet model, we used only **ghost modules**, replacing the 1×1 squeeze and the 3×3 expand convolutions. With the GhostNet model solution, we only adapted the grouped pointwise convolution to the first 1×1 convolution of ghost modules. From Table 2 we can see that applying the shuffle mechanism to ghost modules beats the accuracy achieved without grouped 1×1 convolutions while reducing the parameters and the FLOAT number. With **SE connections** we improved the accuracy of the models with the expected increase in size.

3.3 Post-training Compression

Post-training compression, we proceeded on three different ResNet models (unmodified baseline (No. 1). and models No. 11 and No. 13). Table 3 shows the effects of iterative unstructured pruning with amount 0.2 and norm L1. In each iteration, the model was fine-tuned with a small learning rate. We ran the baseline model pruning much longer than the other two because its larger number of model parameters allowed us to prune the model to greater sparsity. With the progressive pruning, we can observe how its accuracy changes depending on the number of unpruned parameters.

Post-training static quantization in PyTorch required changing the model code and feeding validation data into the model to determine the distribution of values. At the input, we placed 2 layers for quantizing the input data and dequantizing the response at the output. Adding and concatenation also required argument dequantization before they were performed. We performed quantization on both pruned (with an accuracy above 90% and a similar number of non-zero parameters, i.e., No. 11 with a sparsity of 79.0% and No. 13 with 48.8%; for a No. 1, we chose the version with the highest sparsity) and unpruned models. With quantization applied, all selected models decreased their accuracy as expected. For the pruned models, they decreased to 89.77% (baseline), 89.68% (No. 11), and 85.76% (No. 13), while for the unpruned models, they dropped to 94.07%, 91.49%, and 87.11%, respectively. Hardware vendors explicitly allow 8-bit data (after quantization) to be processed faster than 32-bit data (before quantization). Smaller models consume less power and memory, which benefits edge device deployments.

Table 3 Pruning technique for ResNet-101 trained on CIFAR-10

Sparsity [%]	No. 1—baseline		No. 11		No. 13	
	Non-zeroed	Acc. [%]	Non-zeroed	Acc. [%]	Non-zeroed	Acc. [%]
0.0	42 512 970	94.33	400 115	92.43	150 715	90.82
20.0	33 930 189	94.34	304 678	92.51	105 273	90.86
36.0	27 148 247	94.36	243 742	92.41	84 218	90.51
48.8	21 722 694	94.36	194 994	91.82	67 374	90.11
59.0	17 382 251	94.39	155 995	91.79	53 899	89.85
67.2	13 909 897	94.32	124 796	91.68	43 119	89.03
73.8	11 132 014	94.36	99 837	91.21	34 495	87.85
79.0	8 909 707	94.11	79 870	90.59	27 596	87.03
83.2	7 131 862	94.16	63 896	90.19	22 077	84.88
86.5	5 709 586	93.95	51 117	89.28	17 662	81.98
89.2	4 571 765	93.72	40 894	88.73	14 130	80.02
91.4	3 661 508	93.37	–	–	–	–
99.0	411 432	91.53	–	–	–	–

4 Summary

The aim of this work was to evaluate the impact of architecture compression techniques for CNN models that have been developed over the years, which we mentioned in Sect. 2. We also studied the applicability of these methods by applying them to different models: ResNet-101 and VGG-19. All of them result in a model that is smaller (in terms of the number of parameters and size in device memory) and has accelerated computations (reduced FLOPS).

It is difficult to point to a universal solution to this optimization problem. Fire modules have proven unsuitable for newer architecture, such as ResNet. Based on our research results, we consider high-level building blocks, especially ghost bottlenecks, to be a good start toward the goal of CNN compression. After high-level optimization proceeded, we noticed the great potential in lower-level modifications, such as the popular depthwise separable convolution or shuffle mechanism. The same goes for the GAP, which we recommend in most cases. However, we recommend carefully optimizing crucial parts of architectures like the classifier. For more complex problems, completely removing the last fully connected layers can significantly reduce accuracy. We can highlight configuration No. 11 (Table 1), which with 106 times fewer parameters, underperformed the baseline model by less than 2 pp in accuracy. Pruning has proven effective even for the most architecture-optimized models, while quantization may be appropriate for models that are not drastically compressed—otherwise, losses may be too high.

References

1. Khan, A., et al.: A survey of the recent architectures of deep convolutional neural networks. Artif. Intell. Rev. **53** (2020). 10 Artur Sobolewski and Kamil Szyc
2. Zhang, X., et al.: Shufflenet: an extremely efficient convolutional neural network for mobile devices. In: Proceedings of the IEEE Conference on Computer Vision and Pattern Recognition (2018)
3. Howard, A., et al.: Searching for mobilenetv3. In: Proceedings of the IEEE/CVF International Conference on Computer Vision (2019)
4. Simonyan, K., Zisserman, A.: Very deep convolutional networks for large-scale image recognition (2014). arXiv:1409.1556
5. He, K., et al.: Deep residual learning for image recognition. In: Proceedings of the IEEE Conference on Computer Vision and Pattern Recognition (2016)
6. Szegedy, C., et al.: Rethinking the inception architecture for computer vision. In: Proceedings of the IEEE Conference on Computer Vision and Pattern Recognition (2016)
7. Chollet, F.: Xception: deep learning with depthwise separable convolutions. In: Proceedings of the IEEE Conference on Computer Vision and Pattern Recognition (2017)
8. Hu, J., Shen, L., Sun, G.: Squeeze-and-excitation networks. In: Proceedings of the IEEE Conference on Computer Vision and Pattern Recognition (2018)
9. Liang, T., et al.: Pruning and quantization for deep neural network acceleration: a survey. Neurocomputing **461** (2021)
10. P Véstias, M.: A survey of convolutional neural networks on edge with reconfigurable computing. Algorithms **12**(8) (2019)
11. Qassim, H., Verma, A., Feinzimer, D.: Compressed residual-VGG16 CNN model for big data places image recognition. In: 2018 IEEE 8th Annual Computing and Communication Workshop and Conference (CCWC). IEEE (2018)
12. Alzubaidi, L., et al.: Review of deep learning: concepts, CNN architectures, challenges, applications, future directions. J. Big Data **8** (2021)
13. Iandola, F.N., et al.: SqueezeNet: AlexNet-level accuracy with 50x fewer parameters and! 0.5 MB model size (2016). arXiv:1602.07360
14. Lin, M., Chen, Q., Yan, S.: Network in network (2013). arXiv:1312.4400
15. Sandler, M., et al.: Mobilenetv2: inverted residuals and linear bottlenecks. In: Proceedings of the IEEE Conference on Computer Vision and Pattern Recognition (2018)
16. Han, K., et al.: Ghostnet: more features from cheap operations. In: Proceedings of the IEEE/CVF Conference on Computer Vision and Pattern Recognition (2020)

Scheduling Resource to Deploy Monitors in Automated Driving Systems

Peng Su, Tianyu Fan, and Dejiu Chen

1 Introduction

Automated driving systems (ADS) consist of different tasks for operating the ego-vehicle. Basically, these tasks can be divided into two steps [6, 10]: (1) **Scene Understanding**: With the help of multiple sensors (e.g., LiDAR, camera, IMU), the ADS receives different data to perceive environments. Deep learning architectures are widely used for processing these sensory signals. By interpreting and fusing these data, the DNNs provide a general and intuitive context (e.g., relative distances, geometry locations) to downstream components of ADS. (2) **Decision-making and Planning**: According to outputs of DNN modules, the ADS maps its location and makes proper driving policies. With these steps, the ADS accomplishes a workflow to perceive the environment and control the ego vehicle. In this workflow, achieving a high precision of environmental data is critical to ensure safety of ADS. However, faults and uncertainties (e.g., image soiling, soft errors) across different sensors and DNNs prevent the ADS from making accurate decisions [4, 8, 11]. One solution is to use Conditional Monitors (CM) [3, 7, 15] to detect faults in these processes. Such monitors can be used in safety mechanisms for mitigating and avoiding accidents. To maintain the best performance for detecting faults, these monitors are usually sophisticated. For example, Variational Autoencoders are used in [7] to detect soiling cameras. In [15], a supervised learning method is combined with Kullback-Leibler divergence to detect anomalous data in the DNNs. In [3], an exhaustiveness search is used for enumerating all possible faulty states in the perception modules of ADS. Regardless of the real-time performance of these approaches, these complex architectures occupy extensive data and memory space of ADS hardware (Fig. 1), leading to execution latency of the ADS workflow. However, considering the integration

P. Su · T. Fan · D. Chen (✉)
KTH Royal Institute of Technology, 100 44 Stockholm, Sweden
e-mail: chendj@kth.se

P. Su
e-mail: pensu@kth.se

T. Fan
e-mail: tianyuf@kth.se

W. Zamojski et al. (eds.), *Dependable Computer Systems and Networks*, Lecture Notes in Networks and Systems 737, https://doi.org/10.1007/978-3-031-37720-4_26

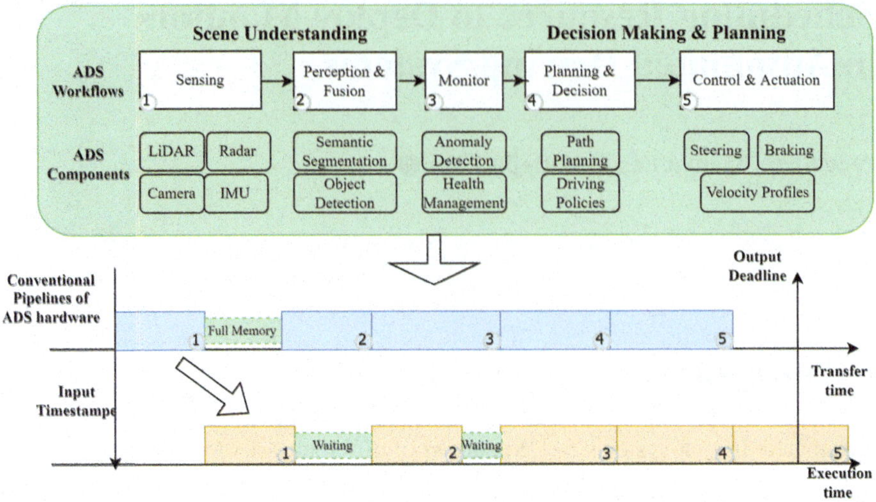

Fig. 1 A pipeline of an ADS workflow with a conditional monitor task. The memory of ADS hardware is full after loading the DNN architectures from disks (e.g., The program code with well-trained parameters of the YOLOv5 is around 1GB), leading to the GPU/CPU wait for executing tasks. In the conventional ADS workflow, such latency is acceptable because the workflows meet a deadline. However, additional transfer/execution time of the monitor leads to the workflow missing the deadline

and flexibility, the ADS software frameworks (e.g., Autoware) and hardware platforms (e.g., NVIDIA Jetson) are indifferent to resource management. Such conflicts challenge deployment of conditional monitors in the ADS.

This paper proposes a method for overcoming some of these challenges. The monitors prioritized detect critical data from the perception modules to avoid and mitigate accidents, whereas Responsibility-Sensitive Safety (RSS) is used to quantify the risk of workflows under different operational situations. According to the time cost of the monitors, a predictable dynamic scheduling ensures that the automated driving hardware executes the workflows in time. As a case study, we select YOLOv5 as a scene understanding module of the car-following module to detect objects [12]. Then, we list current monitors [16] and measure their execution time. Next, we summarize the time consumption of different monitors under the conventional pipeline and our scheme. Compared with the conventional method, our scheme reduces 43.7% of execution time per cycle.

2 Background

2.1 Dynamic Memory Scheduling in the ADS

Current ADS architectures cause data and memory management problems from the following aspects: (1) Entire computing units (e.g., CPU and GPU) of current ADS hardware platforms (e.g., NVIDIA Jetson series, Intel OpenVINO) share unified and limited memories, resulting in memory schedules unrelated to the risk of current workflows. (2) ADS software frameworks (e.g., Autoware [9], Apollo [13]) running on these platforms do not provide a guarantee of predictability and memory management. To solve these issues, current methods in [17, 18] focus on optimizing the input/output efficiency to improve memory management performance. However, these methods ignore the availability to access the memory and operational systems of the platforms, resulting in difficulties to apply their approaches to the current ADS platforms and frameworks. A predictable resource management method is proposed in [1] to reserve the memory space for the ADS workflow. However, to preserve memories of all tasks in the ADS workflows, such a scheduling approach is time-consuming, causing a scalability problem for deploying it in the autonomous embedded hardware.

2.2 Responsibility-Sensitive Safety

The Responsibility-Sensitive Safety model [14] is a formal method to describe the safety distances of the ADS. The RSS models analyze the maneuvers of the ego-vehicle from longitudinal and lateral geometries (e.g., the longitudinal and lateral distances d^{long}, d^{lat} in Fig. 2). The RSS calculates the minimum safe longitudinal $d^{long}_{\min}$ and lateral $d^{lat}_{\min}$ distances as follows:

$$d^{long}_{\min} = \left[v_r \rho + \frac{1}{2} \alpha_{\max} \rho^2 + \frac{v_r + \rho \alpha_{\max}{}^2}{2 \beta_{\min}} + \frac{v_f{}^2}{2 \beta_{\max}} \right] \tag{1}$$

$$d^{lat}_{\min} = \mu + \left[\frac{v_1 + v_{1,\rho}}{2} \rho + \frac{v_{1,\rho}}{2 \alpha^{lat}_{\min}} - \left(\frac{v_2 + v_{2,\rho}}{2} \rho + \frac{v_{2,\rho}}{2 \alpha^{lat}_{\min}} \right) \right] \tag{2}$$

where v_r and v_f are the longitudinal velocities of the rear and the front vehicles, respectively. α and β denote the front vehicle acceleration and the rear vehicle brake. ρ refers to the reaction time before the rear vehicle starts braking. v_1 and v_2 represent the lateral speed of the vehicles. The RSS models define $v_{1,\rho} = v_1 + \rho \alpha^{lat}_{\min}$ and $v_{2,\rho} = v_2 - \rho \alpha^{lat}_{\min}$. The safety distances depend on the velocities, acceleration/deceleration of the two vehicles and their reaction time.

3 Methodology

The motivation of our work is to solve the conflict between the limited resource of ADS hardware and the efficiency of conditional monitors. Therefore, we summarize the research question: To deploy a monitor in the ADS, what strategies should be used to offset additional time consumption caused by resource insufficiency? To answer this question, we propose the following steps:

3.1 Risk Evaluation of Perception Data

A multi-camera perception system is widely used in the ADS to receive data from multiple directions (Fig. 2) [19]. In the conventional ADS workflow, these sensory data are processed by the perception modules sequentially, leading to a risk of the workflows to exceed the deadlines. Such overdue situations may cause accidents. To process and monitor sensory data in time, we propose a risk evaluation of perception data. With the help of the RSS, we select critical data related to operational situations. Then monitors prioritized inspect these data. We create a task set to map the image data c and other sensors data l (e.g., Radars, IMU) at any timestamp k. The task set T_k is defined as follows:

$$T_k = \left\{ \left(c_k^1, l_k^1\right), \ldots, \left(c_k^i, l_k^i\right) \right\} \tag{3}$$

where i refers to the objects detected from cameras and other sensors. $c^{\{1,\ldots,i\}}$ is used to image processes such as object detection and semantic segmentation. $r^{\{1,\ldots,i\}}$ concludes the information (e.g., relative distance and velocity from longitudinal and latent aspects) between the detected objects and the ego vehicle to evaluate the risk of operational situations. According to the RSS rules in Eqs. 1 and 2. We define the risk of each r^i as follows:

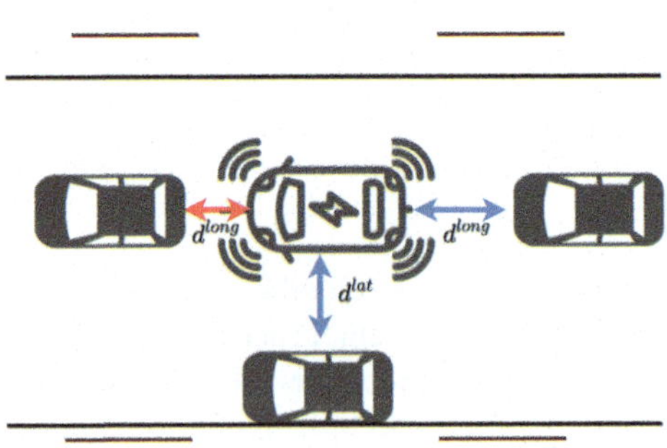

Fig. 2 A bird-eye view of a driving scenario. Ego vehicle receives sensory data from multiple directions. The red arrow indicates the critical distance which will violate the longitudinal safety distance of RSS (Eq. 1). This distance has a high risk and the image from the front camera needs to be prioritized monitored

$$r_k^{i,long} = \frac{[2v_r t_c + \alpha_{\max} t_c^2 - (2v_f t_c + \beta_{\max} t_c^2)]}{d_{\min}^{long}} \tag{4}$$

$$r_k^{i,lat} = \frac{\left[\frac{v_1 + v_{1,t_c}}{2} t_c - \frac{v_2 + v_{2,t_c}}{2} t_c\right]}{d_{\min}^{lat}} \tag{5}$$

t_c is the Worst Case Execution Time (WCET) of the task set to avoid accidents. We define the overall risk $r_k^i = r_k^{i,long} + r_k^{i,lat}$. This metric indicates that with a higher relative velocity, the ego vehicle contains higher risks. To prioritize detect possible faults in the task set T_k, we map a risk set $R_k = \{r_k^n, \ldots, r_k^m\}$ with an image set $C_k = \{c_k^n, \ldots, c_k^m\}$. Compared with the original task set T_k, the monitors spend less time to inspect data from C_k.

3.2 *Dynamic Scheduling for Reserving Resources*

As we mentioned in Sect. 2.1, the sporadic workflow of the ADS frameworks leads to an unpredictable output. We propose dynamic scheduling of executing the workflow by utilizing the free interval among workflows. We modify the Heijunka and ReScue model [1] to reserve resources for executing the monitor modules. By utilizing the free interval among task set cycles, we calculate the starting timestamps s_T for executing ADS workflows and monitors:

$$s_T = t_1 - \left(\frac{m}{v_m}\right) + n_T \cdot t_{\min} + t_M \cdot n_{C_k} \tag{6}$$

where s_T represents a dynamic timestamp for starting to load ADS data and modules from disks. t_1 is the timestamp that depends on the real-world time k and the execution cycle Δ for a task set. m represents the memory space for a task set T and v_m is the memory Input/Output (I/O) speed. $t_{\min}$ refers to the shortest execution time in one task set T_k and n_T is the number of modules in the ADS workflow, t_M refers to the execution time of the monitor per image, and n_{C_k} is the image amount of C_k.

Except for the variables k, m, C_k, which are related to the current task set and time, there are also other variables, which can be measured from the ADS running history. This lightweight approach arranges the execution time of the ADS workflow (Fig. 3), avoiding the complexity of accessing the memory of the integrated ADS platforms and frameworks. Such scheduling ensures the ADS monitors inspect critical data. The entire process is summarized in Algorithm 1.

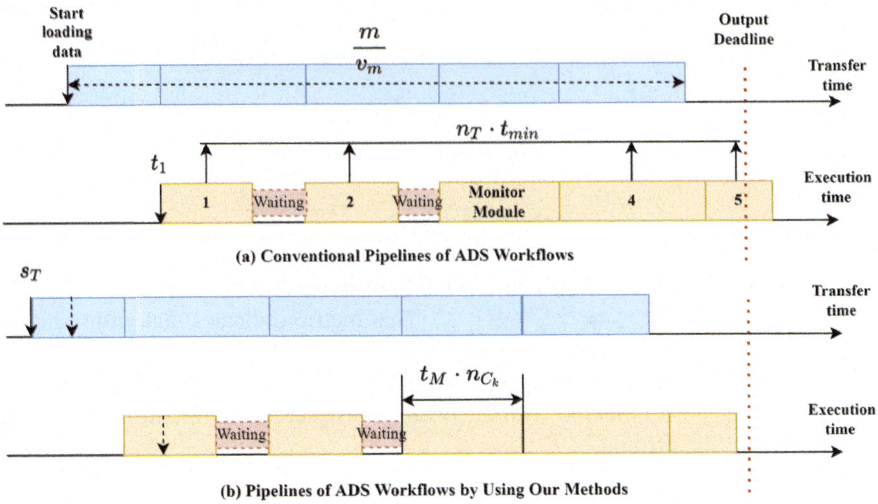

Fig. 3 A comparison between the conventional ADS workflow with our method. The reservation time s_T is calculated by Eq. 6. Even if the monitor inspects the same data as the conventional method, our approach still meets the deadline

---

Algorithm 1: Dynamic Scheduling with Risk Evaluation

Input: Task set$T_k = \{(c_k^1, l_k^1), \ldots, (c_k^i, l_k^j)\}$, timestamp k,
Output: Timestamp to load ADS tasks s_T

Initialize C_k, $R_k = \emptyset$;
for *each data* (c_k^i, l_k^j) *in* T_k **do**
  r_k^i = getRisks(c_k^i, l_k^j); ▷ Risk evaluation of the current task set T_k
  Add r_k^i into set R_k;
  if r_k^i *in* R_k > *threshold* **then**
    Map r_k^i to c_k^i; ▷ Selection of the critical image data
    Add c_k^i into set C_k;
$s_T = getTimstamp(C_k, T_k, k)$ ▷ Get the starting timestamp according to Eq. 6.
return s_T;

---

4 Case Study

We select the car-following system [2] as a case study to evaluate our data scheme for deploying monitors. In this system, we use YOLOv5 to detect objects from front cameras. Once the YOLOv5 classifies the front objects as vehicles, the ADS uses data from radars and LiDars to analyze relative distance and velocity. Next, the speed control process is initialized to follow the front vehicle within a safe distance (Fig. 4). However, when this DNN meets faults, it cannot detect objects or output a wrong classification. In this case, monitors of soiling and anomaly detection can be used for supervising the image quality and the DNN functionality. In our experiment, we use the KITTI dataset [5] to train YOLOv5. We select different monitors, including

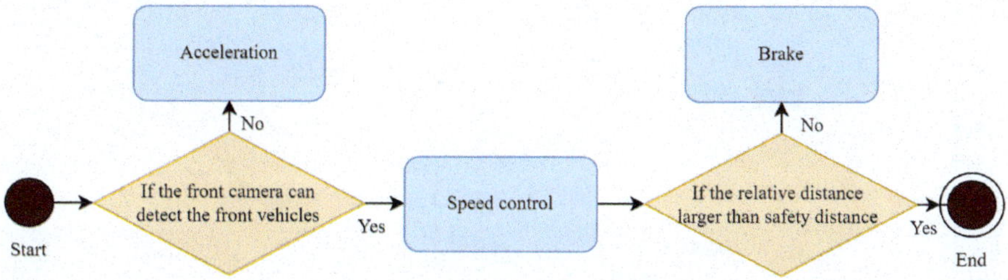

Fig. 4 A functional overview of the car-following produce based on UML Activity Diagram

AutoEndocer (AE), AutoEndocer-Gaussian Mixture Models (AE-GMM), Fisher's Exact Test (FET), Likelihood Ratios (LRT), Maximum Mean Discrepancy (MMD), to detect anomalies in the images. Next, we measure the time consumption of these monitors. Finally, we evaluate the improvement of our method compared with the conventional pipelines of ADS workflows.

4.1 *Timing Analysis of the ADS Modules and Workflows*

We deploy the well-trained perception and monitor modules in the car-following system. By setting a semaphore for the timestamps of loading and executing the tasks, we record the time consumption of these modules. Compared with the execution time of perception modules, most of monitors spend more time to detect faults from images (Fig. 5a). To measure the time cost of ADS workflows, we choose a typical ADS with 6 cameras [19] , which refer to any task sets $T_k = \{(c_k^1, l_k^1), \ldots, (c_k^i, l_k^i)\}, i = 6$. In Fig. 5b, we measure the time cycle of loading and executing the ADS workflows with the monitors. Without any scheduling methods, the time consumption to meet deadlines increases from 155 to 190 ms. Such increased cost indicates that these monitors could be non-trivial in the car-following system.

4.2 *Deploying Monitors Based on the Proposed Data Scheme*

By using the dynamic scheduling method, the results in Fig. 5b indicate that the ADS workflows spend $180ms$. After adding the monitors, these time cycles still increase by 20%. Meanwhile, by decreasing the detected images of the monitors, the risk evaluation method shows similar results in Fig. 5b.

To further optimize the time cycles of ADS workflows, we combine the dynamic scheduling with the risk evaluation. Compared with the conventional ADS workflows, the execution time cycle increases by 6.7%. The main reason for this significant improvement is that we select critical data from the task set, reducing the

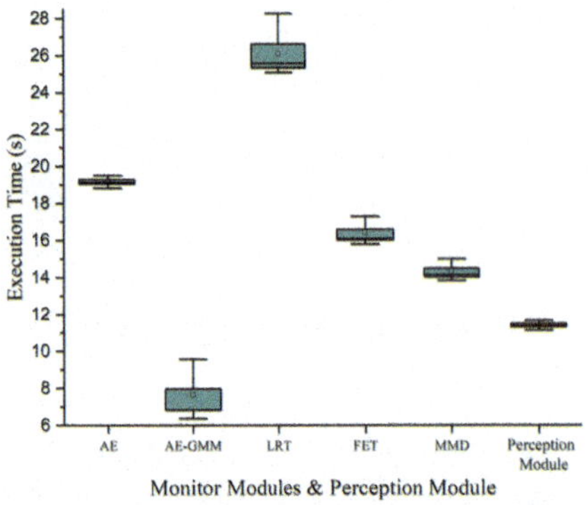

(a) Time consumption of different monitors and the perception module

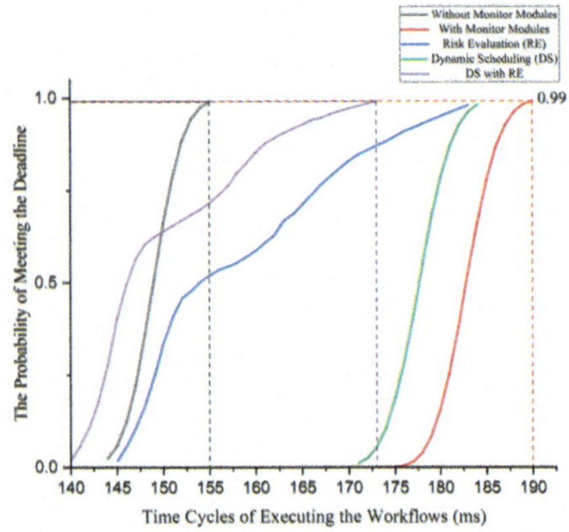

(b) Time cycle with different scheduling methods

Fig. 5 Some results of experiments. In **a**, we evaluate the time consumption of different methods for one task set by iterating 10,000 cycles. In **b**, we illustrate the time cycle with different scheduling methods of deploying LRT in the ADS

loading and executing time of the monitors. Meanwhile, the dynamic scheduling reserves sufficient resources for the ADS workflows, reducing the waiting time for loading and executing the ADS tasks.

To prove the robustness of our approach, we use different monitor modules to test our scheme further (Table 1). We define the decrease rate R_d as follows:

$$R_d = \frac{\left(\bar{t}_M - \bar{t}_O\right) - \left(\bar{t}_{M,A} - \bar{t}_O\right)}{\bar{t}_M - \bar{t}_O} \tag{7}$$

where $\bar{t_O}$ refers to the average time cycles of conventional ADS workflows without monitors. $\bar{t_M}$ refers to the average time cost for the conventional pipeline of ADS workflows with monitor modules. $\bar{t}_{M,A}$ refers to the average time of an ADS workflow with monitor modules by using our scheme. By comparing the increased time of the conventional pipelines with our scheme's, R_d quantifies improvements in the time cycles. From Table 1, we conclude that our method has a better impact on a more time-consuming monitor module. For example, the average time consumption of LRT and AE are 26 and 19 ms, while their decrease rates are 51.4 and 48.1%. Using our scheme, the ADS workflows with monitors spend less time meeting the deadlines.

5 Conclusion and Future Work

In this paper, we propose a data scheme by using dynamic scheduling with RSS. Compared with current methods, our approach has the following benefits: (1) The scheme can deploy in any integrated ADS hardware platforms since our method does

Table 1 Time consumption of different monitors with conventional pipelines and our method by iterating 10,000 cycles of ADS workflows

Monitor name	$\bar{t_O}$	$\bar{t_M}$ (ms)	$\bar{t_{M,A}}$ (ms)	R_d (%)
AE	155	182	169	48.1
AE-GMM		164	161	33.4
LRT		190	172	51.4
FET		179	168	45.8
MMD		175	167	40.0
Average		178	167	43.7

not need to access the memory architectures of these platforms. (2) By scheduling the tasks of the ADS workflows, our method is compatible for current ADS software frameworks. (3) We define the ADS risk metrics of ADS workflows, improving the ADS framework's predictability and reliability. (4) The results of our method indicate that the scheme is robust for data-centric conditional monitors, especially the DNN-based architectures that involve high dimensional data (e.g., pixel-wise data).

We plan to further exploit the scheme's performance by integrating our scheme with open-source ADS frameworks and simulation tools (e.g., Autoware, CARLA). In our current work, the selections of task sets (T and C) are based on multi-sensors. Such a hypothesis has discrepancies in real-life cases. To close our work to the ADS products, where some only equip cameras as perception systems, we consider refining our scheme by purely relying on image data.

Acknowledgements This work is supported by 1. KTH Royal Institute of Technology with the industrial research project ADinSOS (2019065006); and 2. the Swedish government agency for innovation systems (VINNOVA) with the cooperative research project Trust-E (Ref: 2020-05117) within the programme EUREKA EURIPIDES.

References

1. Bateni, S., Liu, C.: Predictable data-driven resource management: an implementation using autoware on autonomous platforms. In: 2019 IEEE Real-Time Systems Symposium (RTSS), pp. 339–352. IEEE (2019)
2. Claussmann, L., Revilloud, M., Gruyer, D., Glaser, S.: A review of motion planning for highway autonomous driving. IEEE Trans. Intell. Transp. Syst. **21**(5), 1826–1848 (2019)
3. Dreossi, T., Fremont, D.J., Ghosh, S., Kim, E., Ravanbakhsh, H., Vazquez-Chanlatte, M., Seshia, S.A.: Verifai: A toolkit for the formal design and analysis of artificial intelligence-based systems. In: International Conference on Computer Aided Verification, pp. 432–442. Springer (2019)
4. Dreossi, T., Jha, S., Seshia, S.A.: Semantic adversarial deep learning. In: International Conference on Computer Aided Verification, pp. 3–26. Springer (2018)
5. Geiger, A., Lenz, P., Stiller, C., Urtasun, R.: Vision meets robotics: the kitti dataset. Int. J. Robot. Res. **32**(11), 1231–1237 (2013)

6. Grigorescu, S., Trasnea, B., Cocias, T., Macesanu, G.: A survey of deep learning techniques for autonomous driving. J. Field Robot. **37**(3), 362–386 (2020)
7. Hartsell, C., Ramakrishna, S., Dubey, A., Stojcsics, D., Mahadevan, N., Karsai, G.: Resonate: a runtime risk assessment framework for autonomous systems. In: 2021 International Symposium on Software Engineering for Adaptive and Self-Managing Systems (SEAMS), pp. 118–129. IEEE (2021)
8. Ibrahim, Y., Wang, H., Liu, J., Wei, J., Chen, L., Rech, P., Adam, K., Guo, G.: Soft errors in DNN accelerators: a comprehensive review. Microelectron. Reliab. **115**, 113969 (2020)
9. Kato, S., Tokunaga, S., Maruyama, Y., Maeda, S., Hirabayashi, M., Kitsukawa, Y., Monrroy, A., Ando, T., Fujii, Y., Azumi, T.: Autoware on board: enabling autonomous vehicles with embedded systems. In: 2018 ACM/IEEE 9th International Conference on Cyber-Physical Systems (ICCPS), pp. 287–296. IEEE (2018)
10. Kiran, B.R., Sobh, I., Talpaert, V., Mannion, P., Al Sallab, A.A., Yogamani, S., Pérez, P.: Deep reinforcement learning for autonomous driving: a survey. IEEE Trans. Intell. Transp. Syst. (2021)
11. Li, G., Hari, S.K.S., Sullivan, M., Tsai, T., Pattabiraman, K., Emer, J., Keckler, S.W.: Understanding error propagation in deep learning neural network (DNN) accelerators and applications. In: Proceedings of the International Conference for High Performance Computing, Networking, Storage and Analysis, pp. 1–12 (2017)
12. Ning, J., Wang, J.: Automatic driving scene target detection algorithm based on improved yolov5 network. In: 2022 International Conference on Computer Network, Electronic and Automation (ICCNEA), pp. 218–222. IEEE (2022)
13. Raju, V.M., Gupta, V., Lomate, S.: Performance of open autonomous vehicle platforms: autoware and Apollo. In: 2019 IEEE 5th International Conference for Convergence in Technology (I2CT), pp. 1–5. IEEE (2019)
14. Shalev-Shwartz, S., Shammah, S., Shashua, A.: On a formal model of safe and scalable self-driving cars (2017). arXiv:1708.06374
15. Su, P., Chen, D.: Using fault injection for the training of functions to detect soft errors of dnns in automotive vehicles. In: International Conference on Dependability and Complex Systems, pp. 308–318. Springer (2022)
16. Van Looveren, A., Klaise, J., Vacanti, G., Cobb, O., Scillitoe, A., Samoilescu, R., Athorne, A.: Alibi detect: algorithms for outlier, adversarial and drift detection (2019). https://github.com/SeldonIO/alibi-detect
17. Wan, B., Li, X., Zhang, B., Zhou, K., Luo, H., Wang, C., Chen, X., Zhou, X.: A predictable servant-based execution model for safety-critical systems. In: 2017 IEEE International Symposium on Parallel and Distributed Processing with Applications (ISPA/IUCC), pp. 892–896. IEEE (2017)
18. Wan, B., Luo, H., Zhou, K., Li, X., Wang, C., Chen, X., Zhou, X.: A time-aware programming framework for constructing predictable real-time systems. In: 2017 IEEE 19th International Conference on High Performance Computing and Communications, pp. 578–585. IEEE (2017)
19. Yogamani, S., Hughes, C., Horgan, J., Sistu, G., Varley, P., O'Dea, D., Uricár, M., Milz, S., Simon, M., Amende, K., et al.: Woodscape: a multi-task, multi-camera fisheye dataset for autonomous driving. In: Proceedings of the IEEE/CVF International Conference on Computer Vision, pp. 9308–9318 (2019)

Power Analysis of BLAKE3 Pipelined Implementations in FPGA Devices

Jarosław Sugier

1 Introduction

Cryptographic hash functions are ubiquitous in contemporary IT security infrastructure: they are used for variety of tasks from key generation, through digital signatures and authentication schemes, to construction of stream ciphers. Most of the current state-of-the-art algorithms of this type originated from the SHA competition started in 2007 by the National Institute of Standards and Technology, which, in order to overcome then-latest advances in cryptanalysis, called for development of the new hash standard with entirely fresh approach to construction of the algorithm. The three ciphers investigated in this work−BLAKE3, Keccak, and BLAKE2−all came from the same background and are rooted in this contest. Evaluation of the candidates with regard to their efficient (i.e. fast) implementation in hardware was an important criterion during the contest (see, e.g., [6]), but the power analysis, being an aspect not as easy in estimation, was not always taken into account just as well.

The aim of this paper was to evaluate the dynamic power losses in FPGA implementations of BLAKE3 cipher organized in iterative and internally pipelined architectures. The calculations were done by the implementation tools using the most accurate approach, i.e. with switching activities of all internal signals computed by the timing simulations of the final, fully placed and routed designs. As an implementation platform, commercial Spartan-7 devices from Xilinx were selected. The results compare power requirements of the iterative architecture of BLAKE3 with analogous implementations of BLAKE2 (its direct predecessor) and Keccak (the core of the SHA-3 standard), as well as illustrate changes in power characteristics which were caused by pipelining the core logic. Six BLAKE3 architectures (the basic iterative one plus five pipelined variants with up to 6 pipeline stages) along

J. Sugier (✉)
Department of Computer Engineering, Wrocław University of Science and Technology, 11/17 Janiszewskiego Str, 50-372 Wrocław, Poland
e-mail: jaroslaw.sugier@pwr.edu.pl

W. Zamojski et al. (eds.), *Dependable Computer Systems and Networks*, Lecture Notes in Networks and Systems 737, https://doi.org/10.1007/978-3-031-37720-4_27

with non-pipelined iterative organizations of BLAKE2 and Keccak algorithms made the test suite of 8 cipher modules. To create a foundation for neutral comparison, all of them were tested with the same basic and simple I/O system and implemented on the same FPGA platform.

The identical set of ciphers/architectures was used in the previous study [11] that evaluated their maximum speed of operation and effective throughput; this work can be seen as an extension of those results, which enhances the comparison with a new perspective of power analysis. The text is organized as follows. The algorithms included in the study and the specifics of their hardware implementations are briefly introduced in Chap. 2, while methodology and results of power analysis are presented in Chap. 3. Evaluation of the results with regard to the structure of the losses, the effects of pipelining, and the differences between the three ciphers are included in Chap. 4, just before conclusions in Chap. 5.

2 The Algorithms and Their Organizations in Hardware

The original version of the BLAKE cipher ([1]) was submitted as a contender to the NIST contest and was among the best 5 algorithms selected for final consideration. Although this cipher ultimately lost the contest to Keccak, it was acclaimed in all stages of evaluation for excellent cryptographic strength and great performance in software implementations. For its advantages (and also due to slower operation of the new SHA-3 algorithms in software), it quickly became popular in the cryptographic community. In 2012 the authors modified their original submission announcing BLAKE2 ([2]) and it soon found application in various data processing or communication systems, from RAR file archive format through Linux kernel utilities to key derivation functions. BLAKE3 is a recent proposition from 2020 ([9]) with further modifications that simplified the processing and led to an additional increase in the effective hashing speed. The newest version quickly became considered as an option for fast yet secure hash generation, e.g., in blockchain transaction protocols ([10]) or automatic key generation ([5]).

The two variants of the cipher considered in this paper—BLAKE2 and BLAKE3—use very similar round organization which can be seen in Fig. 1 presenting as an example architectures of the standard iterative implementations. The processing consists in transforming the 512-bit state in eight instances $G_0 \div G_7$ of the so-called quarter-round functions. An individual function transforms 128 bits (4×32b state words) using also two message words as an additional input; in each round the state goes through a cascade of $4 + 4$ such modules.

There are two differences between BLAKE3 and its predecessor that are important for this analysis: (a) the number of round N_R is reduced from 10 to 7, and (b) while in BLAKE2 a set of ten permutations σ_r decided what message words M were used by G_i modules in the r-th round, in the new algorithm assignment of these words is constant but they are permuted between rounds. Especially the latter modification simplified the hardware remarkably by eliminating costly 16:1, 32-bit

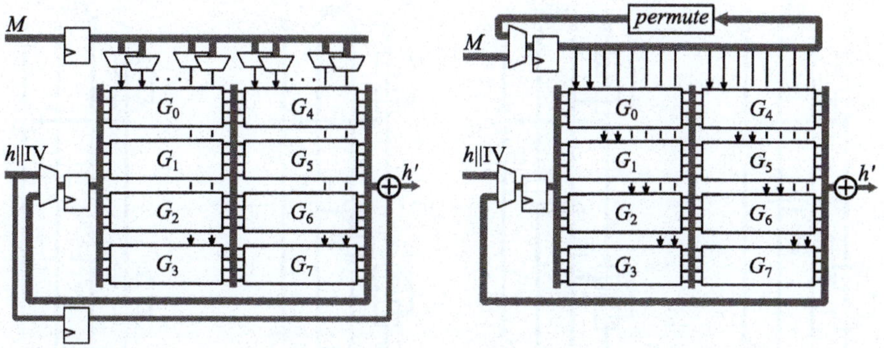

Fig. 1 Organizations of the basic iterative architecture in BLAKE2 (left) and BLAKE3 (right), as proposed in [11]

wide multiplexers (two per each G_i instance). The problems with the implementation of these multiplexers were discussed already in [12] and, as the results published in [11] revealed, their elimination could lead to a more than 40% reduction in the occupancy of the FPGA array.

The internal organization of the Keccak compression function ([3, 8]) is simpler: its large 1600-bit state is organized as an array of 5×5 words, 64-bit each, and the round processing consists in applying 5 specific transformations $N_R = 24$ times. Since the transformations operate on various parts of the state (columns, rows or planes of the $5 \times 5 \times 64$ cube of bits) it is impossible to visualize them in a dataflow diagram as simple as Fig. 1. It is worth noting, though, that in Keccak the message bits are only loaded as a part of the state vector during its initialization before the first round (and later not used) while in BLAKE* they remain active as input parameters in all the rounds, forming a separate 256-bit path running along that of the 512-bit state.

This work includes analysis of six different implementations of BLAKE3 in hardware. Taking a typical iterative organization with one round instance and the state propagated through N_R times in a loop, the higher-throughput variants were derived by pipelining the round and processing multiple data chunks in parallel. The basic iterative (non-pipelined) case will be denoted as "X1" and the pipelined variants as "P*k*", with *k* indicating the total number of pipeline stages inside the round (thus also the number of parallel data streams being hashed concurrently). The set of pipelined variants consisted of P2, P4 and three versions of P6 architectures. The two other algorithms included for comparison—BLAKE2 and Keccak—were implemented in the basic iterative organization X1.

Creating pipelined BLAKE3 cases required decisions about location of stage boundaries within the data path of the round. It was obvious that a natural place for such a boundary was the crossing between two successive G_i blocks, and this made the P2 variant. The remaining architectures P4 and P6 required inner splitting of the G_i logic into, respectively, two or three pipeline stages. The discussion of such a division which considered the structure of internal logic was included in [11] and

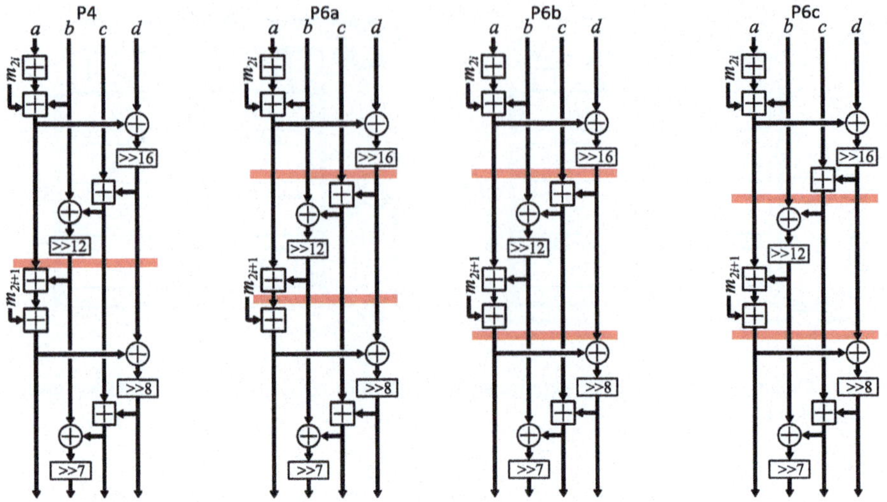

Fig. 2 Internal data paths of the BLAKE quarter-round and proposed stage boundaries for BLAKE3 architectures with 4 or 6 pipeline stages (after performance studies in [11])

ended with three variants denoted as P6a, P6b, and P6c. The divisions of the internal G_i paths for all architectures in which it was needed are presented in Fig. 2.

3 Power Calculations

3.1 Methodology

Implementation. All cipher cores were defined with parallel input and output ports that occupied a considerable number of bits. For example, BLAKE3 would read 512b of the message plus 256b of the hash chain value and 128b of other parameters, and would produce 256b of the output. For power analysis, a complete and realistic device had to be implemented in a physical FPGA device, so the cores were equipped with basic I/O buffers (Serial-In Parallel-Out on the input and Parallel-In Serial-Out on the output) for iterative loading of the input and unloading the results in 32b chunks. The buffers reduced the number of required package pins to a reasonable number and, although consumed some number of registers, had negligible impact on operational and power characteristics of actual cryptographic cores, which were primarily (and heavily) orientated on combinatorial logic.

The designs were then synthesized and implemented by Xilinx Vivado software for the Spartan-7 devices ([14]): all BLAKE3 designs were fitted on the smallest xc7s6cpga196-2 chip, while the BLAKE2 and Keccak X1 cases, being too large, had to be implemented in the second chip on the lineup, xc7s15cpga196-2. The smallest possible devices and the 7 Series budget family were chosen intentionally. The designs were implemented for their maximum frequencies, which were determined in [11], with some margin to help in stable implementation.

Simulation. To evaluate power consumption, special test scenarios were prepared for each device which kept the hash engine running continuously—that is, started a new batch of calculations as soon as the previous one had ended. This was possible, first, by assigning a test vector as an initial value of the input SIPO register so that no data loading phase was necessary and the simulation began instantly with hashing. Such an assignment was done in the HDL source files and, strictly speaking, was a part of the implementation. Secondly, because the pipelined units should be loaded with different input vector every clock cycle at the start (to fill the pipeline) and there were no clock cycles to serially load their values, the next vectors were generated by shifting some random pattern from the input pins to the SIPO register. This did not provide entirely new input for each pipeline stage, as it would be desired for simulating the maximum switching activity of the signals (thus evaluating the maximum power consumption), but the state bits in the adjacent stages soon became sufficiently different, thanks to the cipher's avalanche effect.

With such a test and a design implemented as in the previous stage, a post-route timing simulation was run in order to produce a switching activity file in the SAIF format ([15]). Each implementation was tested with several operating frequencies, and for each one a separate testbench was prepared and a SAIF file generated.

Power calculations. The SAIF file provided switching parameters for all signals in the FPGA array as they had been evaluated in simulation—and these data were the foundation of computations done by the Vivado Power Analysis tool. The calculations were executed for standard values of supply voltages but with maximum process setting, i.e. the power specification of the device was set to the worst-case values. The analysis of this paper is based on the results from power reports generated in this way, which included, among others, total on-chip power, its dynamic and device static parts, junction temperature, etc. ([15]).

Further discussion is limited exclusively to the analysis of the reported dynamic power and omits the static component. The static power represents losses caused by leakage currents depending (for a particular device) on process, voltage, and temperature—and not on the activity caused by the programmed bitstream. As a result, the static power is closely related to heat transport and cooling conditions (ambient temperature and airflow, heat sink, board thermal resistivity, etc.) which are completely independent from the operation of the programmed cipher core. Thus, in order to remove the aspects unrelated to the algorithm under study, the analysis will be limited exclusively to the components of the dynamic power, although if, for example, anyone would need to estimate the final device temperature, then the elements of static device power would have to be taken into account as well as other associated circumstances.

3.2 Results

Table 1 summarizes the parameters of the implemented FPGA designs which were the subject of further power analysis. The first 4 rows describe occupancy of the array (i.e. design size): number of used slices (nodes of the array), LUTs (function generators), registers, and dedicated multiplexers. The last type of resources is used exclusively in the BLAKE2 core, where it is required to multiplex the message words between G_i modules (as noted in Chap. 2). The sizes are visualized in Fig. 3 where one can compare array occupancy presented as a number of slices, and density of their utilization as an average number of LUTs per slice.

The second part of the table gives requested and actual (estimated for the implemented circuit) clock frequency, and also describes the longest propagation path that determined actual F_{clk}: percentage of its delay which came from routing and the number of its logic levels with included carry primitives.

Table 1 Size and speed of the eight implementations that were examined in power analysis

	X1	P2	P4	P6a	P6b	P6c	BL2 (X1)	Kecc. (X1)
Slices	761	892	1054	1150	1199	1207	1247	1397
LUTs	2541	2607	2779	3473	3458	3491	4362	4623
Registers	2184	3206	4743	5804	5774	5783	2437	4815
MUXes	0	0	0	0	0	0	1536	0
Target F_{clk} [MHz]	40,0	90,0	140,0	190,0	190,0	170,0	40,0	200,0
Actual F_{clk} [MHz]	44,4	91,9	146,6	194,2	197,2	172,0	40,6	215,5
Route delay	56%	49%	51%	42%	54%	52%	56%	88%
Logic levels	45	23	14	12	10	11	45	1
(inlc. Carry)	(33)	(18)	(10)	(9)	(7)	(8)	(30)	(0)

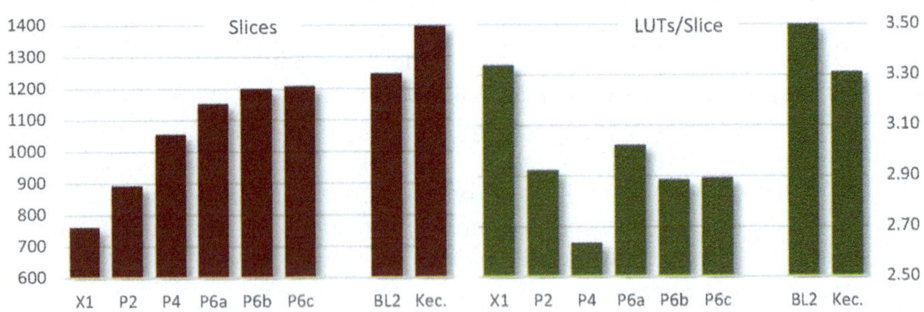

Fig. 3 The 8 implementations: total sizes as numbers of occupied slices (left) and their utilization as average numbers of LUTs per slice (right)

The discussion of size versus speed parameters of these implementations was the subject of study in [11]. In summary: (a) removal of message multiplexing in BLAKE3 led to a remarkable size reduction (somewhat reversed by pipelining, but less than the number of pipeline stages could suggest), but only moderate increase in speed, (b) the Keccak largest size should be related to a significantly wider state path (1600b vs. 512 + 256b), (c) the SHA-3 hash still boast the fastest hardware operation thanks to the best mapping of round transformations to function generators (just 1 LUT element in the longest path), and (d) 32-bit adders which are common for both versions of BLAKE impair their implementations because they prevent full utilization of the extended, 6-input LUTs available in the array.

The effects of power analysis—the topic of this paper—are finally presented in Table 2. The numbers were calculated for each implementation running with its requested (target) operating frequency. The total dynamic power consists of parts attributed to clock distribution networks, slices (split into LUT generators, carry elements, registers and MUXes), routing and I/O blocks. For each design such computations were repeated with other (reduced) frequencies in order to evaluate the dynamic power as a function of frequency, and the results are shown in Fig. 4, with both axes in logarithmic scale. The figure allows one to compare algorithms functioning in different frequency ranges but also confirms the generic principle in estimation of digital power losses, which states that such losses are linear with regard to F_{clk}. For each design, its plot ends with the maximum frequency but, of course, it can be extended linearly towards lower values.

Because the dynamic power is indeed a linear function of the clock frequency, additional metrics can be calculated and they are included as the last two rows in Table 2. The parameter P_{MHz} binds the power with operating frequency (mW/MHz) just as in any digital circuit and is visualized in the left part of Fig. 5. The right part of this figure or the last row of the table gives E_h—the total energy consumed by computation of one hash output (μJ/hash). The latter value can also be interpreted as the dynamic power required to reach a specific hashing throughput in millions of hashes per second (W/Mhps). As is easy to see, these two metrics have the same distribution of relative values among all versions of the BLAKE3 algorithm, whereas the differences in BLAKE2 and Keccak come from different numbers of rounds (hence from different speeds of output expressed in hashes per clock cycle) in those algorithms.

4 Evaluation of Results

4.1 *Structure of Power Losses*

Figure 6 presents power details as they are given in Table 2, i.e., divided into different resources. To make the graph readable the two smallest components—the I/O blocks and dedicated multiplexers—were combined in the "Other" category, but their values

Table 2 Dynamic power (values given in Watts) calculated by the tools for the implementations operating with their maximal frequencies

	X1 40 MHz	P2 90 MHz	P4 140 MHz	P6a 190 MHz	P6b 190 MHz	P6c 170 MHz	BL2 40 MHz	Kecc. 200 MHz
Total	9,219	0,931	0,401	0,413	0,373	0,326	9,214	1,801
Clocks	0,002	0,010	0,020	0,033	0,034	0,031	0,003	0,030
Slice logic	3,330	0,357	0,144	0,146	0,137	0,121	3,394	0,674
LUT	2,969	0,303	0,118	0,112	0,107	0,091	3,029	0,671
Carry	0,361	0,051	0,022	0,026	0,023	0,022	0,366	0,000
Register	0,001	0,001	0,004	0,008	0,008	0,007	0,001	0,003
MUXes	0,000	0,000	0,000	0,000	0,000	0,000	0,001	0,000
Routing	5,884	0,560	0,231	0,228	0,194	0,168	5,816	1,092
I/O	0,001	0,003	0,007	0,007	0,007	0,007	0,001	0,004
Efficiency metrics								
P_{MHz} (mW/MHz)	230	10,3	2,87	2,18	1,96	1,92	230	9,01
E_h (μJ/hash)	1,61	0,072	0,020	0,015	0,014	0,013	2,30	0,216

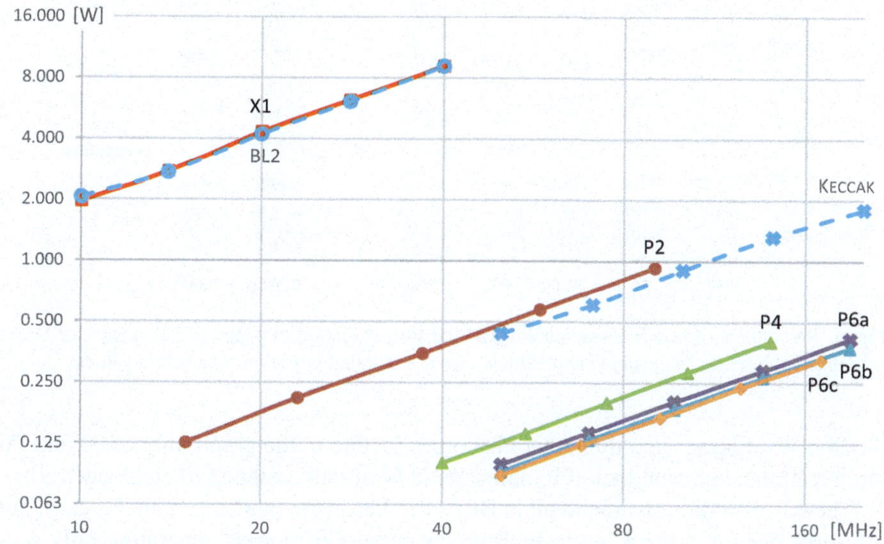

Fig. 4 Dynamic power of the designs as functions of clock frequency

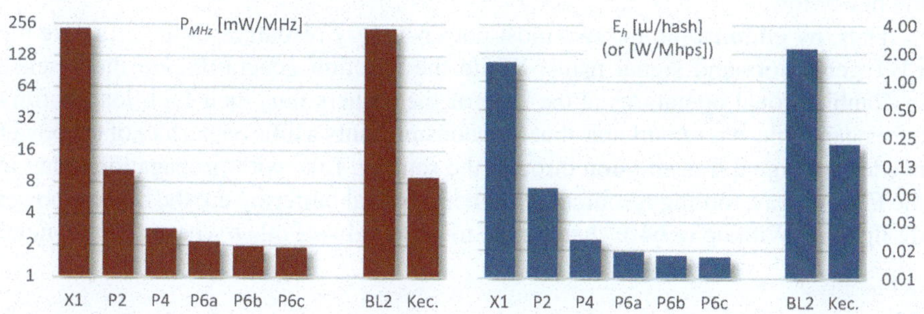

Fig. 5 Power efficiency metrics on logarithmic scale: dynamic power dissipated per MHz of operating frequency on the left and energy consumed for one hash calculation (which is also the power dissipated per million hash calculations per second), on the right

were anyway negligible: the I/O activity in the test scenarios was purposely kept minimal and, more importantly, the MUXes in BLAKE2 core turned out to be very efficient generating just 0,001mW of losses despite such a significant size of the message multiplexing (as indicated by the difference in size of BLAKE2 versus BLAKE3).

Also clock distribution networks are of minor importance here as only in the most pipelined cases, when the number of registers they must drive is above 5000 and F_{clk} approaches 200 MHz, they consume close to 10% of the total power. Speaking of registers, their power efficiency is very good in this application: they generate at most

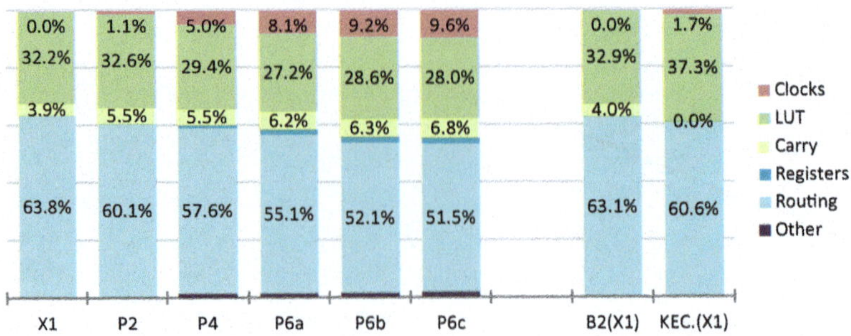

Fig. 6 Percentage of power losses generated by different resources of the FPGA array (for better readability the series "Registers" and "Other" are not labelled with their negligible values)

0,008 mW of losses and their share becomes visible in the graph only in P4 and P6 organizations, but even then it remains about ¼ of that assigned to clock networks.

The 32-bit adders are abundant in BLAKE* hardware (and absent in Keccak), but the carry primitives turn out to be little consumers of power, generating only 4 ÷ 7% of the losses. Similarly to MUXes, building the adders with dedicated primitives and connections within slices (and not with the general configurable routing) keeps the power low.

Thus, by elimination, the two most power-hungry resources of the array are the LUT generators and signal paths. While the function generators are the essence of combinational operations at the core of the ciphers thus their high load is fully understandable, it is remarkable that they consume only a little over a half of the power needed for signal transmission through the routing. Like with propagation delay in the FPGA array, routing resources with their internal buffering are the biggest power dissipators. With up to 64% share in the total they dwarf other types of component.

4.2 The Effects of Pipelining in BLAKE3

Massive reductions in dynamic power caused by pipelining is the most striking observation from a comparison of all BLAKE3 organizations. Already dividing the X1 round into two pipeline stages reduced the E_h energy from 1,61 to 0,072 μJ/hash, i.e. by 95%, and adding another two stages in the P4 case resulted in a further 62% reduction down to 0,020 μJ/hash; the P6 cases need on average 30% less E_h energy than the P4 one.

Possible power reductions caused by the introduction of a pipeline in any digital circuit are well known, especially on FPGA platforms ([4, 7, 13]), and it turns out that the realization of the BLAKE core is a striking example of this effect. The mechanism is rooted in elimination of *signal glitches*—transient and intermittent values on an output of a combinatorial gate which appear before it settles to the final (correct) value. Such transients occur when different inputs do not arrive to the gate synchronically due to (a) unequal delays on their propagation paths and (b) different levels of logic they traverse. The former factor is an essential concern of all FPGA implementations, where transmission lines must be formed with numerous configurable segments of various delays, while the latter one is unavoidable in realization of complex and involved combinatorial functions. The BLAKE round, being essentially a module with $512 + 256$ input bits and 512b output with *very* complicated and purposely irregular internal structure, is extremely glitch-prone—and the resulting switching activities on the long routing lines with parasitic capacitive loads generate a lot of power losses. The pipelining eliminates glitches thanks to two effects: (a) by decomposition of the combinatorial propagation into pipeline segments, the number of logic levels is reduced (according to the Table 1 from 45 in X1 down to $10 \div 12$ in P6*) hence dissimilarities in input delays are smaller, and (b) the pipeline registers block propagation of the glitches between adjacent stages so they are not further reproduced in an avalanche effect.

An inspection of the SAIF files (which were generated by simulation with number of transitions for all internal signals in the array) illustrates the scale of the problem in the examined designs. While the top 5% most active signals in the P6* cores had on average about 5 transitions per one clock cycle, in the X1 designs this number was 1380 and 1270—such a huge amount of glitches *per one clock cycle* explains the generation of so high power losses. In the iterative (not pipelined) implementation of Keccak, on the other hand, this amount was approximately 10.

It is also worth noting the contribution to the glitching which came from the 32-bit adders (6 per each G_i module) appearing in the BLAKE transformations. Because in the FPGA arrays adders are implemented in the most elementary ripple-carry organization, sequential carry propagation in the ripple chain generates extra transient values on the output. While such an organization is sufficiently fast and itself is not especially power-hungry (as Fig. 6 shows), the glitching generated in this way is transferred to the signal routes where it adds to their power losses.

Finally, on another note: when comparing the three options of the 6-stage pipelined architectures (P6a $\div$ c), the results of this analysis deliver a more visible differentiation than the one which came from performance evaluation in [11]. While the comparison in that paper could not find a clear winner, here the case P6c has the best E_h metrics, with P6b worse by nearly 3%, but P6a—by a noticeable 14%. The decomposition of the G_i paths in the (now the worst) P6a organization was the most natural one with an even distribution of the 6 adders (2 per each stage), but it turned out that the concern raised in [11]—separating two consecutive adders by a stage boundary—indeed indicated a possible implementation issue, but more in the power than in the speed aspect. The variant P6c was the worst one with regard to the maximum clock achievable in the performance-oriented implementation but here, when the timing constraints were not so tight, it turned out to be the most power-efficient.

4.3 BLAKE3 Versus BLAKE2 and Keccak

It is a noteworthy finding that the dynamic power of the two BLAKE versions turned out to be virtually identical: the raw power dissipation as a function of frequency (W/MHz) in both circuits was equal, and the change in their E_h metric is a purely arithmetical effect of different number of rounds N_R. Peculiarity comes from the fact that such a close equivalence was achieved for two modules of so different size, with the BLAKE3 being approx. 40% smaller in both slices and LUTs. In [11] the low resource utilization was seen as a notable advantage of the new cipher, but it is now completely invisible in the power parameters.

The size difference came from elimination of the 32-bit wide, 16-input multiplexers from the message schedule of BLAKE3 (see Chap. 2) but it turned out that this massive amount of logic, first, dissipated little power despite its size (because of utilization of internal routing and dedicated MUX primitives within a slice), and, secondly, it generated very little glitches thanks to equalized propagation paths. For a multiplexer of any size, its internal structure is essentially a regular and balanced binary tree, and—only if its implementation does not overuse general, configurable routing resources—this balance can be preserved in the FPGA array. Therefore, despite the 40% variance in size, the calculated power parameters of the two BLAKE designs in Table 2 remain practically the same, within $\pm$ 2% deviation.

Finally, the results allow for comparison of the new hash with Keccak. The SHA-3 core is not plagued with glitches as much as BLAKE*: with single LUT elements in the combinatorial paths and without the need to include 32-bit adders, the average number of transitions in 5% most active signals is about 10 rather than 1300. As a result, even though routing in this design also contributes 60% of the power, its total value is significantly lower, leading to 7 times better E_h parameter (μJ/hash) and even 25 times higher Watt per megahertz efficiency—and this with the occupancy of the array higher by 80%.

5 Conclusions

The paper discussed dynamic power losses of different FPGA implementations of BLAKE3, BLAKE2, and Keccak hash functions, based on their calculation by the implementation tools from timing simulation of complete, fully routed designs. Using the same platform and evaluation methodology, the total of 8 different designs could be compared in a uniform and fair method. The results indicated essential importance of signal activities, particularly glitches, which in these ciphers determine total dynamic power more than, e.g., their size and amount of used logic resources.

Overall, the exceptional level of signal glitches in a full (unpipelined) BLAKE* round makes its X1 organization power-wise virtually impractical, at least for higher clock frequencies. Even if pipelining is not needed for an increase in the hashing throughput, it can be applied solely for power savings, unless a power budget of the entire design can accommodate high losses generated by the cipher core. It should be noted, though, that in this comparison the efficiency characteristics for the pipelined variants were computed assuming their optimal working conditions, i.e. the pipeline filled in 100% and producing new hash on average every N_R clock cycles (which comes from k results every $k \cdot N_R$ cycles, for a k-stage pipeline). If the operating environment cannot provide enough data on time to fill the pipeline, the efficiency and the difference would be proportionally smaller.

Another remark should be made about validity of the power calculation models in case of glitches as intense as those found in this study. The (most accurate) power evaluation method offered by the tools is based on the simulated SAIF data which provide just one number for each signal giving its total transitions over the whole simulation period, without differentiating between transient glitches and stable levels. If the glitches appear more than thousand times in one clock cycle (as in the extreme cases considered here), typical models for calculation of current consumption based on recharging a capacitive load may need refinement.

References

1. Aumasson, JP., Henzen, L., Meier, W., Phan, R.C.-W.: SHA-3 proposal BLAKE, version 1.3. (2010). https://www.aumasson.jp/blake/blake.pdf. Last accessed Mar 2023
2. Aumasson, J.P., Neves, S., Wilcox-O'Hearn, Z., Winnerlein, C.: BLAKE2: Simpler, smaller, fast as MD5. In: Jacobson, M., Locasto, M., Mohassel, P., Safavi-Naini, R. (eds.) Applied cryptography and network security. ACNS 2013. Lecture notes in computer science, vol 7954, pp. 119–135. Springer, Berlin, Heidelberg (2013). https://doi.org/10.1007/978-3-642-38980-1_8
3. Bertoni, G., Daemen, J., Peeters, M., Van Assche, G.: The Keccak reference, SHA-3 competition (round 3), 2011. https://keccak.team/papers.html. Last accessed March 2023
4. Boemo, E., Oliver, J., Caffarena, G.: Tracking the pipelining-power rule along the FPGA technical literature. In: 10th FPGA world Conference—Academic proceedings 2013, FPGAworld (2013). https://doi.org/10.1145/2513683.2513692
5. Ciocan, I. T., Kelesidis, E. A., Maimuţ, D., Morogan, L.: A modified argon2i using a tweaked variant of Blake3. In: 2021 26th IEEE Asia-Pacific Conference on Communications (APCC). Kuala Lumpur, Malaysia, pp. 271–274. IEEE Xplore (2021). https://doi.org/10.1109/APCC49754.2021.9609933
6. Gaj, K., Kaps J.P., Amirineni, V., Rogawski, M., Homsirikamol, E., Brewster, B.Y.: ATHENa—Automated tool for hardware evaluation: Toward fair and comprehensive benchmarking of cryptographic hardware using FPGAs. In: 20th International conference on field programmable logic and applications, pp. 414–421. Milano, Italy (2010). https://doi.org/10.1109/FPL.2010.86
7. Grover, N., Soni, M.K.: Reduction of power consumption in FPGAs—An overview. Int. J. Inf. Eng. Electron. Bus. **4**(5), 50–69 (2012). https://doi.org/10.5815/ijieeb.2012.05.07
8. National Institute of Standards and Technology: SHA-3 Standard: Permutation-based hash and extendable-output functions (FIPS 202). https://csrc.nist.gov/publications/detail/fips/202/final. Last accessed March 2023. https://doi.org/10.6028/NIST.FIPS.202
9. O'Connor, J., Aumasson, J. P., Neves, S., Wilcox-O'Hearn, Z.: BLAKE3: one function, fast everywhere. Real World Crypto 2020 (lightning talk). (2020). Available at https://github.com/BLAKE3-team/BLAKE3-specs/blob/master/blake3.pdf. Last accessed March 2023

10. Sinha, S., Anand, S., K, K. P.: Improving smart contract transaction performance in hyperledger fabric. In: 2021 Emerging trends in industry 4.0 (ETI 4.0), pp. 1–6. IEEE Xplore (2021). https://doi.org/10.1109/ETI4.051663.2021.9619202
11. Sugier, J.: FPGA implementations of BLAKE3 compression function with intra-round pipelining. In: Zamojski, W., Mazurkiewicz, J., Sugier, J., Walkowiak, T., Kacprzyk, J. (eds) New advances in dependability of networks and systems. DepCoS-RELCOMEX 2022. Lecture notes in networks and systems, vol 484. Springer, Cham. https://doi.org/10.1007/978-3-031-06746-4_31
12. Sugier, J.: Simplifying FPGA implementations of BLAKE hash algorithm with block memory resources. Procedia Eng. **178**, 33–41 (2017). https://doi.org/10.1016/j.proeng.2017.01.057
13. Wilton, S.J.E., Ang, S.S., Luk, W.: The impact of pipelining on energy per operation in field-programmable gate arrays. In: Becker, J., Platzner, M., Vernalde, S. (eds) Field programmable logic and application. FPL 2004. Lecture notes in computer science, vol 3203. Springer, Berlin, Heidelberg (2004). https://doi.org/10.1007/978-3-540-30117-2_73
14. Xilinx, Inc.: 7 Series FPGAs data sheet: Overview. DS180.PDF. www.xilinx.com. Last accessed Mar 2023
15. Xilinx, Inc.: Vivado design suite user guide: Power analysis and optimization. UG907.PDF. www.xilinx.com. Last accessed Mar 2023

Deep Learning ECG Signal Analysis: Description and Preliminary Results

Mateusz Surowiec, Piotr Ciskowski, Kondrad Kluwak, and Łukasz Jeleń

1 Introduction

According to the World Health Organization (WHO), cardiovascular disease is the leading cause of death worldwide. It is estimated that they cause about 19.9 million deaths a year.

This paper describes a system that can comply with one of the most important goals of the World Health Organization in relation to cardiovascular diseases, namely to ensure appropriate treatment and prevent premature deaths. Our framework seeks the best representation of the ECG signal, which is often performed in the diagnosis of all primary health facilities.

As reported by Lyon et al. [9], ECG signals can be an important step in predicting cardiovascular disease in diagnosis. Electrocardiography is a clinical examination procedure that can record electrical heart activity. In this recording, it is easy to observe abnormal heart function for a trained eye. They can be caused by many circumstances and have no direct impact on patients' health. From the literature [3, 4, 7] it can be seen that ECG measurements can not only be used to determine heart disease, but also have much broader applications. One of the most popular uses of ECG recordings is the classification of emotions and emotions [4, 10]. Here, popular emotions such as anger, sadness, and so on are classified. Agrafioti et al. [1] reported that their framework can classify the valence of ECG shapes with 89% accuracy. Selvaraj et al. [10], on the other hand, reported an average accuracy of 82.88% in emotion classification. They also showed that "neutral" emotions have the highest accuracy at which they can reach 92.87%.

M. Surowiec · P. Ciskowski · Ł. Jeleń (✉)
Department of Computer Engineering, Wrocław University of Science and Technology, Wybrzeże Wyspiańskiego 27, 50-370 Wrocław, Poland
e-mail: lukasz.jelen@pwr.edu.pl

P. Ciskowski
e-mail: piotr.ciskowski@pwr.edu.pl

K. Kluwak
Department of Control Systems and Mechatronics, Wrocław University of Science and Technology, Wybrzeże Wyspiańskiego 27, 50-370 Wrocław, Poland
e-mail: konrad.kluwak@pwr.edu.pl

W. Zamojski et al. (eds.), *Dependable Computer Systems and Networks*, Lecture Notes in Networks and Systems 737, https://doi.org/10.1007/978-3-031-37720-4_28

In 2018, Liu et al. described an additional accelerometer for activity tracking and a wearable ECG monitoring device. The authors believe that the system can achieve a total accuracy of 96.92% when tested with 13 volunteers aged 5–68 years. They also claim that the reported results can be used to monitor user activity. This method could start a possible use of ECG monitoring for medical support systems. One of the most disadvantaged fields seems to be communication with people who have difficulties generating or understanding spoken or written languages. Communication with these people is a very difficult task, leading to the introduction of an augmented and alternative communication system (AAC). AAC requires a person with disabilities to learn a specific way of communication, so if activity recognition is based on the explicit wearable, it will be easier to perform communication tasks. In this paper, we take the opportunity to describe a deep learning architecture that can be used for ECG signal analysis. These algorithms can also be used to detect activity and emotions in supplementary and alternative communications. In addition, we present preliminary results of the analysis and conclude promising conclusions on implementing a complete AAC framework.

2 ECG Signals Measurements

The electrical activity of the heart is the result of a variety of physiological phenomena. The aim of electrocardiography is to record a voltage-to-time graph to represent this activity. The measurement is carried out with sensors attached to various places in the human body. The first attempt to obtain electronic heartbeat records dates back to the end of the 19th century. In 1895, Willem Einthoven proposed the use of a PQRST notation of the peak of the ECG signal, and today we can distinguish different methods of recording the ECG signal.

The most commonly used clinical standard for medical diagnosis is the 12 lead system. The system uses 10 electrodes placed on the arms, legs, and chest. Other standards use fewer electrodes, for example, 3, 4, or 6 electrodes (Einthhoven, Goldberg and Wilson Methods). In addition, various ECG measurements are applied to patients with different diagnosis needs, such as the most common 12-lead rest measurement, and long-term exercise and Holter electrocardiography. Preprocessing of electrical signals (including filters) is an integral part of all kinds of measurement [8]. The results of ECG measurements are schematically shown in Fig. 1, including the most popular R–R interval measurement and all the above-mentioned peaks. This measurement is used primarily for the analysis of heart rate variability (HR) and heart rate variability (HRV). In many popular reusable devices, the R–R interval is the only calculation parameter based on reading from an optical sensor. Many other measurements are used to analyze entire electrical signals. These measurements include: SDNN—Standard Deviation of Normal to Normal (R-R) intervals, RMSSD—square Root of the Mean from the Sum of Squares of Differences between adjacent NN intervals, PNN50—Percentage of NN50 adjacent NN intervals larger than 50ms, SDANN—Standard Deviation of the 5-minute average NN intervals,

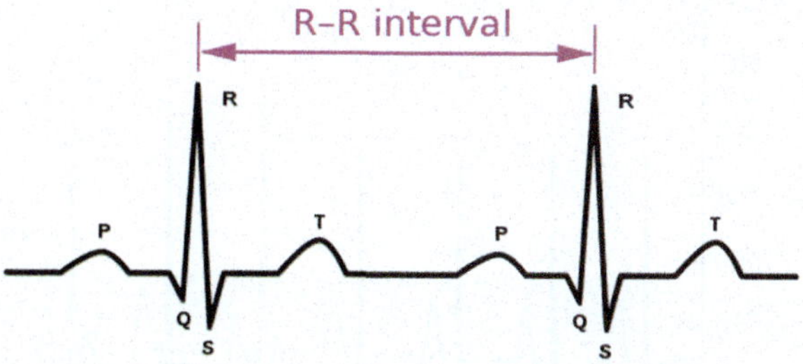

Fig. 1 One cycle of heart activity, with the R-R interval, as well as other peaks indicated, taken from [2]

SDSD—Standard Deviation of Successive Differences, TP—total power, HF, LF, VLF—high–, low– and very low–frequency power, as well as Poincare plots.

This research used a Polar H10 sensor. It is a popular wearable sports device that is widely available, as well as a reliable method of measuring heart rates. It has been studied in several studies [5, 6] and has been tested against Holter medical devices. These studies showed that in healthy subjects, Polar Sensors showed the same accuracy for low- and moderate-intensity activities and even higher accuracy for intense activities. Most studies comparing different wearable devices have validated methods based solely on measurements and HR calculations of the R–R interval, e.g. [6]. In our approach, we use all the data from the ECG signal, and the Polar H10 sensor is the only device that can accurately record this signal through the available API. We have developed a mobile application to collect ECG data from the Polar H10 device. An example of a raw signal recorded with this application is shown in Fig. 2. The use of the Polar Sensor allows access to complete and reliable ECG recordings. Therefore, we can calculate and analyze not only HR or HRV but also all the measures mentioned above. In addition, we can use raw one- or two-dimensional signals as input to deep neural networks. The data collected by the H10 sensor can be used for various applications such as heart disease diagnosis, reliable blood pressure estimation, and emotion detection in alternative communication systems.

3 Deep Neural Networks in ECG Signal Analysis

The purpose of this study was to propose the best architecture to manage future measurements. Furthermore, to perform the analysis mentioned above, we searched for a signal representation that would be best suited to the task of distinguishing various ECG signals, namely one-dimensional (vector) or two-dimensional (image). Two well-known data sets were used to compare, namely *MIT-BIH Normal Sinus Rhythm Databse* and *MIT-BIH Arrhythmia Database*. The first contains long-term recordings of signals from 18 healthy patients, while the second contains half-hour recordings of signals from 48 patients with different types of arrhythmia. Signals were downloaded and decoded using the *Python waveform-database* tool from Physionet's *wfdb* library. Additionally, all signals were converted 128 Hz at the same

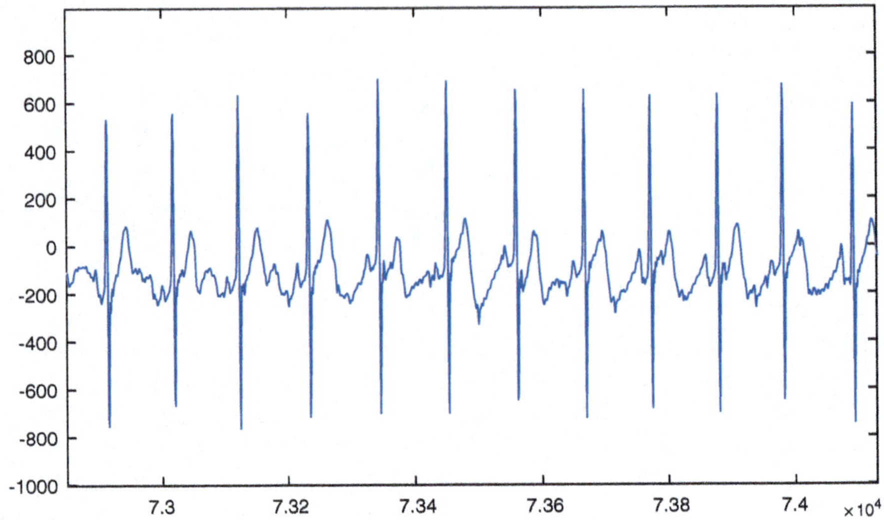

Fig. 2 An example of a raw ECG signal measured by the Polar H10 sensor

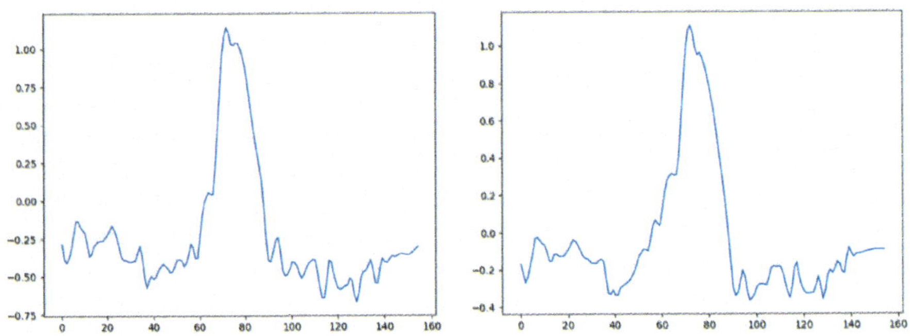

Fig. 3 ECG signal before and after preprocessing

frequency, filtered with the Butterworth filter 0.1–54 Hz to avoid major signal failures, and converted with wavelet transformations to eliminate high-frequency noise. An example of a pre- and post-preprocessing cycle is shown Fig. 3.

To detect R spikes and select consecutive heart cycles, a classical Pan-Tompkins algorithm is applied to the data. For healthy patients, 2,000 cycles were chosen and for diseased patients, 1,000 cycles were chosen. To avoid excessive errors due to physical measurement errors in the initial and final stages of the process, only the middle stage cycles of the recording are used. The length of each cycle was set to 160 measurements. Since all measurements come from different devices, each cycle is normalized to the 0–1 range.

One of the main purposes of the study was to compare the representation of signals, so data was prepared as 160-digit vectors or images with 160×30 resolution. An

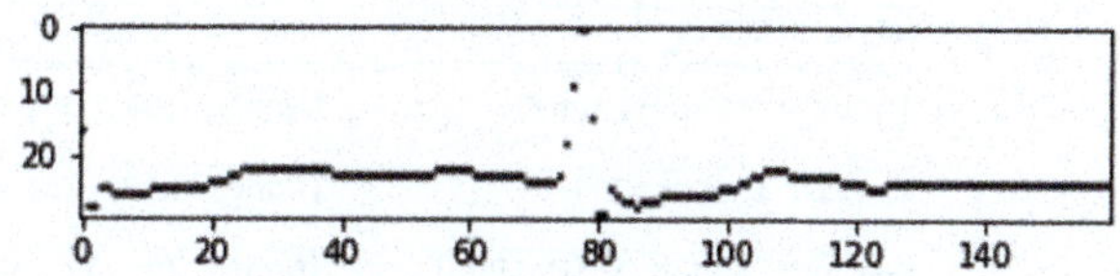

Fig. 4 Example of a two-dimensional ECG signal

example of two-dimensional ECG signals for convolutional networks is presented in Fig. 4.

During training the data sets were randomly divided as:

1. **Sample-oriented**—here a training and testing samples were prepared. Each patient was directed to one of these groups with all his samples. The training group was later divided into a training set and a validation set.
2. **Patient-oriented**—in this approach, all samples are randomly divided into three subsets: training, validating and testing. The number of samples used in these three groups was 55, 15 and 20 thousand, respectively.

The number of samples in both groups was similar. The second method has the advantage of having the same proportion of healthy and rhythmic signals in all subgroups, while the first sample can be similar in training and test data.

4 Comparison of Neural Network Architectures and Performance

As mentioned earlier, the main objective of the study was to determine the best classification scheme. Therefore, we compare the following neural network architectures:

1. **convolutional networks**—these networks were the first architectures examined in our research. The network provides a two-dimensional representation of the ECG signal. The simple version of this network, with only two convolutional layers and large filter sizes, achieved high accuracy of over 98% when tested with the sample-oriented method. Here, a complete random distribution of training, verification, and testing data is performed. When tested with patient-oriented methods, the accuracy dropped to about 76–86%. In further analysis, networks with additional convolution layers were used with fewer filter kernel sizes and an increased number of epochs. We have tested several models but the best accuracy of 90% was achieved for the architecture presented in Fig. 5 when a patient–oriented testing dataset was used.
2. **recurrent networks**—these networks use LSTM layers that receive one-dimensional ECG signals. Initially, two simple LSTM-layer architectures were used to test sample-oriented sets that allowed to achieve the accuracy of 97%. For patient-oriented data sets, this architecture reached about 85% accuracy. The results can be noted as similar to those for the convolution network, but the training time was much longer and the loss was high. The attempts to add a two-sided

Layer (type)	Output Shape	Param #
input_1 (InputLayer)	(None, 160, 30, 1)	0
conv2d_1 (Conv2D)	(None, 140, 25, 62)	7874
max_pooling2d_1 (MaxPooling2	(None, 70, 12, 62)	0
conv2d_2 (Conv2D)	(None, 67, 11, 128)	63616
max_pooling2d_2 (MaxPooling2	(None, 33, 5, 128)	0
conv2d_3 (Conv2D)	(None, 30, 2, 128)	262272
flatten_1 (Flatten)	(None, 7680)	0
dropout_1 (Dropout)	(None, 7680)	0
dense_1 (Dense)	(None, 256)	1966336
dense_2 (Dense)	(None, 128)	32896
dense_3 (Dense)	(None, 2)	258

Total params: 2,333,252
Trainable params: 2,333,252
Non-trainable params: 0

Fig. 5 Structure of the final convolutional network

layer at the front and a dense layer at the end of the network have not improved the results. A final architecture of a recurrent network is presented in Fig. 6.

3. **hybrid networks**—The next step was to replace the first two layers of the previously analyzed network with one-dimensional convolutions and convert two-dimensional images into one-dimensional vectors. LSTM layers are used to analyze the resulting vector. This operation improved accuracy by 90% and reduced training time by 30%. However, by replacing a single-dimensional convolution layer (for vector operations) with a two-dimensional layer (for image operations), accuracy dropped to 72–75% in patient-oriented test sets and the calculation time increased significantly. The architectures of one-dimensional and two-dimensional hybrid networks are presented in Figs. 7 and 8.
4. **double–input networks**—are the last architecture taken into consideration in this study. It uses convolutional and recurrent networks for classification, and its structure is described in Fig. 9. It is based on the best one- and two-dimensional networks obtained previously and is built with two branches, each of which analyzes the ECG signal differently. The first branch is composed of a one-dimensional lateral layer, followed by a vector signal, and a LSTM layer. The second branch is made up of two-dimensional convolutional layers. The output of the two branches is concatenated and followed by a dense layer.
In our experiments, we conducted several training analyzes that showed similar results. Although the best performance of this architecture was obtained for the train and validation sets (99.9% and 99.2% respectively), the accuracy for

```
Layer (type)                 Output Shape              Param #
=================================================================
input_1 (InputLayer)         (None, 160, 1)            0
_________________________________________________________________
bidirectional_1 (Bidirection (None, 160, 32)           2304
_________________________________________________________________
bidirectional_2 (Bidirection (None, 160, 64)           16640
_________________________________________________________________
lstm_3 (LSTM)                (None, 160, 128)          98816
_________________________________________________________________
lstm_4 (LSTM)                (None, 64)                49408
_________________________________________________________________
dense_1 (Dense)              (None, 64)                4160
_________________________________________________________________
dense_2 (Dense)              (None, 32)                2080
_________________________________________________________________
dense_3 (Dense)              (None, 1)                 33
=================================================================
Total params: 173,441
Trainable params: 173,441
Non-trainable params: 0
_________________________________________________________________
```

Fig. 6 Structure of the final recurrent network using one- and two-directional LSTM layers

Table 1 The best results for each architecture

Model	Loss	Accuracy (%)
1. **convolutional 2d**	**0.33**	**90.1**
2. recurrent 1d (only LSTM)	0.59	85.9
3. **hybrid (conv1d+LSTM)**	**0.28**	**92.2**
4. hybrid (conv2d+LSTM)	0.77	75.8
5. double-input	0.59	85.3
6. **double–input (pretrained)**	**0.30**	**93.6**

testing set remained at 80–85%. The obvious advantage of this structure is a short training time, measured in epochs (30) and in time (45s for each epoch). The last architecture modification was to use pretrained networks in both branches of double input networks. In this study, we use the vector network, which achieves 90% accuracy in the first branch and 92% accuracy in the repeat neural network of the second branch (asymmetric convolution). Training of the two branches was blocked, so only the last layer of thickness could be adapted. Due to the freezing of a large part of the network, only short training was given. This was sufficient to achieve the highest accuracy of 93.6% and the lowest loss of 0.30.

The results achieved for all tested architectures are summarized in Table 1 with the best results highlighted in bold. From the table, it can be seen that the networks with convolutional layers achieved high accuracies (above 90%). Furthermore, the best accuracy was noted for a mixed model for which both input types (one-dimensional vectors and two-dimensional images) were used and where both branches were pre-trained separately.

```
Layer (type)                 Output Shape              Param #
=================================================================
input_1 (InputLayer)         (None, 160, 1)            0
_________________________________________________________________
conv1d_1 (Conv1D)            (None, 156, 64)           384
_________________________________________________________________
conv1d_2 (Conv1D)            (None, 152, 128)          41088
_________________________________________________________________
dropout_1 (Dropout)          (None, 152, 128)          0
_________________________________________________________________
lstm_1 (LSTM)                (None, 152, 256)          394240
_________________________________________________________________
lstm_2 (LSTM)                (None, 128)               197120
_________________________________________________________________
dense_1 (Dense)              (None, 64)                8256
_________________________________________________________________
dense_2 (Dense)              (None, 32)                2080
_________________________________________________________________
dense_3 (Dense)              (None, 1)                 33
=================================================================
Total params: 643,201
Trainable params: 643,201
Non-trainable params: 0
_________________________________________________________________
```

Fig. 7 Structure of the final hybrid network using one-dimensional convolutional layers

```
Layer (type)                 Output Shape              Param #
=================================================================
input_1 (InputLayer)         (None, 160, 30, 1)        0
_________________________________________________________________
conv2d_1 (Conv2D)            (None, 140, 25, 64)       8128
_________________________________________________________________
max_pooling2d_1 (MaxPooling2 (None, 70, 12, 64)        0
_________________________________________________________________
conv2d_2 (Conv2D)            (None, 67, 11, 128)       65664
_________________________________________________________________
max_pooling2d_2 (MaxPooling2 (None, 33, 5, 128)        0
_________________________________________________________________
conv2d_3 (Conv2D)            (None, 30, 2, 128)        262272
_________________________________________________________________
max_pooling2d_3 (MaxPooling2 (None, 15, 1, 128)        0
_________________________________________________________________
flatten_1 (Flatten)          (None, 1920)              0
_________________________________________________________________
dropout_1 (Dropout)          (None, 1920)              0
_________________________________________________________________
reshape_1 (Reshape)          (None, 1920, 1)           0
_________________________________________________________________
lstm_1 (LSTM)                (None, 1920, 64)          16896
_________________________________________________________________
lstm_2 (LSTM)                (None, 32)                12416
_________________________________________________________________
dense_1 (Dense)              (None, 128)               4224
_________________________________________________________________
dense_2 (Dense)              (None, 64)                8256
_________________________________________________________________
dense_3 (Dense)              (None, 2)                 130
=================================================================
Total params: 377,986
Trainable params: 377,986
Non-trainable params: 0
_________________________________________________________________
```

Fig. 8 Structure of the final hybrid network using two-dimensional convolutional layers

Layer (type)	Output Shape	Param #	Connected to
bin (InputLayer)	(None, 160, 30, 1)	0	
conv2d_1 (Conv2D)	(None, 140, 25, 64)	8128	bin[0][0]
float (InputLayer)	(None, 160, 1)	0	
max_pooling2d_1 (MaxPooling2D)	(None, 70, 12, 64)	0	conv2d_1[0][0]
conv1d_1 (Conv1D)	(None, 156, 64)	384	float[0][0]
conv2d_2 (Conv2D)	(None, 67, 11, 128)	65664	max_pooling2d_1[0][0]
max_pooling1d_1 (MaxPooling1D)	(None, 78, 64)	0	conv1d_1[0][0]
max_pooling2d_2 (MaxPooling2D)	(None, 33, 5, 128)	0	conv2d_2[0][0]
conv1d_2 (Conv1D)	(None, 74, 128)	41088	max_pooling1d_1[0][0]
conv2d_3 (Conv2D)	(None, 30, 2, 128)	262272	max_pooling2d_2[0][0]
max_pooling1d_2 (MaxPooling1D)	(None, 37, 128)	0	conv1d_2[0][0]
max_pooling2d_3 (MaxPooling2D)	(None, 15, 1, 128)	0	conv2d_3[0][0]
dropout_2 (Dropout)	(None, 37, 128)	0	max_pooling1d_2[0][0]
flatten_1 (Flatten)	(None, 1920)	0	max_pooling2d_3[0][0]
lstm_1 (LSTM)	(None, 37, 256)	394240	dropout_2[0][0]
dropout_1 (Dropout)	(None, 1920)	0	flatten_1[0][0]
lstm_2 (LSTM)	(None, 128)	197120	lstm_1[0][0]
concatenate_1 (Concatenate)	(None, 2048)	0	dropout_1[0][0] lstm_2[0][0]
dense_1 (Dense)	(None, 256)	524544	concatenate_1[0][0]
dense_2 (Dense)	(None, 128)	32896	dense_1[0][0]
dense_3 (Dense)	(None, 2)	258	dense_2[0][0]

```
Total params: 1,526,594
Trainable params: 1,526,594
Non-trainable params: 0
```

Fig. 9 Structure of the final double input network

5 Conclusions and Future Work

This paper presents several architectures for ECG data analysis. From Table 1 it can be seen that the best architecture was observed for the double input network, achieving 93.6% accuracy. Other architectures also provide good classification performance. The obtained results were very optimistic and we came to the conclusion that the proposed methodology is applicable to the classification of ECG data and could be further used in wearable devices.

The conclusions and results of this study allow us to investigate real-time data analysis using wearable devices to predict certain cardiovascular conditions. In future research, algorithms can be developed to improve communication between patients with communication contact disorder, depending on the patient's emotional state.

References

1. Agrafioti, F., Hatzinakos, D., Anderson, A.K.: ECG pattern analysis for emotion detection. IEEE Trans. Affect. Comput. **3**(1), 102–115 (2012). https://doi.org/10.1109/T-AFFC.2011.28
2. Atkielski, A.: Schematic diagram of normal sinus rhythm for a human heart as seen on ECG, two periods forming a RR-interval (2009). https://en.wikipedia.org/wiki/File:ECG-RRinterval.svg. Wikipedia Online; Accessed 19 Feb. 2021
3. Brás, S., Ferreira, J., Soares, S., Pinho, A.: Biometric and emotion identification: an ECG compression based method. Front. Psychol. **9**, 467 (2018). https://doi.org/10.3389/fpsyg.2018.00467
4. Bulagang, A., Weng, N., Mountstephens, J., Teo, J.: A review of recent approaches for emotion classification using electrocardiography and electrodermography signals. Inform. Med. Unlocked **20**, 100363 (2020)
5. Etiwy, M., Akhrass, Z., Gillinov, L., Alashi, A., Wang, R., Blackburn, G., Gillinov, S., Phelan, D., Gillinov, A., Houghtaling, P., Javadikasgari, H., Desai, M.: Accuracy of wearable heart rate monitors in cardiac rehabilitation. Cardiovasc. Diagn. Ther. **9** (May 2019). https://doi.org/10.21037/cdt.2019.04.08
6. Gilgen-Ammann, R., Schweizer, T., Wyss, T.: RR interval signal quality of a heart rate monitor and an ECG Holter at rest and during exercise. Eur. J. Appl. Physiol. **119**, 1525–1532 (2019). https://doi.org/10.1007/s00421-019-04142-5
7. Hatzinakos, D., Agrafioti, F., Anderson, A.K.: ECG pattern analysis for emotion detection. IEEE Trans. Affect. Comput. **3**(1), 102–115 (2012)
8. Kligfield, P., Gettes, L.S., Bailey, J.J., Childers, R., Deal, B.J., Hancock, E.W., van Herpen, G., Kors, J.A., Macfarlane, P., Mirvis, D.M., Pahlm, O., Rautaharju, P., Wagner, G.S.: Recommendations for the standardization and interpretation of the electrocardiogram. Part i. Circulation **115**, 1306–1324 (2007)
9. Lyon, A., Mincholé, A., Martinez, J., Laguna, P., Rodriguez, B.: Computational techniques for ECG analysis and interpretation in light of their contribution to medical advances. J. R. Soc. Interface **15**(138) (2018)
10. Selvaraj, J., Murugappan, M., Wan, K., Yaacob, S.: Classification of emotional states from electrocardiogram signals: a non-linear approach based on Hurst. Biomed. Eng. Online **12**(1), 44 (2013). https://doi.org/10.1186/1475-925X-12-44

Deployment of Deep Models in NLP Infrastructure

Tomasz Walkowiak

1 Introduction

Artificial Intelligence (AI) is no longer just the domain of scientific research but is a component of many commercial applications. Thus, a problem of effective implementation of AI methods as components of applications is emerging. Currently, state-of-the-art AI models (in the areas of NLP and image processing) are typically built around the transformer architecture [15]. It includes such architectures as BERT [2], T5 [13, 20], and GPT-2 [12] and their modifications.

The application area of theses models is very wide, including such tasks as text classification [18], sentence embeddings [14], key phrase extraction [9], punctuation restoration [10], and devulgarization [4].

CLARIN-PL[1] is the Polish node of the CLARIN ERIC research infrastructure. The CLARIN-PL Language Technology Center [11] (LTC) is a platform that supports the digital research paradigm in language data analysis. As such, it must implement the latest NLP technologies, such as deep language models. In this paper, we outline how these models have been implemented in the CLARIN-PL infrastructure and the problems encountered during their development.

The paper is organized as follows. Section 2 describes the CLARIN-PL LTC architecture. It is followed by a discussion of how to implement deep models. Section 4 presents the resulting solution for implementing deep model services. Section 5 demonstrates the use of these services within the CLARIN-PL infrastructure and presents an example application from the area of semantic similarity analysis.

[1]https://clarin-pl.eu/.
[2]https://clarin.eu.

T. Walkowiak (✉)
Faculty of Information and Communication Technology, Wroclaw University of Science and Technology, Wroclaw, Poland
e-mail: tomasz.walkowiak@pwr.edu.pl

W. Zamojski et al. (eds.), *Dependable Computer Systems and Networks*, Lecture Notes in Networks and Systems 737, https://doi.org/10.1007/978-3-031-37720-4_29

2 CLARIN-PL Language Technology Center

One of the goals of CLARIN-PL LTC [11] is to provide a technological infrastructure for social sciences and humanities. CLARIN-PL provides a complete NLP processing pipeline for Polish as well as for various types of NLP applications (e.g., those available at https://ws.clarin-pl.eu/). The development of CLARIN-PL LTC required the design of an effective way of processing texts using various language tools. These tools are developed in different programming environments (such as C++, Java, Python, and R). Moreover, many of them require large models and the time needed to load a model is much longer than the time needed to process a single text file. The solution is to run a language tool as a service with the necessary models or data preloaded in memory [16]. That is why CLARIN-PL Language Center [11] is based on the micro-service architecture [8], where each service runs its own process and communicates with each other via lightweight mechanisms for sending messages (start, end and progress of tasks) and a shared file system for input and output data for language tools. For message communication, we used the AMQP[2] protocol and the open source RabbitMQ[3] broker for message queuing. The queuing system acts as a load balancer, allowing for easy horizontal scaling. We did not use AMQP for input/output data exchange (we used a shared file system) because these data (files) could be of large size, i.e. several GB, and would block the messaging system. The programmer only needs to write a single function that transforms input data and stores it in a network storage visible to all other services that need to access it. The components that form processing chains (langauge tools implemented as microservices) are orchestrated by a Language Processing Management Notation [17]. This allows for much greater efficiency through the ability to parallelize operations. For example, for the processing chain shown in Fig. 1, each file in the input corpus (zip file) could be processed in parallel (depending on available system resources) by a ppipeline of language tools `any2txt` | `embedder`, and then the results are merged (directive `dir`) and processed in a single pipe by `mds`.

As a result, LTC runs a configurable chain of language tools with robust and parallel processing on demand. We now have more than 100 different tools implemented in the system.

```
filezip(path)|any2txt|
embedder({"type":"sbert-paraphrase-multilingual-mpnet-base-v2"}|
dir|mds("method":"mds")
```

Fig. 1 A LPMN [17] statement defining a semantic similarity analysis process (see Chap. 5)

[2] https://www.amqp.org/.

[3] https://www.rabbitmq.com/.

3 Deep Models Deployment

As mentioned in the introduction, modern language tools are based on extensive use of rransformers-based language models. Unfortunately, the number of available graphics cards in CLARIN-PL LTC, which are the primary means of running deep models, is limited. Therefore, it was necessary to extend the infrastructure with the effective use of GPU cards. To do this, it was necessary to analyze the methods of implementing transformer based neural models.

One of the more popular repositories of publicly available transformer-based language models is Huggingface. It supports a Transformers library that allows models to be read, inferred, or taught, and includes support for all the popular Transformers models in NLP, such as BERT [2], T5 [13], RoBERTa [6], BART [5], and GPT-2 [12]. Implementations are available in three of the most popular deep learning frameworks: PyTorch, TensorFlow and JAX. In addition, models can be transformed into ONNX, an open-source, machine-independent format that is widely used for sharing neural network models [1]. This raises the question of which runtime implementation to use. Deep models can run on both GPU and CPU. It is obvious that GPU gives much better performance than CPU, but the question is how much better. The question is really important because GPU is very often a limited resource. Another important aspect in the case of deep models is the size of the batch to be used for inference. To answer these questions, a number of tests were performed.

We compared the inference time and memory usage for PyTorch and the ONNX implementation running on CPU and GPU [19]. In the tests, we used AMD Ryzen 9 3950X and NVIDIA GeForce RTX 3090 graphics cards. We used the multilingual Sentence BERT model[4] and tested the longest sequences (512 tokens). All tests were repeated 100 times and the results were averaged. The results are shown in Tables 1 and 2.

The results show that the ONNX implementation outperforms PyTorch in terms of inference time for both CPU and GPU cases. However, the difference is much more significant in the CPU case. Furthermore, the results show that using batches speeds up the average inference time, but the efficiency of using batches is low. Therefore, the batch size must be chosen carefully. Moreover, using batches for CPU is only justified if we have more CPU cores than the batch size, otherwise there is no average speedup.

Table 1 Sentence BERT inference time per task in ms for different deployments on the GPU. The reported time is a time to process one text, i.e., in case of batches, it is a time to process one batch divided by the size of a batch

Type	PyTorch			ONNX		
Batch size	1	4	8	1	4	8
Processing time [ms]	9.6	7.4	6.5	8.1	7.2	6.4

[4] https://huggingface.co/sentence-transformers/paraphrase-multilingual-mpnet-base-v2.

Table 2 Sentence BERT inference time per task in ms for different deployments on the CPU. The reported time is a time to process one text, i.e., in case of batches, it is a time to process one batch divided by the size of a batch

Type	PyTorch			ONNX		
Batch size	1	4	8	1	4	8
1 core	3,103	3,116	3071	968	1,006	1,103
2 cores	1,597	1,611	1640	527	521	532
4 cores	848	858	858	298	293	290
8 cores	536	502	498	202	187	184
16 cores	415	360	341	177	158	152

4 Deep Model Service

According to the results of the experiments described in the previous section, a library has been implemented that allows deep models to be used in CLARIN-PL LTC using the ONNX format. Currently, CLARIN-PL uses such models as Senetence-BERT [14] to generate embeddings of short text passages and T5 [13] to determine keywords [9] and devulgarize texts [4]. In all these cases, the input data is text of limited length. This is related to the fact that BERT or T5 models can process text of limited size (typically 512 tokens). While the output data is embedding vector (with 512 or 768 dimensionality), a list of keywords (up to 5) or text (of similar size as the input). According to the CLARIN-PL LTC architecture, we have based the communication on AMQP, but in contrast to the language tool services described in Sect. 2, the input and output data are also sent via the AMQP protocol, since they are always of limited size.

The results reported in the previous section suggest the use of batches. To enable this, we used a feature of AMQP that allows to read more than one message (if available) from the task queue and process these tasks (texts) at once (as a batch).

The HuggingFace Optimum[5] library has been used to convert PyTorch models to ONNX format. For Sentence-BERT models, the library does not convert the pooling layer or the additional dense layer often used in this class of networks. Therefore, these additional layers are additionally stored in PyTorch format. T5 models often contain additional parameters and prompts, so an additional configuration file is used that contains information not represented in ONNX format.

In order to save memory on the graphics card, the solution was enriched with a mechanism of automatic memory release. That is, if the model is not used for a certain time, it is removed from memory and reloaded at the next use. In the current state, we have deployed four models (see Table 3 for a list of them). The first two are Sentence-BERT one, the next two are T5. Table 3 shows the observed memory usage and load time. The experiments were performed on NVIDIA Tesla A100 GPU. We

[5] https://huggingface.co/docs/optimum.

Table 3 Time of loading and sage of memory by service in unloaded and loaded state for GPU

Deep model	Loading	Memory usage [GB]	
	Time [s]	Unloaded	Loaded
sbert-distiluse-base-multilingual-cased-v1 [14]	0.5	0.22	0.86
sbert-paraphrase-multilingual-mpnet-base-v2 [14]	1.7	0.4	1.0
t5-voicelab-vlt5-base-keywords [9]	1.4	1.2	4.3
t5-DEPOTxT5-base [4]	1.2	1.0	4.4

can notice that the mechanism of freeing the memory works and the usage of GPU memory is 4 times smaller if models are unloaded.

CLARIN-PL LTC is deployed in Kubernetess,[6] which allows one pod or container per GPU, i.e., containers and pods do not share GPUs. Since we have many more models than available GPUs, we need to implement multiple deep models in a pod. To do this, we add functionality to the library that allows multiple ONNX-based models to run as separate processes in a container.

5 Semantic Similarity of Texts

Books or articles are much larger than the limit of 512 tokens for the Transformer models. Therefore, if we want to process such data using language deep models, it is necessary to divide it into fragments and send each fragment to the deep model and then combine the results. For this purpose we have developed a CLARIN-PL service called *embedder*,[7] i.e., embedder divides input text into passages of fixed length (in words), calls the deep model (e.g.,Sentence-BERT) for each chunk, and merges the results from the deep model into a JSONL file. To communicate with the deep model services (described in the previous section), a special class (called Sentence) has been developed. It allows sending text and non-blocking waiting for results. Its use in the embedder service is shown in Fig. 2.

The embedder service could be used for document similarity visualization, where the goal is to represent documents as points on a 2D plane. Documents that are more similar should be closer together in the plot than objects that are different [3]. The multidimensional passage embeddings obtained by embedder could be projected onto the 2D plane by the *mds*[8] service using the UMAP [7] method.

The results of the similarity visualization obtained by the LPMN statement shown in Fig. 1 are given in Fig. 3. The input corpus consisted of six books by Polish authors.

[6] https://kubernetes.io/.

[7] https://wiki.clarin-pl.eu/nlpservices/list/embedder.

[8] https://wiki.clarin-pl.eu/nlpservices/list/mds.

```
sentence = None
embeddings = None
for chunk in chunker.chunks():
    sentences.append(chunk)
    if len(sentences) >= limit:
        if sentence is not None:
            embeddings = sentence.get_results()
        sentence = Sentence(sentences, type)
        sentence.run(blocking=False)
        if embeddings is not None:
            save(out, previous_sentences, embeddings)
        previous_sentences = sentences
        sentences = []
```

Fig. 2 Code in Python illustrating the communication with CLARIN-PL deep model services

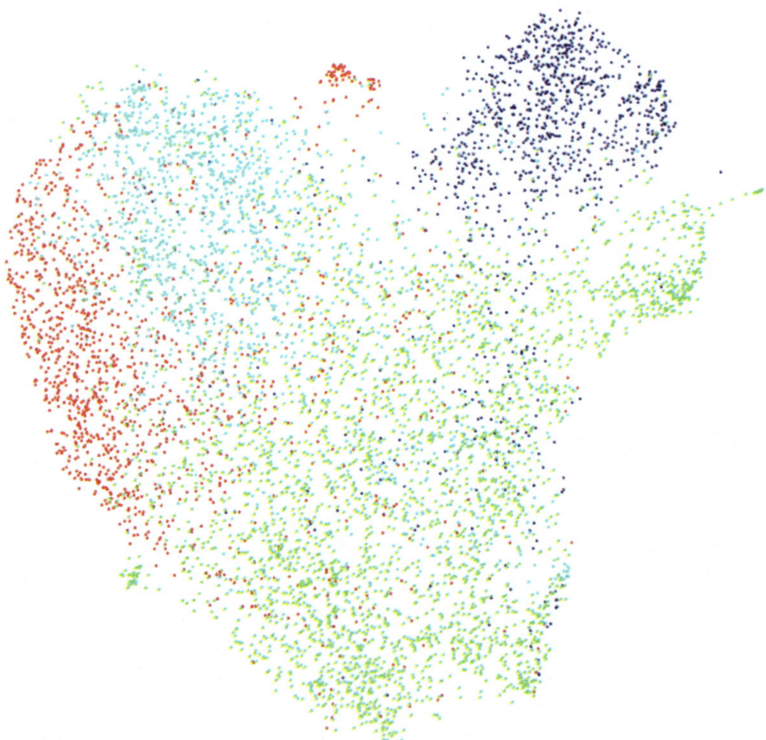

Fig. 3 Similarity visualization of passages from books by Eliza Orzeszkowa (green dots) and Bolesław Prus ("Emancypantki" in red, "Faraon" in dark blue, and "Lalka" in light blue). 2D placements were obtained using the UMAP [7] method applied to embeddings from paraphrase-multilingual-mpnet-base-v2 Sentence BERT model

There were three books by Eliza Orzeszkowa (each passage is marked by a green dot) and three by Bolesław Prus (marked in three different colors: red, dark and light blue). It can be noticed that passages from Orzeszkowa are placed in different areas than those from Prus. And that passages from each book of Prus are concentrated in different areas of the final plot. It shows that each book by Prus has different thematic areas (they do not overlap in the diagram) and, moreover, different from the books of Orzeszkowa.

6 Conclusions

We have shown how deep language models are implemented in the CLARIN-PL Language Technology Center. The solution is based on the ONNX format and AMQP for inter service communication. It allows an effective and scalable use of deep models within CLARIN-PL services. Furthermore, we illustrated the obtained solutions with the results of the similarity visualization of literary texts. Similar analyses could be easily performed using the developed NLP infrastructure and is available for users by a web interface.[9]

Acknowledgements Financed by the European Regional Development Fund as a part of the 2014–2020 Smart Growth Operational Programme, CLARIN—Common Language Resources and Technology Infrastructure, project no. POIR.04.02.00-00C002/19.

References

1. Bai, J., Lu, F., Zhang, K., et al.: Onnx: open neural network exchange (2019). https://github.com/onnx/onnx
2. Devlin, J., Chang, M.W., Lee, K., Toutanova, K.: Bert: Pre-training of deep bidirectional transformers for language understanding (2018). arXiv:1810.04805
3. Gniewkowski, M., Walkowiak, T.: Assessment of document similarity visualisation methods. In: Vetulani, Z., Paroubek, P., Kubis, M. (eds.) Human Language Technology. Challenges for Computer Science and Linguistics, pp. 348–363. Springer International Publishing, Cham (2022)
4. Klamra, C., Wojdyga, G., Żurowski, S., Rosalska, P., Kozłowska, M., Ogrodniczuk, M.: Devulgarization of polish texts using pre-trained language models. In: Groen, D., de Mulatier, C., Paszynski, M., Krzhizhanovskaya, V.V., Dongarra, J.J., Sloot, P.M.A. (eds.) Computational Science-ICCS 2022, pp. 49–55. Springer International Publishing, Cham (2022)
5. Lewis, M., Liu, Y., Goyal, N., Ghazvininejad, M., Mohamed, A., Levy, O., Stoyanov, V., Zettlemoyer, L.: BART: denoising sequence-to-sequence pre-training for natural language generation, translation, and comprehension. In: Proceedings of the 58th Annual Meeting of the Association for Computational Linguistics, pp. 7871–7880. Association for Computational Linguistics (Jul. 2020). https://aclanthology.org/2020.acl-main.703

[9] https://services.clarin-pl.eu.

6. Liu, Y., Ott, M., Goyal, N., Du, J., Joshi, M., Chen, D., Levy, O., Lewis, M., Zettlemoyer, L., Stoyanov, V.: Roberta: a robustly optimized BERT pretraining approach (2019). http://arxiv.org/abs/1907.11692
7. McInnes, L., Healy, J., Melville, J.: Umap: uniform manifold approximation and projection for dimension reduction (2020)
8. Newman, S.: Monolith to Microservices: Evolutionary Patterns to Transform Your Monolith. O'Reilly Media (2019)
9. Pęzik, P., Mikołajczyk, A., Wawrzyński, A., Nitoń, B., Ogrodniczuk, M.: Keyword extraction from short texts with a text-to-text transfer transformer. In: Szczerbicki, E., Wojtkiewicz, K., Nguyen, S.V., Pietranik, M., Krótkiewicz, M. (eds.) Recent Challenges in Intelligent Information and Database Systems, pp. 530–542. Springer Nature Singapore, Singapore (2022)
10. Pogoda, M., Walkowiak, T.: Comprehensive punctuation restoration for English and Polish. In: Findings of the Association for Computational Linguistics: EMNLP 2021, pp. 4610–4619. Association for Computational Linguistics, Punta Cana, Dominican Republic (Nov. 2021). https://aclanthology.org/2021.findings-emnlp.393
11. Pol, M., Walkowiak, T., Piasecki, M.: Towards CLARIN-PL LTC digital research platform for: depositing, processing, analyzing and visualizing language data. In: Kabashkin, I., Yatskiv, I., Prentkovskis, O. (eds.) Reliability and Statistics in Transportation and Communication, pp. 485–494. Springer International Publishing, Cham (2018)
12. Radford, A., Wu, J., Child, R., Luan, D., Amodei, D., Sutskever, I.: Language Models are Unsupervised Multitask Learners (2019). https://www.semanticscholar.org/paper/Language-Models-are-Unsupervised-Multitask-Learners-Radford-Wu/9405cc0d6169988371b2755e573cc28650d14dfe
13. Raffel, C., Shazeer, N., Roberts, A., Lee, K., Narang, S., Matena, M., Zhou, Y., Li, W., Liu, P.J.: Exploring the limits of transfer learning with a unified text-to-text transformer. J. Mach. Learn. Res. **21**, 67 (2020). https://jmlr.csail.mit.edu/papers/v21/20-074.html, id/No 140
14. Reimers, N., Gurevych, I.: Sentence-Bert: sentence embeddings using Siamese Bert-networks. In: Proceedings of the 2019 Conference on Empirical Methods in Natural Language Processing. Association for Computational Linguistics (Nov. 2019). http://arxiv.org/abs/1908.10084
15. Vaswani, A., Shazeer, N., Parmar, N., Uszkoreit, J., Jones, L., Gomez, A.N., Kaiser, L.U., Polosukhin, I.: Attention is all you need. In: Guyon, I., Luxburg, U.V., Bengio, S., Wallach, H., Fergus, R., Vishwanathan, S., Garnett, R. (eds.) Advances in Neural Information Processing Systems, vol. 30. Curran Associates, Inc. (2017)
16. Walkowiak, T.: Web based engine for processing and clustering of polish texts. In: Zamojski, W., Mazurkiewicz, J., Sugier, J., Walkowiak, T., Kacprzyk, J. (eds.) Theory and Engineering of Complex Systems and Dependability, pp. 515–522. Springer International Publishing, Cham (2015)
17. Walkowiak, T.: Language processing modelling notation–orchestration of NLP microservices. In: Zamojski, W., Mazurkiewicz, J., Sugier, J., Walkowiak, T., Kacprzyk, J. (eds.) Advances in Dependability Engineering of Complex Systems-Proceedings of the 12th International Conference on Dependability and Complex Systems DepCoS-RELCOMEX, July 2–6, 2017, Brunów, Poland. Advances in Intelligent Systems and Computing, vol. 582, pp. 464–473. Springer (2017). https://doi.org/10.1007/978-3-319-59415-6_44
18. Walkowiak, T.: Subject classification of texts in polish-from TF-IDF to transformers. In: Zamojski, W., Mazurkiewicz, J., Sugier, J., Walkowiak, T., Kacprzyk, J. (eds.) Theory and Engineering of Dependable Computer Systems and Networks, pp. 457–465. Springer International Publishing, Cham (2021)
19. Walkowiak, T., Mazurkiewicz, J., Sugier, J., Śliwiński, P.: Performance analysis of intelligent agents in complex event processing systems. Saf. Reliab. Syst. Process. Summer Saf. Reliab. Semin. **2022**, 195–204 (2022)
20. Xue, L., Constant, N., Roberts, A., Kale, M., Al-Rfou, R., Siddhant, A., Barua, A., Raffel, C.: mT5: a massively multilingual pre-trained text-to-text transformer. In: Proceedings of the 2021 Conference of the North American Chapter of the Association for Computational Linguistics: Human Language Technologies, pp. 483–498. Association for Computational Linguistics (Jun. 2021). https://aclanthology.org/2021.naacl-main.41

Analysis of Handwritten Texts to Detect Selected Psychological Characteristics of a Person

Marek Woda and Grzegorz Oliwa

1 Introduction

This research touches on digital image processing, deep machine learning and *graphology*, which is a pseudoscience that enables the analysis of handwritten texts to detect certain psychological characteristics of the individuals who authored them.

Graphology is a pseudo-science that involves the analysis of writing from a psychological point of view, and therefore aims to identify links between writing style and personality traits. There are several factors that influence the specific conclusions drawn by graphologists when analysing a person's handwriting, and these are generally not consistent with one another, as many graphology experts look for certain traits in different places. Among these are: slanting of the letters, the intensity of pressure, the way in which bonds are formed between individual letters, the distance between letters, words and even lines, the style of writing.

Due to the limited time available, this paper focuses only on the last of the above, i.e. how a person writes individual letters. During template graphological analysis, the entire text is generally considered (except in special cases which are extremely rare), so it would be inappropriate to subject specific letters to verification one by one. Therefore, the intermediate steps described in the previous subsection are required, as without this the work would be flawed at its very foundation.

Although the subject of graphology is dealt with daily by many experts qualified in this field, it has not been possible in recent years to reliably and in-depth study of its effectiveness. Developing a universal model is an extremely difficult task, but nevertheless much of the graphology community agrees on certain, often extreme issues. These include, for example, that it is perfectly possible to identify individuals affected by neurodegenerative diseases. As with electrocardiograms, it is necessary

M. Woda (✉) · G. Oliwa
Department of Computer Engineering, Wroclaw University of Technology, Janiszewskiego 11-17, 50-372 Wrocław, Poland
e-mail: marek.woda@pwr.edu.pl

W. Zamojski et al. (eds.), *Dependable Computer Systems and Networks*, Lecture Notes in Networks and Systems 737, https://doi.org/10.1007/978-3-031-37720-4_30

to define the norm beforehand and then look for any deviations from it to identify selected problems.

In the research itself, two parts can be distinguished. The first contains information on the technology and algorithms used to realize the final software, which enables a full graphological analysis based on the selected models. It also touches on the efficiency of the extraction of the constituent results at the various stages of the work, which make up the whole. The second part of the work, on the other hand, deals with issues such as the method of surveying the subjects, the methodology of conducting the research, as well as the interpretation of the results and drawing conclusions. It also describes the strengths and weaknesses of the approach used and suggests directions for further development of the work described. It is based on research conducted by qualified and recognized graphologists and psychologists, so that although its foundations are based on ambiguous and sometimes even controversial issues, the reliability of the features extracted from handwritten texts is, by definition, acceptable to the wider graphological community.

2 Related Work

This subsection is devoted to a discussion of the three most important—from the point of view of the discussed work—literature items, based on which an algorithm for inferring the characteristics of individual people, whose handwriting was subjected to graphological analysis, was created. The first, and much more substantive than the others, was [1]. It covers many different aspects of psychology and graphology and includes specific examples, which makes the individual issues easy to understand, even for a beginner. The work itself is divided into thematic segments devoted to the characteristics of writing, on what graphology is, and many others which, due to the scope of the research, were not discussed in depth. They have, however, had a significant impact. In his book, the author tries to identify factors that influence people's levels of intelligence, honesty, secretiveness, introversion, or extroversion, among others. However, he does not juxtapose the individual personality traits on a contrasting basis, which makes the precise determination of the consequences of the occurrences of the different writing styles not a simple task.

The second important item [2] claims that graphology has not changed in a in a significant way for last decades. Moreover, extracts from this book are relatively often used as a canvass for certain theories (directly or indirectly). For this reason, it was decided to use it to create a model to detect the specific personality traits of the individuals studied. However, it was written in a similar way to [1]. At times, it resembles a collection of randomly scattered theories and assumptions about handwriting, rather than a structured guide on how to approach graphological analysis correctly and reliably.

Both above books are based on many years of research through experts in psychology and graphology, which was a key factor in their selection. To this day, however, the field that is the subject of the thesis described above has not received satisfactory research that would enable it to stop being called a *pseudoscience*, as no one has been able to unequivocally state and let alone evaluate its effectiveness. For this reason, it was not possible to use existing research and it was decided to create two models that were different from each other. For the purposes of this study, they were named after the initials of the authors of the books on which they were based–*DA* [1] and *KAMR* [2]. For this purpose, however, it was necessary to find a common part for both items mentioned earlier. Based on this, a questionnaire was created, discussed in more detail in Chap. 4 on the methodology of data collection. To simplify the level of complexity of the engineering part as much as possible, it was decided to choose a limited set of characters as the said common part, rather than the individual writing features and the conclusions drawn from them. In both the first and second abovementioned works on which the models were based, differing implications of certain facts were mostly indicated. These themselves, however, were relatively straightforward to establish by means of a survey of the respondents.

The third piece of literature, which served as a complement to the conclusions drawn from the previous two, is [5]. The work itself saw the light of day as early as 1991, but to this day–like the aforementioned papers–it is one of the among publications by psychologists and graphologists around the world. The authors do not focus on a single language. Instead, they try to identify more universal phenomena that may be common to a certain, selected family of them–but they nevertheless favor English and Chinese. Counter-intuitively, the information contained in this item was very helpful, as the people interviewed included people from central Europe and Australia (born and raised there since birth). There were 42 and 8 people respectively, making a total of 50 writing samples and completed questionnaires. This fact was not used at the graphological analysis software development stage, although it served as some indication in drawing conclusions.

An unquestionably essential book for anyone wishing to delve into the subject of graphology [8]. It treats how to correctly interpret the most popular phenomena occurring in writing, seemingly based on the positions mentioned above. Compared to them, however, it is much less specific. Equally interesting is the research presented in [3]. It is said to be groundbreaking in the application of neural networks to tasks such as graphology.

3 System Design

The initial stage was to obtain the results of a questionnaire from the subjects, available at (tinyurl.com/5b7mk38c), together with samples of their writing. On its basis, a personality analysis is carried out, the course of which, as well as the information obtained as a result, is described in Chap. 4. The handwriting samples, on the other hand, undergo digital processing (filtering, thresholding, segmentation) and then the graphological analysis proper. The information obtained from both the images and the questionnaires is compared with each other, so that results can be obtained on the effectiveness of the final software. The creation of a character recognition model is crucial from the point of view of the quality of the graphological analyzer, as it determines whether individual characters are examined from the correct angle. Information about it is provided in Sect. 4.1, on the selection of optimal parameters for *Deep Neural Network* (DNN).

These processes are presented in Fig. 1. All the intermediates included, i.e. individual character segments (1), labelled characters (2) and the results of the graphological analysis (3) propagate their own errors. These, in turn, if not eliminated early enough, are amplified. For this reason, it is important to properly understand the relevance of each (Fig. 2).

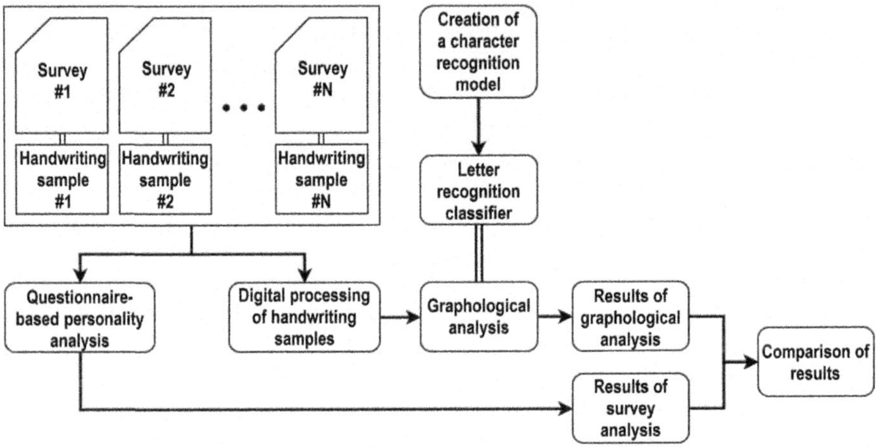

Fig. 1 Research breakdown

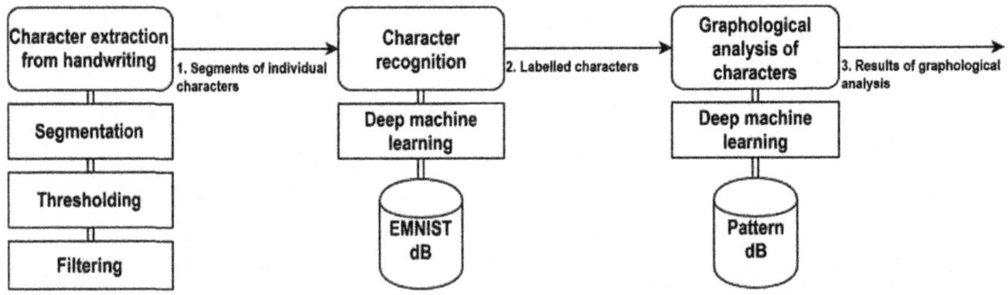

Fig. 2 Graphological analysis processes

4 Research and Its Methodology

It was decided to approach the whole process in a manner similar to *supervised machine learning*. First, information had to be collected on the psychological characteristics of specific individuals. This was done in two ways:

- by drawing up a questionnaire with a fuzzy response scale (*definitely yes, almost certainly yes, rather yes, rather not, almost certainly not, definitely not*), and each was assigned the value specified in Table 3.
- by taking a writing sample with a predetermined content common to all respondents.

The questionnaire consists of 50 questions prepared in such a way that they are paired against each other. This is one of the most popular methods indicated in [9, 10, 12], through which it is possible to determine the error imposed on the answers of the interviewee. This was a necessary condition for the accurate determination of personality traits, as some people tend to under- or overestimate their qualities. The questions were developed based on the traits sought and its aim is to establish these as best as possible. These are listed and categorised below. The use of a fuzzy response scale in it, was introduced because of the logarithmic function used in the deep machine learning process.

Each writing sample is initially digitally processed. This makes it possible to undergo a proper graphological analysis of each letter included in the text. Stripped of Polish diacritical marks, the text originally read *Ochoczy kot skacze przez wysoki mur bo go lubi* (in eng. *an eager cat jumps over a high wall because it likes it*). This somewhat humorous, albeit short and entirely sufficient phrase remained the subject of research until the very end. At the sample collection stage, it was possible to collect exactly 50 pairs of questionnaires and handwriting fragments.

As mentioned above, such an exceptionally short text subjected to graphological analysis was sufficient, as it contained all the letters examined in the work in question. Furthermore, some of them were repeated in juxtaposition with different characters, so it was possible to extract them from the text for multiple joins (Table 1).

Each of the above-mentioned traits could be established by an appropriate interpretation of the survey responses. However, their juxtaposition is not balanced, meaning that some characteristics can be determined based on a very low number of responses, while others unfortunately cannot. This translated into the complexity of the algorithm and the voluminousness of the code (Table 2).

As can be seen, some characteristics are difficult to describe unambiguously. The total coefficient of occurrence of a trait, is equal to the ratio of the sum of points for each answer given (according to Table 3) to the number of related questions. For example, based on the letter a in the DA model, we can identify traits such as a constant sense of pressure and low self-confidence. Assuming that the respondent would answer questions 8, and 19 with a *definitely yes* and 36 and 43 with a *definitely not*, the total probability of these qualities would be set at 50%. However, this is a special case because, as reported in the literature, these are highly correlated characteristics.

Table 1 List of characters assessed

Char	Features sought in DA model	Features sought in KAMR model
a	A constant sense of pressure, low self-confidence	A tendency to change one's mind frequently, to be inspired by others, to improvise constantly
b	Ability to make new friends, responsibility, honesty and integrity	Scepticism, need for manipulation, desire to be in the centre of attention
g	Nervousness, tendency to worry, avoidance of responsibility	Acting only with the approval of others, generosity, openness, petulance
h	Tendency to lie to others, high self-control, preference for living alone	Taking quick and decisive action, low intelligence
k	Agreeableness, curiosity, self-control, egocentrism	Constant sense of pressure, blind stubbornness, gruffness
o	Deceitfulness, treachery, tendency to change one's mind frequently	High extrovertedness, willingness to talk to others, arrogance, taking criticism badly
t	Tendency to aggression, high level of self-confidence, eccentricity	Willingness to be aggressive, high levels of self-confidence, eccentricity
u	Constant rush, introverted lifestyle, frequent reliance on others	Lack of ability to make new friends, low self-confidence, nervousness
y	Openness to people and change, high level of intelligence, ruthlessness	Tendency to worry, critical evaluation of own actions

Table 2 Relationship between survey questions and personality traits

Char	Numbers of questions relevant to the analysis	
	DA	KAMR
a	8, 19, 36, 43	3, 4, 20, 30, 34
b	2, 4, 7, 11, 17, 27, 36, 39, 43, 46	7, 12, 23, 26, 31, 38, 45, 48
g	10, 18, 42, 49	11, 15, 22, 24, 26, 40, 43, 45, 46
h	12, 16, 26, 27, 31, 40, 41, 44	1, 9, 18, 20, 21, 28, 33, 48
k	4, 13, 22, 27, 31, 36, 41, 44, 49	3, 8, 14, 28, 42, 44
o	2, 5, 10, 12, 15, 16, 24, 25, 26, 31, 38, 40, 45	7, 25, 29, 35, 38, 39, 41, 44, 47, 50
t	3, 6, 14, 18, 33, 44, 48	3, 6, 14, 18, 33, 44, 48
u	10, 15, 19, 21, 27, 37	8, 10, 15, 18, 21, 22, 27, 32, 40
y	23, 34, 47, 50	8, 42, 43, 49

Table 3 Points allocated for survey responses

Selected answer	Score
Definitely yes	1.00
Almost certainly yes	0.85
Rather yes	0.65
Rather not	0.35
Almost certainly not	0.15
Definitely not	0.00

In most cases, a given characteristic is indicated by a subset of the answers given, which is implemented as part of the algorithm. A scoring scheme drawn up in this way will always result in a value in the interval [0, 1], so that it is possible to translate the obtained value into a deep neural network score using a sigmoidal function in the process of graphological analysis.

4.1 Selection of Optimal Parameters in Deep Machine Learning

The most important parameters used in the deep machine learning process, included: number of hidden layers and the number of nodes in each layer, activation function used, and number of epochs allocated for fine-tuning.

Initially, there was only one network, aiming to classify characters from the *EMNIST* database as efficiently as possible. Through experiments and a grid search for the optimum solution, it was found that it works best for 5 layers, where each successive layer is defined by a monotonically non-increasing number of nodes—208, 156, 104, 52 and 26, respectively. The first (input) layer naturally has 784 nodes, as the images from the *EMNIST* database have a resolution of 28 × 28 pixels. Such information is treated as a vector with a length equal to the product of the width and height of each image file. The last layer has 26 nodes, as there are 26 membership classes (one for each letter of the English alphabet). Then, using a sigmoidal function and a top-down number of iterations (epochs) of 5000, an efficiency of 92% was achieved.

By replacing the multilayer perceptron model with a convolutional variant with an interleave of 5, it was possible to increase efficiency by as much as 6% to a level of 98%. It was decided to use such a trained network for the remainder of this thesis to identify handwritten letters. An attempt to replace the logistic activation function with a unipolar threshold variant, ended in great failure. Such a restrictive function was wrong much more often, misclassifying letters. The success rate of the DNN acting on it was successively 90% for the multilayer perceptron and 94% for the convolutional network.

The next stage of the work was to detect specific features contained in the letter. Initially, this was done in a primitive way, using the OpenCV library's built-in algorithms for detecting edges, circles and the like, but over time it became apparent that a separate neural network could perform the same task faster and just as effectively (or even better). By means of cross-validation, it was decided to collect all segments containing handwritten letters that met the condition of occurrence of specific patterns, and then 80% of these were allocated to learning. The remaining 20% acted as a training set. The same exercise was repeated for 10 different groups, making the results much more reliable. In case there were not enough patterns, they were manually added. As it turned out, the efficiency of such a network was surprisingly high for the DA model, and considerably worse for the KAMR model from the

Table 4 Summary of the best neural network parameters

Rank	Activation function	Hidden layers	Max. effectiveness (%)
1	Sigmoid	[208–156–104–52–26]	59
2	Sigmoid	[200–150–100–50 - 26]	57
3	Sigmoid	[150–100–75–50–26]	52
4	Sigmoid	[250–200–10–200–150–100–50–26]	51
5	Threshold unipolar	[208–156–104–52–26]	47

following subsection. As before—the attempt to use a thresholded unipolar activation function, only ended up worsening the results. Moreover, in the case of this network, in addition to its high efficiency, the logistic variant had another function—it allowed a stochastic interpretation of the results. The presence of a given trait could equally well be read off from the output of the network as the probability of its occurrence, which translated into an increased number of inferences and a much greater capacity for analysis. A summary of the most effective parameters of the final neural network, is presented in Table 4.

A fixed number of epochs of 5,000 is not included in the table above because, as research and literature [4, 7–9] have shown—as this value increases, the network performs increasingly efficiently up to a certain threshold, above which further development is negligible. All values in the last column refer to the DA model. Regardless of the selected parameters, the effectiveness of the KAMR model was no more than 24%. Further research to determine the reason for this was not possible, due to the extremely computationally expensive duration of the deep neural network algorithm (model formation). However, it is possible that the error related to the part concerning the way the respondents were surveyed. Another possibility is that the algorithm was highly effective, but the hypotheses put forward by the author of the paper based on which the KAMR model was created were incorrect.

As mentioned above, the DA model has an astonishingly high efficiency. For the best set of parameters determined, an efficiency of 47% was achieved. However, the extent to which the indicated personality traits were determined correct requires careful discussion and analysis. A summary of these is presented in the following subsections, which are devoted to each individual model examined.

4.2 DA Model

The model was most effective in diagnosing personality traits classified by respondents as *rather yes*, at 69%. This result means that the majority of questions defining a given trait were answered in this way (i.e. corresponding to reality) (Table 5).

Table 5 Coverage of DNN algorithm results with survey responses (DA/KAMR)

Survey response	*Recognition effectiveness*	
	DA	*KAMR*
definitely yes	12%	5%
almost certainly yes	40%	21%
rather yes	69%	33%
rather not	51%	15%
almost certainly not	42%	47%
definitely not	66%	25%

By far the least successful in interpreting the characteristics were those classified into the group of responses identified as *definitely yes.* Full agreement occurred in only 12% of cases. However, the situation is different in the opposite case −when trying to exclude a given characteristic. For the DA model, the effectiveness in doing so was as high as 66%. Based on the other results, it can also be seen that the efficiency is very promising, as it exceeds 40%. An analysis of the results in terms of the percentage recognition of the features contained in the letters is presented in Table 6.

Evidently, the algorithm was most effective at identifying features resulting from the interpretation of the o, t and u characters. This fact is not particularly surprising, as the peculiarity of their notation allows for a relatively simple detection of anomalies associated with them.

The results discussed earlier were clarified based on the use of range values, marked as default thresholds (Table 7). They mean nothing more than the occurrence of a number in the interval [0, 1] at the output of the neural network, which was translated into the corresponding threshold value. For the purposes of the study, however, it was decided to also test the effectiveness of the DA model for slightly underestimated and overestimated ranges.

These values were chosen experimentally, but as it turned out, the results were very interesting for both the DA model in question and the KAMR described in the subsection below. Another approach to this study could be to use the extended and narrowed interval techniques. Both options would significantly affect an algorithm that was not developed with this in mind. The first would have introduced an ambiguity, involving a feature belonging to more classes. The second, on the other hand, would have forced the introduction of a completely new class, meaning that

Table 6 Recognition of character features for DA and KAMR models

Character	*Correct interpretation*	
	DA	*KAMR*
a	44%	29%
b	40%	14%
g	37%	11%
h	45%	12%
k	43%	22%
o	58%	26%
t	66%	66%
u	52%	19%
y	41%	20%

Table 7 Values of the thresholds tested

Description of characteristic	Default threshold (%)	Underrated threshold (%)	Overrated threshold (%)
Definitely yes	85–100	80–100	90–100
Almost certainly yes	70–84	65–79	75–89
Rather yes	50–69	45–64	55–74
Rather not	30–49	25–44	35–54
Almost certainly not	15–29	10–24	20–34
Definitely not	0–14	0–9	0–19

the feature would not belong to any of the options listed in Table 7 (there would have been a possibility of not matching them) (Fig. 3).

As can be seen, and as seems natural, the lowering of the threshold results in a significant reduction in the effectiveness of the algorithm for answers classified into the *definitely not* group. It thus has a smaller range, but thanks to this, it can be seen that, in this particular case, as much as 17% of the effectiveness was between 10 and 14%. A similar conclusion can be drawn for the membership of the *definitely yes* group. In the remaining cases, it is not possible to clearly determine what percentage of effectiveness was gained and what was lost, as only the sum of these can be seen (Fig. 4).

When examining each of the characters separately, it can be seen that changing the threshold had a stronger effect on some of them than on others. Certainly, broadening the thresholds would allow for a much higher efficiency in interpreting the features of the letter *k*. The situation is less favourable for the letters *g* or *t*. This may indicate a defect in the model indicating the selected features, or an imperfection in the algorithm itself.

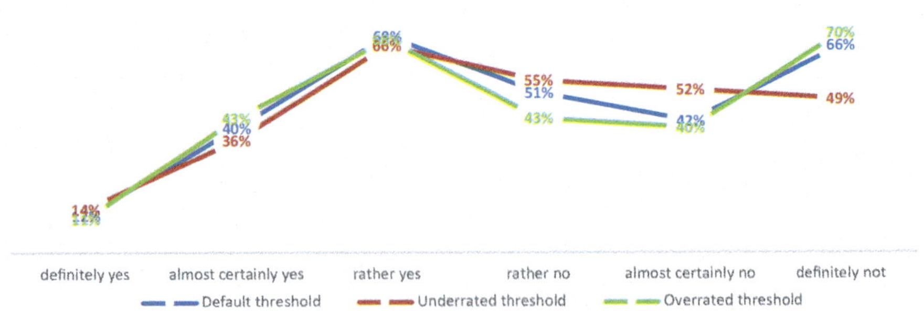

Fig. 3 Effect of changing the threshold for classifying a feature on coverage with respondents' answers in the DA model

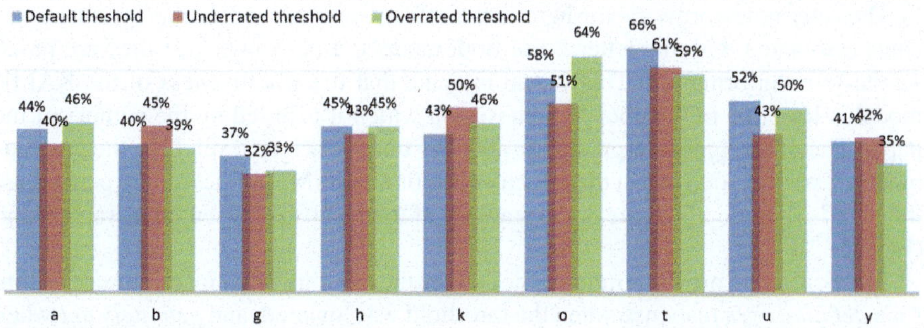

Fig. 4 Effectiveness of DA in recognizing sign features for different thresholds

4.3 KAMR Model

In the case of this model, the total efficiency was much lower (almost twice) and did not exceed 24%. It coped best with answers classified as *almost certainly not*, and the worst—as in the DA model—defined as certain ones. All other results were much worse, which makes it not a good idea to include this model in predictions on a larger scale. However, it would be much better to detect differences based on which it achieves greater effectiveness in responses classified as *almost certainly not* and implement them in the DA model, creating a new hybrid variant.

As can be seen (Table 5), the model in question has an incredibly low efficiency in virtually all possible response groups. Compared to the DA model, it can perform up to three times worse, depending on the case. An analysis of the results in terms of the percentage recognition of the features contained in the letters is presented in Table 8.

Table 8 Effectiveness of the DA and KAMR models in interpreting responses for different thresholds

Description of characteristic	Default threshold		Underrated threshold		Overrated threshold	
	DA (%)	KAMR (%)	DA (%)	KAMR (%)	DA (%)	KAMR (%)
Definitely yes	12	5	14	9	11	4
Almost certainly yes	40	21	36	19	43	25
Rather yes	69	33	66	29	68	34
Rather not	51	15	55	17	43	10
Almost certainly not	42	47	52	51	40	35
Definitely no	66	25	49	14	70	32
Average	47	24	45	23	46	23

The only noteworthy factor here is the effectiveness in indicating the characteristics associated with the letter *t*. For both models, it is identical, as they are based on the same information. This would indicate that the effectiveness of the KAMR model is low, due to the poor hypotheses on which it is based by the authors of the paper. The other signs interpreted are at a very unsatisfactory level. Like the previous model, it was decided to test the effectiveness of the KAMR model for the percentage thresholds specified in Table 8. A summary of the results is presented in Table 9 and (Fig. 5).

Identical to the previous model, one can see the obvious behaviour of the algorithm for a *definitely yes* response when the threshold was lowered, and *definitely not* when it was moved upwards. Unlike the case discussed earlier, however, one can see a much larger deviation from the default range for the lowered threshold. This is a welcome development, as it allows one to infer that it is not worth modifying the threshold for the *rather not* and *almost certainly not* responses. In other cases, the difference is too

Table 9 Effectiveness of models in recognising character features for different thresholds

Character	Default threshold		Underrated threshold		Overrated threshold	
	DA (%)	KAMR (%)	DA (%)	KAMR (%)	DA (%)	KAMR (%)
A	44	29	40	27	46	33
B	40	14	45	14	39	16
G	37	11	32	10	33	13
H	45	12	43	12	45	8
K	43	22	50	25	46	17
O	58	26	51	29	64	22
T	66	66	61	56	59	61
U	52	19	43	15	50	16
Y	41	20	42	23	35	25
Average	47	24	45	23	46	23

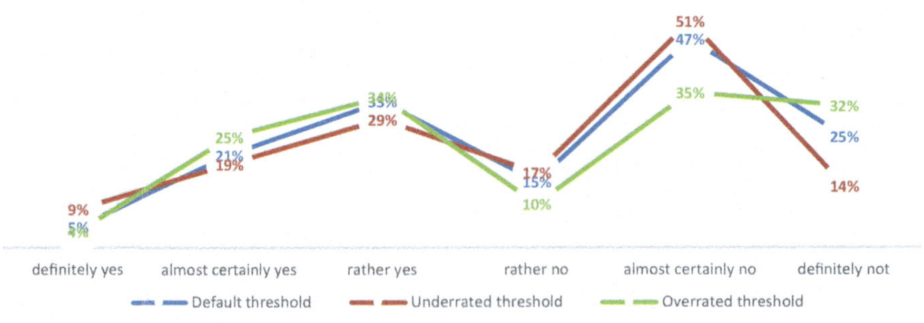

Fig. 5 Effect of changing the threshold for classifying a feature on coverage with respondents' answers in the KAMR model

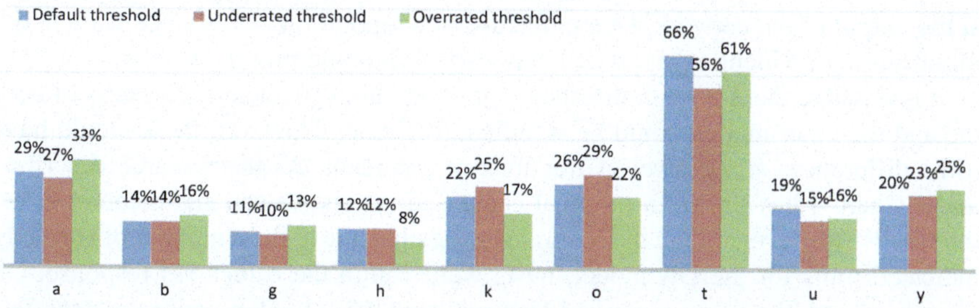

Fig. 6 Effectiveness of KAMR in recognizing sign features for different thresholds

small to draw far-reaching conclusions. Due to the low effectiveness of this model, the results obtained at this stage were not very satisfactory. Ultimately, a significant qualitative leap was expected, which would have disproved the theory of the model's low effectiveness and blamed the poor choice of initial parameters. Unfortunately, however, it is most likely that it was the conclusions drawn by the authors of the paper on which the KAMR model was based that proved to have little validity (Fig. 6).

Unfortunately, based on this identical study carried out within individual characters, the exact same thing was shown. The conclusions drawn come down to the fact that the KAMR model is not very effective. It does not guarantee high quality in indicating personality traits based on handwriting.

5 Summary

The use of the DA model yielded much better results than initially expected. With its unexpectedly high efficiency of 47%, it can be concluded that the use of deep neural networks in combination with graphology makes a great deal of sense indeed. If indicating simple personality traits such as openness and honesty is possible, then the more complex ones, for example introversion and extroversion, are also possible. Unfortunately, the same cannot be said based on the KAMR model. Its incomparably inferior—almost twice as effective at 24%—only highlights the problems facing graphology. Very often, these are merely hypotheses and assumptions without coverage in reality. It should be emphasized that the study was carried out only for a small group of people from central Europe and Australia. It is possible that the results could be completely different if they were repeated for people from America, Africa, or Asia [5]. The use of an in-house *Deep Neural Network* proved to be time-consuming yet translated into satisfactory results. Flexibility in the choice of activation functions allowed the better one to emerge, which turned out to be the sigmoidal (logistic) one. Interpreting the network in a stochastic way, allowed the selection of thresholds within which the results were considered acceptable. As it turned out, it is a much better idea to categories the results based on the value obtained at the output of the network. Due to the use of a 6-point fuzzy response scale in the initial survey, an identical approach was also used to interpret the results.

It is possible that for other threshold values in Table 10, the effectiveness of both the and the other model, could be significantly higher. However, these would have to be differences at the level of fractions of a percent. As the research described under Chap. 4 has shown, a minimal change in thresholds can significantly affect the results. However, taking the best ranges together, the highest number of correctly indicated traits for the DA model, 69%, were within the *rather yes* response. On the other hand, the worst identified features were those hidden under the *definitely yes* answer. Their total recognition was only 12%. For the KAMR model, on the other hand, traits interpreted as very weak were most effectively recognized. Like its predecessor, it also did worst with features classified as certain. Their values were 47 and 5%, respectively.

Based on the above information, the DA model seems to be better and, as mentioned earlier–it is possible to develop methods that successfully indicate personality traits. It should be emphasized that the questionnaire–although based on reliable sources of information on topics bordering on psychology and graphology–was created by an amateur. Perhaps the key element in the work in question would be precisely to draw up a new, much more precise test, which would more effectively extract information from people about their personality traits. If professional research is carried out on the topics raised in literature about the potential applications of the work under discussion, then use of very similar systems in medicine and forensics seems to make a lot of sense [6]. It is possible that it would revolutionize current approaches to preventive testing and would also contribute to significant improvements in security.

As for suggested direction for further work the first and probably most important element that the current software would need to be supplemented with is an extension of the classifier's ability to indicate a much larger number of features. This is currently relatively small, which wastes a huge amount of information. Naturally, to obtain an answer to the question of whether their classification has been implemented correctly, it would also be necessary to enlarge the pool of questions appearing in the survey. It would be advisable to entrust the creation of this to an expert in this field. Only if the software obtained in this way is successful, the next step would be to create a dedicated module, analyzing the writing in more detailed way.

Table 10 Method of translating graphological analysis results into survey results

Survey response	Interpretation of characteristic	Suggested value of DNN threshold (%)
Definitely yes	Certain	85–100
Almost certainly yes	Very strong	70–84
Rather yes	Strong	50–69
Rather not	Weak	30–49
Almost certainly not	Very weak	15–29
Definitely not	None	0–14

References

1. Antony, D.J., Cap, O.F.M.: Personality profile through handwriting analysis. In: Anugraha (Tamil Nadu Capuchin Institute for Counselling, Psychotherapy and Research) Nochiodaipatti Post Dindigul-624 003 Tamil Nadu, India Tel: 0451–2550100, 2550324, 255083 Email: anugrahacap@ eth. net. Anugraha Publications, (2008)
2. Amend, K.K., Ruiz, M.S.: Handwriting analysis: The complete basic book. Red Wheel/Weiser, (2000). ISBN: 978–0–87877–050–2
3. Champa, H.N., AnandaKumar, K.R.: Artificial neural network for human behavior prediction through handwriting analysis. Int. J. Comput. Appl. **2**(2), 36–41 (2010). Last accessed 17 Jan 2023. https://pdfs.semanticscholar.org/9962/3bfbc10270210cc2a4caeda6ae98c49644b2.pdf
4. Halilaj, E., Rajagopal, A., Fiterau, M., Hicks, J.L., Hastie, T.J., Delp, S.L.: Machine learning in human movement biomechanics: Best practices, common pitfalls, and new opportunities. J. Biomech. **81**, 1–11 (2018)
5. Kao, H.S., Hoosain, R., Van Galen, G.P. eds.: Graphonomics: Contemporary research in handwriting. Elsevier, (1986). ISBN: 978–0–444–70047–6
6. Matozza F., Ortiz A.: Levy D.: Handwriting analysis in cancer patients clinical-radiological and graphological correlation. Last accessed 17 Jan 2023. https://www.slideshare.net/frankmatozza/HRA-ADOBE-english-presentation
7. Mehta, Y., Majumder, N., Gelbukh, A., Cambria, E.: Recent trends in deep learning based personality detection. Artif. Intell. Rev. **53**(4), 2313–2339 (2020)
8. Obrębska, A.: Sekrety charakteru zawarte w piśmie. Jak wykonać ekspertyzę grafologiczną. Praktyczny podręcznik. Wydawnictwo Medium, Łódź, (2004). ISBN: 978-83-87025-38-0
9. Samek, W., Montavon, G., Lapuschkin, S., Anders, C.J., Müller, K.R.: Explaining deep neural networks and beyond: A review of methods and applications. Proc. IEEE **109**(3), 247–278 (2021)
10. Roberts, R., Woodman, T.: Personality and performance: Moving beyond the Big 5. Curr. Opin. Psychol. **16**, 104–108 (2017)
11. Woda, M., Batogowski, J.: Prediction of selected personality traits based on text messages from instant messenger. In: International conference on dependability and complex systems (pp. 672–685). Springer, Cham (2020)
12. Wright, T.: Handwriting Recognition with Artificial Neural Networks and OpenCV. CS488-Senior Capstone, (2012). https://cs.earlham.edu/~twright09/paper.pdf. Last accessed 17 Jan 2023
13. Wójtowicz, A.: Grafologia dla początkujących (praktyczny podręcznik). Wydawnictwo Wolumen, Bytom, (2007). ISBN: 978–83–926158–0–4
14. Zuckerman, M.: Zuckerman-Kuhlman personality questionnaire (ZKPQ): An alternative five-factorial model. Big five assessment, pp.377–396. (2002)

Architecting Cloud-Based Business Software—A Practitioner's Perspective

Andrzej Zalewski and Szymon Kijas

1 Introduction

Cloud-based software is built out of components being executed in a cloud environment. This marks a substantial shift in software architectures and software development, over the last 10–15 years. Today, software architecting focuses on looking for appropriate COTS, coordination of computing performed by independent entities, and integration of independent external services into the system being developed. These changes are a result of developments in software engineering such as microservices, CI/CD, containerization, microservices, non-SQL, cloud computing and popularization of agile software development methodologies [1]. Due to these changes, software evolution can be sped up in response to rapidly changing business requirements.

The purpose of this paper is to study how cloud computing has transformed the practice of architecture decision-making of large business software systems. The following research questions define this goal more precisely:

RQ.1. What are the key architectural problems in cloud-based software development?

RQ.2. What are the most important architectural drivers that shape architecture decision-making in cloud-based software architectures?

RQ.3. How have software architectures and architecture decision-making changed in comparison to the pre-cloud era?

In order to address the issues raised above, we analyses two practical, representative cases of cloud-based software systems which have been developed recently and

A. Zalewski (✉) · S. Kijas
Warsaw University of Technology, Warsaw, Poland
e-mail: a.zalewski@ia.pw.edu.pl

S. Kijas
e-mail: szymon.kijas@pw.edu.pl

W. Zamojski et al. (eds.), *Dependable Computer Systems and Networks*, Lecture Notes in Networks and Systems 737, https://doi.org/10.1007/978-3-031-37720-4_31

are currently being operated. This enables analysis of how software architectures and architecture decision-making have changed in the cloud computing era. It also inspires a number of insights into the logic and nature of these changes.

2 Related Work

Software architecture is commonly perceived nowadays as a series of architectural decisions [2]. They can be documented in the form of text records [3] that comprise, for instance, a problem statement, considered and chosen options, and rationale. In past papers [6] and [7], we identified the set of architectural decisions that define the top-level organization of large-scale systems, which has been referred to as a *System Organization Pattern.* These crucial choices significantly influenced the system's buildability and achievability of key non-functional properties such as performance and reliability.

An extensive list of architecting challenges in cloud-enabled systems and proposals as to how they might be resolved can be found in [5]. They have been divided into four groups: resource management, workflow management, service level agreement compliance, and energy awareness.

3 Research Method

On the basis of two real-world case studies, we have identified how a System Organization Pattern (SOP) introduced in our past papers [6, 7] changed in the cloud computing era if compared to the version of 2007. As both case studies concern business software, the outcomes will concern this type of software.

Additionally, we take a retrospective view of the architecting of these two systems and identify the main drivers that shape architectures of cloud-based software. We compare them to the critical issues identified in [7].

4 Case Studies

4.1 Case No. 1. Payment Messages Exchange System

A Payment Messages Exchange System was developed to enable the exchange of payment-related messages between a Third Party Provider (TPP) and Account Servicing Payment Service Provider (ASPSP)—typically banks, according to PSD2. The system is actually a message broker, which is supposed to aggregate messages coming from various TPPs and direct them to an appropriate ASPSP.

The system is based on hybrid-cloud and Kubernetes container infrastructure. There are pairs of containers, which host microservices that handle communication with ASPSP and message translation. Each pair is devoted to various ASPSPs, which do not use a common standard of API. This means that messages sent/received to/from ASPSPs have to be translated to/from a common format used between our system and TPP.

There is also a separate component called "HUB", which manages internal communication and a container service that ensures access to the encryption keys stored in the "on-premises" part of the system. All the components are automatically scalable, which means that their instances can be automatically created as load soars. Below we study key architectural problems, architectural drivers, chosen solutions and their rationale:

1. **Payment transaction logs storage placement?** *Drivers*: storage cost, regulatory/legal concerns, data security. *Chosen solution*: logs stored "on-premises". *Rationale*: the transaction logs contain personal data and require a large amount of data storage. The data has to be retained for more than a year.
2. **Encryption key placement?** *Drivers*: storage cost, regulatory/legal concerns, data security. *Chosen solution*: Use of HSM module "on-premises"; *Rationale*: as the encryption keys are obtained from TPP they were stored in a HSM device that enables also remote key generation.
3. **How to ensure ease of connecting additional ASPSP** *Drivers:* modifiability. *Chosen solution*: pairs of microservices (connector/translator) for each ASPSP/containers; *Rationale*: template microservices are expected to speed up extensions needed to attach further ASPSPs.
4. **How to cope with peak loads** *Drivers*: insufficient performance or system break-down under peak loads. *Chosen solution*: microservices and cascading calls/composition, autoscaling coordinator; *Rationale*: the expected hugely varying loads of payment messages (e.g. during Black Fridays) a container auto-scalability feature has been adopted.
5. **How to implement message translation effectively**. *Drivers*: bugs in internal communication; bottlenecks in the communications layer; problems with access to the encryption keys. *Chosen solution*: Encryption/Apache Camel integration framework/asynchronous communication/special container connectors to HSM. *Rationale*: variety of APIs delivered by the ASPSP (integration framework), secure access to the encryption keys (container connectors to HSM).
6. **How to implement external communication** *Drivers*: Problems with integration with external systems. *Chosen solution*: REST API compliant with national banking standard. *Rationale*: this minimizes the scope of changes needed to connect other APSPs (the implementation of the national standard varies among APSPS).
7. **How to manage API updates and modifications** *Drivers*: problems with API versioning and integration with external systems. *Chosen* solution: API management. *Rationale*: ability to publish several versions of API simultaneously.

8. **How to achieve unlimited scalability under the regulatory/legal concerns.** *Drivers:* data leak, encryption keys get compromised, licensing violation, violation of legal/regulatory rules. *Chosen solution*: computing model: hybrid cloud and mixed computing (cloud and non-cloud); *Rationale*: a hybrid cloud enables scalability of a system's capacity to varying the load of payment messages."On-premises" storage of sensitive data addresses the issue of data protection and legal/regulatory compliance.

4.2 Case No. 2. Mutual Betting System

The system has been developed for a company that organizes mutual betting for classic live sport events as well as for virtual games. It consists of the following main subsystems: bet registry, archive of bet records, player admittance system, user web, and mobile application front-end. The system does not follow any architectural style, such as microservices. It is rather an ad-hoc orchestration of internal and external entities that address certain functional and legal requirements (incl. security requirements).

The most important architectural problems, considered options, chosen solutions and their rationale are listed below:

1. **How to verify a player's identity** *Drivers*: to maximize access to a system's services, while meeting legal requirements, regulatory acceptance. *Chosen solution*: all available options, namely using electronic identity services, "manual" verification of scanned documents, and automatic verification with external document content recognition services based on a scan of the identification card or passport. *Rationale*: obvious.
2. **How to store client data, their bet and finance history and bet records** *Drivers*: security of bet records, legal issues, regulatory acceptance. *Chosen solution*: replication of bet records from operational cloud-based to an "on-premises" relational data base management system. *Rationale*: combining cloud scalability with legal and security requirements.
3. **How to ensure easy integration of virtual game providers** *Drivers*: modifiability and evolvability of the system in a rapidly changing environment. *Chosen solution*: integrate the vision directly into a web-page using Iframes, while obtaining game results via the REST interface. *Rationale*: simplicity of the solution, popularity of communication protocol.
4. **How to archive bet records** *Drivers*: legal regulations imposing time constraints and security requirements. *Solution*: real-time replication from an operational database onto a specialized archive server, placed "on-premises" because of the legal requirements.

Table 1 SOP Theme: Data management

Arch. drivers	Arch. risk
• Data access type (e.g. transactional, logs, documents, archival data)	• Data loss
• Volume of data	• Data inconsistency
• Data structuralization (structured/unstructured)	• Data access performance
• Data access / search performance	• Cost of data storage beyond assumed limits
• Data security	• Data leak • Unauthorized data modification • Regulatory/legal/internal rules violations • Insufficient storage scalability • Development cost exceeding budget • Private key get compromised

Options:
Database engine: RDBMS, Non-SQL, incl. document DB, Distributed repositories (HADOOP, Datalake), Blockchain;
Data consistency: Data synchronisation, transactions;
Data security: Data encryption, HSM, KeyStore

5 System Organization Pattern for Cloud-Based Systems

The key architectural issues that define architectures of cloud-based systems as well as drivers and risks that influence decision-making are presented in Fig. 1–6. They were identified on the basis of the two case studies (Sect. 4) and architecting retrospection. These decisions define the System Organization Pattern for modern cloud-based business systems. The architectural options considered by the architects are enumerated below each Tables 1, 2, 3, 4, 5 and 6.

6 Discussion and Limitations

Despite the important changes to software technologies in recent years, software systems are still about data input, output, storage, processing and transfer. It is no wonder that a System Organization Pattern of modern cloud-based software systems comprises similar topics to that of 2007. However, many SOP themes have been transformed. We summarize this below:

- The geographical allocation of system entities and data storage has become an important factor in many designs because of regulatory and legal issues, which enforce physical location of data and hardware within a certain area (country,

Table 2 Theme: Data storage location and ownership

Arch. drivers	Arch. risk
• Legal requirements	• Insufficient scalability
• Internal organization's policy	• Unexpectedly high cost of data storage operation
• Agreements with the clients	• Excessive data distribution
• Quality norms and branch regulations	• Regulatory/legal rules violation
• Data storage operation cost	• Potential problems with data extraction from cloud storage on contract termination or services provider change
• Volume of data	• Data leak
• Licensing models of cloud-based software components	• Unauthorized access/data modification
• Security requirements	• Non-optimal/costly licensing model

Architectural options:
Data location: Cloud, On-premises, DB as a service, Hybrid cloud, Mixed (cloud and no cloud)

Table 3 Theme: Functional decomposition style & computing organization

Arch. drivers	Arch. Risk
• Frequency of software releases	• High coupling between system's components hindering its evolution
• Stakeholders' concerns	• Excessive data distribution (e.g. in case of huge microservices networks)
• Level of data distribution	• Poor software modifiability
• Number of integrated external and internal systems	
• Component reuse	
• Use of domain cloud-based COTS	
• Licensing models	

Architectural options:
Decomposition: Containters, Microservices, frontend implemented by client application;
Internal cloud solutions: Cloud libraries, cloud services

union of countries). This also relates to the ownership of data storage infrastructure hardware and software. This was not an issue in SOP'2007, as the entire infrastructure was typically owned by the owner of the system, who was also the owner of the data.

- Decomposition—currently is generally much more fine-grained than in SOP'2007 and often based on composition of COTS. It is rather a question of choosing a functional decomposition style with regard to back-end (microservices, services, monolith) and front-end accessing the latter ones via API. An important aspect here is the choice of appropriate COTS in order to meet the requirements.

Table 4 Theme: Computing organization

Arch. drivers	Arch. risk
• Frequency of software releases	• Operating cost explosion (e.g. in case of unlimited autoscaling)
• Stakeholders' concerns	• To high coupling between system components
• Level of data distribution	• Data dispersed among too many entities
• Number of integrated external and internal components	• Loss of performance under peak loads
• Component reuse	• High cost of maintenance of predeclared cloud infrastructure
• Availability of domain cloud-components	
• Licencing models	

Options:
Compuring organisation style :Microservices and cascading calls/composition, Monolith, Autoscaling Coordinator, Amorfous

Table 5 Theme: Communication

Arch. drivers	Arch. risk
• Number of system's users	• Kryptographic key access problems
• Time characteristics of system load	• Internal communication implementation errors
• Performance requirements	• Communication performance problems
• Security requirements	

Options:
Communication types: Synchronous, Asynchronous;
Communication hardening: Communication encryption, Integration frameworks;

Table 6 Theme: Data input/output organization

Arch. drivers	Arch. risk
• Diversity of communication protocols	• API versioning problems (integracja)
• Number of system's users	• Insufficient performance (e.g. due to suboptimal load-balancing)
• Time characteristics of system load	• Faulty integration with external systems
• Data input performance	
• Data input security	

Options:
Input/output techniques: API (REST, SOAP), Queues;
Input/output management: API management, Load balancing solutions, Encryption/ Decryption

- The data management topic includes former topics of managing and processing distributed data, including transaction management. Our observation is that architects do not pay as much attention to these issues as before.

Between 2007 and 2021, there was a shift with regard to the challenging aspects of system design. The synchronization and transactional processing in a cloud environment is not as challenging as before because the communication between system entities located in the cloud is reliable enough. Therefore, the design challenges currently focus on security, coordination and complexity of the interactions between system components, the cost of system operation, and last but not least selecting domain-based COTS in order to fulfill functional requirements.

Security has become a common issue of significant architectural importance. This has its roots in the structural changes to system architectures. They are much more exposed to external penetration nowadays than before, as system entities execute and exchange information via the Internet. Additionally, what would have been a single entity is currently in many cases decomposed into a set of entities (e.g. microservices). This increases the number of entry points to the system and chances of a successful attack as well as the number of communication paths that have to be secured.

Pricing schemes of cloud-based services, on which a modern system runs, are inherently complex. This together with features such as autoscaling of resources complicates the estimation of system operation costs and becomes an important concern when choosing auto-scaling options.

Cloud-based systems are usually made up of a number of smaller components (e.g. microservices) interacting in order to provide required computations. As these components are easily accessible via APIs or other types of interfaces, the number and complexity of interactions between them may become an important concern hindering the evolution of the system as a whole, similarly to the popular "integration spaghetti".

The above challenges fall mainly in the resource management category proposed in [5]. They are results of observation of practical architecting of business systems, which is an important but not the only category of cloud-based systems.

As case study 2 shows, thanks to the wealth of the available COTS the architecting of cloud-based system may now be focused mainly on meeting functional requirements. This, in turn, enables architecting, in which little attention is paid to the system's structure, as far as functional (and regulatory) requirements are met.

Although we find the two cases studies representative for the broader class of cloud-based business systems, as they reflect the issues common to a broader number of cloud-based systems, the main limitation of the validity of results remains the small number of case studies.

7 Summary and Research Outlook

Software architectures and architectural decision-making has changed considerably in the cloud era, as it focuses mainly on delivering required functionality, ensuring security, operation cost, system evolvability, and meeting regulatory concerns. Further research will comprise:

- Analysis of a larger number of case studies;
- In-depth analysis of architectural issues, choices made by the architects, and their rationale;
- Capturing architectural wisdom that results from the analyzed case studies. This in turn should become a basis for a cloud-based software architecture evaluation method and/or architectural decision-making support.

References

1. Bosch, J.: Speed. Data, and ecosystems: The future of software engineering in IEEE software **33**(01), 82–88 (2016)
2. Jansen, A., Bosch, J.: Software architecture as a set of architectural design decisions. In: Proceedings—5th Working IEEE/IFIP Conference on software architecture, WICSA 2005. pp 109–120. (2005)
3. Harrison, N.B., Avgeriou, P., Zdun, U.: Using patterns to capture architectural decisions. IEEE Softw. **24**, (2007)
4. Wood, W., Zubrow, D., Nord, R., Bass, L.: Risk themes discovered through architecture evaluations in software architecture. In: Working IEEE/IFIP Conference on (p. 1). Mumbai, India, (2007)
5. Chauhan, M. A., Babar, M. A., Benatallah, B.: Architecting cloud-enabled systems: a systematic survey of challenges and solutions. Softw. Pract. Exper., **47**. (2017)
6. Zalewski, A.: Beyond ATAM: Architecture analysis in the development of large scale software systems. In: Lecture Notes in Computer Science (including subseries Lecture Notes in Artificial Intelligence and Lecture Notes in Bioinformatics) (2007)
7. Zalewski, A., Kijas, S.: Beyond ATAM: Early architecture evaluation method for large-scale distributed systems. J. Syst. Softw. **86**, (2013)

Appendix A
Emerging Challenges in Technology-Based Support for Surgical Training (Invited Lecture)

Jerzy W. Rozenblit[1] [https://orcid.org/0000-0002-7348-4128]
and Minsik Hong[1] [https://orcid.org/0000-0002-0431-1127]

[1]The University of Arizona, Tucson, AZ 85,718, USA

jerzyr@arizona.com, mshong@arizona.edu

Abstract: This summary of the presentation describes the emerging challenges and trends in computer-based assistive systems for minimally invasive surgery training.

Keywords: Surgical training, minimally invasive surgery, virtual coaching, haptics

Minimally Invasive Surgical Training Challenges

Our work focuses on the development of methods and systems that can support training of students, residents, and practicing fellows to acquire and improve skills in standard laparoscopy and robotically assisted surgery [1–7]. Such procedures are minimally invasive and significantly reduce recovery time as well as postoperative pain [1, 2]. However, they are difficult to learn due to the lack of depth perception and direct tactile feel (especially when using devices such as the DaVinci robot [3]). We provide an overview of the techniques and supportive systems that we have developed to date and discuss our vision for further advancing computer aided support for surgical training in minimally invasive surgery.

Specifically, we present our concept of using force as a means of providing guidance when executing standard laparoscopy training tasks [4]. We introduce methods for using augmented reality (AR) as the basis for visual navigation in a training environment. AR provides overlays with visual cues (in addition to force-based navigation) to enhance the trainees' visuo-spatial awareness in the surgical field of

W. Zamojski et al. (eds.), *Dependable Computer Systems and Networks*, Lecture Notes in Networks and Systems 737, https://doi.org/10.1007/978-3-031-37720-4

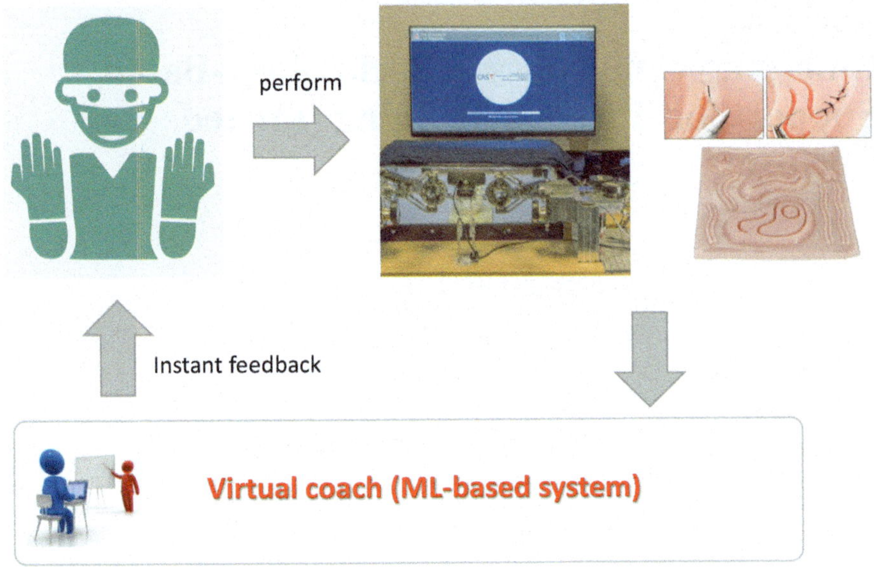

Fig. A.1 A virtual coaching system concept

view [6]. Machine learning models are being developed so that our Computer-Aided Surgical Trainer (CAST) [4] could reinforce learning through progressively more complex surgical exercises.

Our present focus is on advancing the technology, especially for robot-assisted training. Two key challenges in this realm are haptics and a broader area of virtual coaching. When operating with the DaVinci robot, surgeons do not have any sense of tactile sensation and rely on visual cues from an endoscopic camera to discern tissue resistance or pliability. This is a challenge for novices and, therefore, incorporating methods that mimic the feel of surgery is one of our key research objectives.

We also focus on the development of a framework for virtual coaching systems. This is motivated by several factors, among them the need to train surgeons is low resource settings such remote rural areas, and the lack of extensive access to expert surgeons who could devote sufficient time to didactic activities. Our concept of a virtual coaching system is shown in Fig. A.1. In such a system, a trainee would perform an exercise in our CAST system using physical reality setups. The system would then collect data and compute performance metrics in real time and provide instant feedback on how well the student progress, point out deficiencies, and adjust task difficulty based on progress in learning and attendant metrics.

The presentation will address technical challenges and propose solution for building such virtual coaching systems.

References

1. Dankelman, J.: Surgical robots, and other training tools in minimally invasive surgery. In: Systems, Man, and Cybernetics (SMC), 2004 IEEE International Conference on, pp. 2459–2464 (2004)
2. Stylopoulos, S. et al.: Computer-enhanced laparoscopic training system (CELTS): bridging the gap. Surg. Endosc. Other Intervent. Techniq. **18**(5), 782–789 (2004)
3. Intuitive Surgical Homepage. https://www.intuitive.com/en-us. Accessed 21 Mar 2023
4. Rozenblit, J.W. et al.: The computer assisted surgical trainer: design, models, and implementation. In: Proceedings of the 2014 Summer Simulation Multiconference, pp. 211–220 (2014)
5. Napalkova, L., Rozenblit, J.W., Hwang, G., Hamilton, A.J., Suantak, L.: An optimal motion planning method for computer-assisted surgical training. Appl. Soft Comput. **24**, 889–899 (2014)
6. Schreiber, A., Hong, M., Rozenblit, J.W.: Monocular depth estimation using synthetic data for an augmented reality training system in laparoscopic surgery. In: 2021 IEEE International Conference on Systems, Man, and Cybernetics (SMC), pp. 2221–2226 (2021). https://doi.org/10.1109/SMC52423.2021.9658708
7. Hong, M., Rozenblit, J.W.: An adaptive force guidance system for computer guided laparoscopy training. IEEE Trans. Cybernet. **52**(9), 8019–8031 (2022)

About the Speaker

Prof. Jerzy W. Rozenblit
University Distinguished Professor
Raymond J. Oglethorpe Endowed Chair
Department of Electrical and Computer Engineering
Department of Surgery
BIO5 Institute
The University of Arizona

Education

Ph.D. Computer Science, Wayne State University, Detroit, Michigan, 1985.

M.Sc. Computer Science, Wayne State University, Detroit, Michigan, 1983.

M.Sc. Computer Science and Engineering, Technical University of Wroclaw, Poland, 1980.

Academic Experience

University Distinguished Professor, The University of Arizona, Tucson, Arizona, 2009–present.
Raymond J. Oglethorpe Endowed Chair, Department of Electrical and Computer Engineering, The University of Arizona, Tucson, Arizona, 2006–present.
Professor of Surgery, College of Medicine, The University of Arizona, 2007–present.
Professor, BIO5 Institute, The University of Arizona, 2018–present.
Department Head, Department of Electrical and Computer Engineering, The University of Arizona, Tucson, Arizona, 2003–2011.
Professor, Department of Electrical and Computer Engineering, The University of Arizona, Tucson, Arizona, 1998–present.
Associate Professor, Department of Electrical and Computer Engineering, The University of Arizona, Tucson, Arizona, 1992–1998.
Assistant Professor, Department of Electrical and Computer Engineering, The University of Arizona, Tucson, Arizona, 1986–1992.
Visiting Assistant Professor, Department of Computer Science and Engineering, Oakland University, Rochester, Michigan, 1984–1985.
Research Fellow, US Army Research Institute, Ft. Huachuca, Arizona, 1992.

Key Related Publications

J. Nikodem, K.J. Kluwak, D. Zysko, R. Klempous, B. Hrymniak, J. Rozenblit, T. Zelniker, A. Wytyczak-Partyka, M. Bozejko and D. Jagielski, "A Novel Approach for Continuous Hearth Rhythm Monitoring," R. Moreno-Dıaz et al. (Eds.): EUROCAST 2022, Lecture Notes in Computer Science 13789, pp. 512–519, Springer Nature, 2022, 10.1007/978-3-031-25312-6_60
R. Klempous. J.W. Rozenblit, K. Kluwak, J. Nikodem, D. Patkowski, S. Gerus, M. Palczewski, Z. Kielbowicz, and A. Wytyczak-Partyka, "Our Early Experience Concerning an Assessment of Laparoscopy Training Systems," in: *Computer Aided Systems Theory, EuroCast 2017*, Springer, pp. 372–379, January 2018, 10.1007/978-3-319-74727-9_44.
A. Rao, N. Carreon, R. Lysecky, and J. Rozenblit, "FIRE: A Finely Integrated Risk Evaluation Methodology for Life-critical Embedded Systems," *Information,* 2022, 13, 487. 10.3390/info13100487
A.M. Schreiber, M. Hong, J. W. Rozenblit, C. J. Hinkel, S. Kim, and B. R. Lee, "Enhancing Training by Using Mixed Reality Rendering in Laparoscopic Surgery Training," *Journal of Urology,* vol. 206, no. 3, Sept 2021, 10.1097/JU.0000000000001983.01
K. Kluwak, R. Klempous, Z. Chaczko, J.W. Rozenblit, J.W. and M. Kulbacki, "People Lifting Patterns—A Reference Dataset for Practitioners," *Sensors,* 2021, 21, 3142. 10.3390/s21093142
M. Hong and J. W. Rozenblit, "An Adaptive Force Guidance System for Computer-guided Laparoscopy Training," *IEEE Transactions on Cybernetics*, Jan 2021, 10.1109/TCYB.2021.3051837

M. Hong, J. W. Rozenblit, and A. J. Hamilton, "Simulation-based Surgical Training Systems in Laparoscopic Surgery: A Current Review," *Virtual Reality*, 2020, 10.1007/s10055-020-00469-z

Z. Chaczko, R. Klempous, J. Rozenblit, T. Adegbija, Ch. Chiu, K. Kluwak, Cz. Smutnicki "Biomimetic Middleware Design Principles for IoT Infrastructures," *Acta Polytechnica Hungarica*, vol. 17, no. 5, pp. 135–150, 2020.

G. Ditzler, S. Miller, and J. Rozenblit, "Learning What We Don't Care About: Regularization with Sacrificial Functions," *Information Sciences*, vol. 496, pp. 198–211, September 2019.

L. Napalkova, J.W. Rozenblit, G. Hwang, A. Hamilton and L. Suantak, "An Optimal Motion Planning Method for Computer-Assisted Surgical Training," *Applied Soft Computing*, 24(11), pp. 889–899, 2014, 10.1016/j.asoc.2014.08.054.

M. Riojas, C. Feng, A.J. Hamilton and J.W. Rozenblit, "Knowledge Elicitation for Performance Assessment in a Computerized Surgical Training System," *Applied Soft Computing,* 11(4), pp. 3697–3708, 2011.

C. Feng, J.W. Rozenblit, and A.J. Hamilton, "A computerized assessment to compare the impact of standard, stereoscopic and high definition laparoscopic monitor displays on surgical technique," *Surgical Endoscopy*, 24(11), pp. 2743–2748, 2010. 10.1007/s00464-010-1038-6.

R. Klempous, J. Nikodem, K. Kluwak, M. Nikodem, A Kolcz, P. Gawlowski, J.W. Rozenblit, C. Chiu, and M. Olesiak, "Motion Capture Analysis of Supporting Lifting Technique Optimization for Occupational Safety Diagnosis," in: *17*th*Intl. Conference on Computer Aided Systems Theory*, Springer, pp. 313–320, April 2020, 10.1007/978-3-030-45096-9_39.

Z. Chaczko, R. Klempous, J. Rozenblit, C. Chiu, K. Kluwak, and Cz. Smutnicki, "Enabling Design of Middleware for Massive Scale IoT-based Systems," *Proc. of the 23rd IEEE Intl. Conference on Intelligent Engineering Systems,* pp. 219–224, Godollo, Hungary, April 2019.

Appendix B
Neural Computation Methods for the Early Diagnosis and Prognosis of Alzheimer's Disease: An Overview (Invited Lecture)

Carmen Paz Suárez-Araujo[1] [https://orcid.org/0000-0002-8826-0899],
Ylermi Cabrera-León[1] [https://orcid.org/0000-0001-5709-2274],
Pablo Fernández-López[1] [https://orcid.org/0000-0002-2135-6095],
and Patricio García Báez[2] [https://orcid.org/0000-0002-9973-5319]

[1]Instituto Universitario de Ciencias y Tecnologías Cibernéticas, Universidad de Las Palmas de Gran Canaria, Parque Científico Tecnológico, Campus Universitario de Tafira, Las Palmas de Gran Canaria, CN, Spain

[2]Departamento de Ingeniería Informática y de Sistemas, Universidad de La Laguna, Camino San Francisco de Paula, 19, San Cristóbal de La Laguna, CN, Spain

carmenpaz.suarez@ulpgc.es

ylermi.cabrera101@alu.ulpgc.es

pablo.fernandezlopez@ulpgc.es

pgarcia@ull.es

Extended Abstract

More patients with dementia will exist as the elderly population increases and ages. Alzheimer's Disease (AD), the most common dementia, is a neurodegenerative disease that takes years to fully develop. Mild Cognitive Impairment (MCI) has been considered a construct which could be the equivalent to the oligosymptomatic or prodromal stage of AD [3].

Diagnosis and differential diagnosis of AD and MCI are cumbersome problems because both standardized diagnostic criteria and a specific biomarker are unavailable [3]. Among them, computation-based ones have played a major role because of their automatic characteristic. Neural approaches have been reaching very good

W. Zamojski et al. (eds.), *Dependable Computer Systems and Networks*, Lecture Notes in Networks and Systems 737, https://doi.org/10.1007/978-3-031-37720-4

performance results in this area, because of their ability to deal with complex, multi-dimensional, very noisy, and high class overlapping information. Due to this, in the last years approaches based on neural computation, both shallow Artificial Neural Networks (ANNs) and Deep Neural Networks (DNNs), have become commonplace for aiding dementia diagnosis and prognosis.

Researchers have focused their efforts on the two possible AD-related tasks: diagnosis and prognosis. Both cross-sectional and longitudinal studies have been proposed. In turn, these studies can be grouped into neuroimaging-based or non-neuroimaging-based.

Neuroimaging techniques refers to all imaging used to study the structure and function of the central nervous system, most frequently the brain. They are complementary medical tests frequently used for the diagnosis of AD and other dementia.

In works from the last years the following neuroimaging and brain signals techniques were found, sorted in decreasing order of frequency: Structural Magnetic Resonance Imaging (sMRI), Positron Emission Tomography (PET), Functional Magnetic Resonance Imaging (fMRI), Electroencephalography (EEG) and Functional Near-Infrared Spectroscopy (fNIRS).

Multimodal scheme is essentially used to take advantage of synergies among different modalities working together.

The Alzheimer's Disease Neuroimaging Initiative (ADNI) database was the most frequently used, followed by far by private and the Open Access Series of Imaging Studies (OASIS) datasets. Convolutional Neural Networks (CNNs) were the most prevalent neural method, whereas Multilayer Perceptrons (MLPs) were the most common shallow ANNs. Magnetic Resonance Imaging (MRI), especially sMRI, outnumbered the other neuroimaging techniques.

Several non-neuroimaging techniques started being utilized earlier than neuroimaging due to their lower cost and technological complexity. A much wider variety of modalities in this group were found: blood, demographic data, gait and movements of body parts (such as eyes and limbs), genes, neuropsychological tests, and speech.

Once again, the ADNI database was the most popular. Other types of DNNs different to CNNs, such as Recurrent Neural Networks (RNNs), appeared more, especially with speech data. Among shallow ANNs, MLPs and Backpropagation Networks (BPNs) were common, but non-neural methods sometimes outperformed them.

No longitudinal studies and prognosis of AD works based on shallow ANNs were found. Comparatively, the number of works was much lower than cross-sectional ones and those based on non-neuroimaging techniques were quite scarce. Most of them could be divided in two groups of tasks: conversion time prediction, and prognosis or conversion prediction. Once again, ADNI was the most used database.

Although DL-approaches that used neuroimaging are almost always at the top of the list, several solutions have outperformed them, such as those based on multimodality and hybrid, essentially ontogenic, architectures [1, 2].

While AD remains incurable novel modalities and methodologies will be developed and works utilizing multimodality or ensembles of ANNs, possibly DNNs, will be probably the best option and they will possibly arrive earlier.

This work will overview the most recent and outperforming ANN-based approaches to in vivo AD diagnosis and prognosis, managing to show the trends on this hot topic.

Acknowledgments This work was funded by the Consejería de Vicepresidencia Primera y de Obras Públicas, Infraestructuras, Transporte y Movilidad del Cabildo de Gran Canaria under Grant Number "23/2021".

References

1. Cabrera-León, Y., Báez, P.G., Fernández-López, P., Suárez-Araujo, C.P.: Study on mild cognitive impairment and Alzheimer's disease classification using a new ontogenic neural architecture, the supervised reconfigurable growing neural gas. In: 2023 Annual Modeling and Simulation Conference (ANNSIM 2023). Mohawk College, ON, Canada (May 2023)
2. Cabrera-León, Y., Suárez-Araujo, C.P., Fernández-López, P., Báez, P.G., Navarro-Mesa, J.L.: Towards an intelligent computing system for the differential diagnosis of Alzheimer's disease based on MyGNG. Acta Polytechnica Hungarica p. (In Press) (2023)
3. Petersen, R.C., Caracciolo, B., Brayne, C., Gauthier, S., Jelic, V., Fratiglioni, L.: Mild cognitive impairment: a concept in evolution. J. Intern. Med. **275**(3), 214–228 (2014). 10.1111/joim.12190

About the Speaker

Prof. Dr. Carmen Paz Suárez-Araujo

Instituto Universitario de Ciencias y Tecnologías Cibernéticas (IUCTC)

Universidad de Las Palmas de Gran Canaria (ULPGC)

Professor of Computer Sciences and Artificial Intelligence at the Universidad de las Palmas de Gran Canaria (ULPGC). She is BS & MS in Physics and PhD. in Computer Sciences. She is a Number Academic Member of the Canary Real Academy of Sciences, and currently is Head of the Computational Neuroscience Research Division at the Institute of Cybernetics Science and Technology of the ULPGC. She has also been the Director of the research groups "Intelligent Computing, Perception, and Big Data" and "Adaptive Neural Computing and Computational Neuroscience". She has been the Director of the Ph.D. Program of Neural Computing in Natural and Artificial Systems at the same University. She has been Vice-President and President of the Canary Neuroscience Society, and Vice-Rector and Vice-Dean of this University for many years, too.

Her teaching activity has been essentially developed at the ULPGC, in Computer Engineering Degree, and Ph.D. Programs, principally devoted to Cybernetics, Neural Computing, Artificial Perception, and Intelligent Computing. She has been Invited Professor at a broad range of European and American Universities, around 30, and at University of Technology Sydney (UTS).

Her research areas are: (a) Natural and Artificial Neural Networks, Design of New Neural Architectures; (b) Application of Neural Computation in Clinical and Environmental Domains, Biomedicine, Neuroinformatics; (c) Computational Neuroscience and Cognitive Computation: Neural communication models and learning. With more than 140 scientific articles and more than 150 contributions to international and national congresses, she has been Project Leader of international and national Research Grants, head of 34 research seminars, and Supervisor of Doctoral and Master Theses.

Her research work has been awarded to the *Spain Real Doctors Academy*, the *ULPGC*, and to several *International Conferences*. She has also been awarded *The CAN of Sciences 2020 by Regional Canary Government*.

She is a referee of a wide range of international scientific journals and international congresses. She has been invited speaker in more than 40 seminars and conferences.

She is President of the Informatic National Committee at the Spanish National Agency, ANECA, referee of several other international and national Evaluation Agencies.

The manufacturer's authorised representative in the EU is Springer Nature Customer Service Centre GmbH, Europaplatz 3, 69115 Heidelberg, Germany. If you have any concerns regarding our products, please contact ProductSafety@springernature.com

Printed and bound by CPI Group (UK) Ltd, Croydon, CR0 4YY
28/09/2026
02224079-0004